Texts in Logic
Volume 2

Model Based Reasoning in Science and Engineering

Volume 1
Proof Theoretical Coherence
Kosta Dosen and Zoran Petric

Volume 2
Model Based Reasoning in Science and Engineering
Lorenzo Magnani, editor

Texts in Logic Series Editor
Dov Gabbay   dov.gabbay@kcl.ac.uk

Model Based Reasoning in Science and Engineering

Cognitive Science, Epistemology, Logic

edited by
Lorenzo Magnani

© Individual author and King's College 2006. All rights reserved.

ISBN 1-904987-23-0
College Publications
Scientific Director: Dov Gabbay
Managing Director: Jane Spurr
Department of Computer Science
Strand, London WC2R 2LS, UK

Cover design by Richard Fraser, www.avalonarts.co.uk
Printed by Lightning Source, Milton Keynes, UK

All rights reserved. No part of this publication may be reproduced, stored in a retrieval system or transmitted, in any form, or by any means, electronic, mechanical, photocopying, recording or otherwise, without prior permission, in writing, from the publisher.

CONTENTS

Editorial Preface   xi
Lorenzo Magnani

Scientific Reasoning, Discovery Processes and Mechanisms

New Foundations for Geometry and Computation   15
Michael Leyton

Theories Looking for Domains Fact or Fiction? Reversing Structuralist Truth Approximation   33
Theo Kuipers

Scientific Cognition as Model-Based Reasoning   51
Ping Li and Dachao Li

Simulation for the Shift of Paradigm   67
Stanislava Mildeová

The Role of Theoretical Models in the Methodology of Physics   75
Andrés Rivadulla

The *Evolutionary History* of Models as Representational Agents   87
Demetris Portides

On Popper's Logic of Scientific Discovery   107
Atocha Aliseda

Models, Mental Models and Representation

Exemplar-Based Reasoning with the Shortest Derivation **Rens Bod**	119
The Role of Simulation Models in Visual Cognition **Arturo Carsetti**	141
Thought Experiements and Imagery in Expert Protocols **John J. Clement**	151
Technological Thinking and Moral Imagination **Michael E. Gorman**	167
Disembodying Minds, Externalising Minds: How Brains Make Up Creative Scientific Reasoning **Lorenzo Magnani**	185
Analogical Reasoning with Animal Models in Biomedical Research **Cameron Shelley**	203
Cognitive Fictions **Giovanni Tuzet**	215
Cognitive Design Principles: from Cognitive Models to Computer Models **Barbara Tversky, Maneesh Agrawala, Julie Heiser, Paul Lee, Pat Hanrahan, Doantam Phan, Chris Stolte and Marie-Paule Daniel**	227
Cognitive Complexity and the "Supports" of Modeling **Zhikang Wang**	249

Abduction, Creative Inferences and Logical Aspects of Model Based Reasoning

A Diagrammatic Proof Search Procedure as Part of a Formal Approach to Problem Solving **Diderik Batens**	265

Diagrams as Physical Models to Assist in Reasoning **Balakrishnan Chandrasekaran**	285
A Formal Model of Abduction **Dov Gabbay and John Woods**	301
Integrating MMASS with a Hybrid Commonsense Spatial Logic **Stefania Bandini, Alessandro Mosca, Matteo Palmonari and Giuseppe Vizzari**	311
On Abductive Equivalence **Katsumi Inoue and Chiaki Sakama**	333
A Transconsistent Lgoic for Model-Based Reasoning **Joseph E. Brenner**	353
An Inductionless and Default-Based Analysis of Machine Learning Procedures **Edoardo Datteri, Hykel Hosni and Guglielmo Tamburrini**	379
The Pragmatic Logic of Ordered Representations **Helmut Pape**	401
Agent-Based Abduction: Being Rational through Fallacies **Lorenzo Magnani and Elia Belli**	415

Preface

The volume is based on the papers that were presented at the International Conference *Model-Based Reasoning in Science and Engineering: Abduction, Visualization, Simulation* (MBR'04), held at the Collegio Ghislieri, University of Pavia, Pavia, Italy, in December 2004. The previous volume *Model-Based Reasoning in Scientific Discovery*, edited by L. Magnani, N.J. Nersessian, and P. Thagard (Kluwer Academic/Plenum Publishers, New York, 1999; Chinese edition, China Science and Technology Press, Beijing, 2000), was based on the papers presented at the first "model-based reasoning" international conference, held at the same place in December 1998. Other two volumes were based on the papers presented at the second "model-based reasoning" international conference, held at the same place in May 2001: *Model-Based Reasoning. Scientific Discovery, Technological Innovation, Values*, edited by L. Magnani and N.J. Nersessian (Kluwer Academic/Plenum Publishers, New York, 2002) and *Logical and Computational Aspects of Model-Based Reasoning*, edited by L. Magnani, N.J. Nersessian, and C. Pizzi (Kluwer Academic, Dordrecht, 2002).

The presentations given at the Conference explored how scientific thinking uses models and explanatory reasoning to produce creative changes in theories and concepts. Some addressed the problem of model-based reasoning in technology, and stressed the issue of technological innovation.

The study of diagnostic, visual, spatial, analogical, and temporal reasoning has demonstrated that there are many ways of performing intelligent and creative reasoning that cannot be described with the help only of traditional notions of reasoning such as classical logic. Understanding the contribution of modeling practices to discovery and conceptual change in science requires expanding scientific reasoning to include complex forms of creative reasoning that are not always successful and can lead to incorrect solutions. The study of these heuristic ways of reasoning is situated at the crossroads of philosophy, artificial intelligence, cognitive psychology, and logic; that is, at the heart of cognitive science.

There are several key ingredients common to the various forms of model-based reasoning. The term "model" comprises both internal and external representations. The models are intended as interpretations of target physical systems, processes, phenomena, or situations. The models are retrieved or constructed on the basis of potentially satisfying salient constraints of the target domain. Moreover, in the modeling process, various forms of abstraction are used. Evaluation and adaptation take place in light of structural, causal, and/or functional constraints. Model simulation can be used to produce new states and enable evaluation of behaviors and other factors.

The various contributions of the book are written by interdisciplinary researchers who are active in the area of creative reasoning in science and technology: the most recent results and achievements about the topics above are illustrated in detail in the papers.

Very interesting cognitive models and architectures have been built, that represent various model-based reasoning performances, for example, in scientific discovery and in knowledge representation. Several of the papers in this volume aim at increasing *epistemological* knowledge about the role of model-based reasoning in various tasks: the role of new models of spatiality through a new extremely interesting foundation of geometry and computation (M. Leyton); the role of models in the interplay between scientific theories and their domains (T.A. Kuipers), in normal science and scientific revolutions (P. Li and D. Li), and in paradigm shift (S. Mildeová); model-based reasoning in the development of our representations of phenomena in recent physics (A. Rivadulla and D. Portides); the problem of the Popperian context of discovery in creative mechanisms of reasoning (A. Aliseda).

Other papers address fundamental *cognitive* issues related to model-based reasoning: the exemplar-based reasoning as a form of model-based reasoning (R. Bod); the role of simulation in visual cognition (A. Carsetti); the problem of the creative and simulative aspects of thought experiments (J. Clement); model-based reasoning and its interplay with moral imagination in ethical reasoning (M. Gorman); the problem of models in the so-called "disembodiment of mind" and the cognitive importance of external representations and mediators (L. Magnani); the important ethical and epistemological problem of the use of "animal models" and disanalogies (C. Shelley); the so called cognitive fictions as forms non–actually–observable states of things inferred from actually observable ones, like simulations and artificial representations of past or future events (G. Tuzet); the powerful model-based cognitive activity of diagrams and visualizations in several cognitive situations studied by cognitive psychology (B. Tversky, M. Agrawala, J. Heiser, P. Lee, P. Hanrahan); and , finally, the description of the capacity of model-based reasoning to support cognitive complex systems (Z. Wang).

In recent years novel analyses of cognitive "logical" models of model-based reasoning and of the interplay abduction/model-based reasoning/ creative inferences have been undertaken. For example, the recent attention given to diagrammatic reasoning (in logic and also in the case of geometrical aspects) has promoted the creation of many new logical models also related to the representation of problem solving tasks (D. Batens) and to the analysis of diagrams as physical models able to assist various reasoning performances (B. Chandrasekaran); moreover, spatial logics can play an important role in computational modeling for example integrating the so called "multilayered multi–agent situated systems (MMASS) (S. Bandini, A. Mosca, M. Palmonari, G. Vizzari).

Finally, the remaining papers address the analysis of abduction and its role in increasing knowledge about model-based reasoning and creative inferences: the short description a new interesting logical model that interprets abduction as an ignorance–preserving inference also endowed with a non–explanationist character (D. Gabbay and J. Woods); "explainable" and "explanatory" equivalence in first order abduction and computational complexity (K. Inoue and C. Sakama); the relationship between the so-called "transconsistent" logics and model-based reasoning (J. Brenner); the relevance of the anti–inductivist views and the role of a so called "induction-free" logic in learning agents (E. Datteri, H. Hosni,G. Tamburrini); the role of formal order in the analysis of Peirce's pragmatism and his creation of the logic of abduction and the system of a model-based "visual logic" in terms of existential graphs (H. Pape); and finally, the rationality embedded in abduction as a form of fallacious reasoning (L. Magnani and E. Belli).

The editor expresses his appreciation to the members of the Scientific Committee for their suggestions and assistance: Atocha Aliseda, Universidad Nacional Autonoma de Mexico (UNAM), Mexico City, Mexico, Lawrence W. Barsalou, Emory University, Atlanta, GA, USA, Diderik Batens, Universiteit Gent, Ghent, Belgium, Walter Carnielli, CLEHC State University of Campinas, UNICAMP, Campinas, SP, Brazil, Balakrishnan Chandrasekaran, Ohio State University, Columbus, OH, USA, Kenneth D. Forbus, Northwestern University, Evanston, IL, USA, Dov Gabbay, King's College, London, UK, David Gooding, University of Bath, Bath, UK, Mary Hegarty, University of California, Santa Barbara, CA, USA, Theo A.F. Kuipers, University of Groningen, Groningen, The Netherland, Michael Leyton, Rutgers University, New Brunswick, NJ, USA, Li Ping, Sun Yat-sen (Zhongshan) University, Guangzhou, P.R. China, Lorenzo Magnani, University of Pavia, Pavia, Italy and The City University of New York, New York, USA, Claudio Pizzi, University of Siena, Siena, Italy, Qiming Yu, Central University for Nationalities, Bejing, P.R. China, Friedrich Steinle,

Max-Planck-Institut, Berlin, Germany, John Woods, University of British Columbia, Vancouver and King's College, London, UK, Andrea Woody, University of Washington Seattle, WA, USA.

Special thanks to the members of the Local Organizing Committee R. Dossena, E. Gandini, M. Piazza, E. Bardone, and C. Bocchiola, for their contribution in organizing the conference, to R. Dossena and E. Bardone for their contribution in the preparation of this volume, and to the copy-editor L. d'Arrigo. The conference MBR04, and thus indirectly this book, was made possible through the generous financial support of the MIUR (Italian Ministry of the University), University of Pavia, and Fondazione CARIPLO (Cassa di Risparmio delle Provincie Lombarde). Their support is gratefully acknowledged. The preparation of the volume would not have been possible without the contribution of resources and facilities of the Computational Philosophy Laboratory and of the Department of Philosophy, University of Pavia. Also special thanks to Dov M. Gabbay for having promoted the publication of this book in the "Philosophy Series" of King's College Publications, London.

Several papers concerning model-based reasoning deriving from the previous conferences MBR98 and MBR01 can be found in Special Issues of Journals: in *Philosophica*: Abduction and Scientific Discovery, 61(1), 1998, and Analogy and Mental Modeling in Scientific Discovery, 61(2) 1998; in *Foundations of Science*: Model-Based Reasoning in Science: Learning and Discovery, 5(2) 2000, all edited by L. Magnani, N.J. Nersessian, and P. Thagard; in *Foundations of Science*: Abductive Reasoning in Science, 9, 2004, and Model-Based Reasoning: Visual, Analogical, Simulative, 10, 2005; in *Mind and Society*: Scientific Discovery: Model-Based Reasoning, 5(3), 2002, and Commonsense and Scientific Reasoning, 4(2), 2001, all edited by L. Magnani and N.J. Nersessian.

Finally, other related philosophical, epistemological, and cognitive oriented papers deriving from the presentations given at the Conference MBR04 will be be published in four Special Issues of Journals: in *Foundations of Science*: Tracking Irrational Sets: Science, Technology, Ethics, and Model-Based Reasoning in Science and Engineering; and in *Mind and Society*: Scientific Discovery: Model-Based Reasoning, and Model-Based Reasoning, Science, and Information, all edited by L. Magnani.

Lorenzo Magnani, University of Pavia, Italy and Sun Yat-sen University, Canton, P. R. China

Pavia, Italy, February 2006

New Foundations for Geometry and Computation

MICHAEL LEYTON

ABSTRACT. My books have argued that the conventional foundations for geometry (Euclid, Klein, Einstein, etc.) are inappropriate to the computational age, because those foundations define a geometric object is an invariant under applied action, and therefore the object cannot store information about the action. Thus my books have developed new foundations for geometry in which a geometric object is the opposite: a *memory store* for applied action. This means that the mathematical system is completely the opposite from the conventional geometry. For example, in the new system, groups are about asymmetries rather than symmetries, symmetry-breaking is defined by extending groups rather than reducing them, objects are reference-frame bound rather than reference-frame free, etc. Because of the central role of memory storage in the theory, and the way it is analyzed, *the new foundations for geometry are equivalent to new foundations for computation*. This is due to the fact that the new geometric theory consists of two components: (1) Inference rules for the extraction of history from geometric objects; and (2) generative operations which create memory stores. The first of these corresponds to the *reading* operation in the new computational theory, and the second corresponds to the writing operation. These reading and writing operations are far more sophisticated than the reading and writing operations of conventional computation. In the new foundations, the reading and writing operations are systems of inference rules and generative operations based on deep relations between asymmetry and symmetry, elaborated in the theory and embodied in new classes of groups I call symmetry-breaking wreath products.

1 Introduction

In the conventional foundations for geometry (Euclid, Klein, Einstein, etc.) a geometric object is an invariant under applied action. Thus Euclid's geometry concerns the invariants under the Euclidean operations; Klein generalizes this to the invariants of any chosen group; and Einstein's relativity principle defines physics to be concerned with the invariants of transformations between observers' reference frames.

I have argued that there is a severe problem with the definition of a geometric object as an invariant under applied action: The object cannot store information about that action. Computation is founded on the use of memory stores, and in the computational age, memory storage is the premium. The geometry of Euclid, Klein, and Einstein, defeat this.

As a consequence, I embarked on a 30-year project to build up entirely new foundations for geometry – a system I recently completed and published as the book, *A Generative Theory of Shape* (Springer-Verlag, 550 pages). In my new foundations, a geometric object is the direct opposite of what it is in the conventional foundations: It is a *memory store* for applied action. This means that the mathematical system is completely the opposite from the conventional geometry. For example, in the new system, groups are about asymmetries rather than symmetries, symmetry-breaking is defined by extending groups rather than reducing them, objects are reference-frame bound rather than reference-frame free, etc.

Because of the central role of memory storage in the theory, and the way the theory handles it, *the new foundations for geometry are equivalent to new foundations for computation*. This is due to the fact that the new geometric theory consists of two components: (1) Inference rules for the extraction of history from geometric objects; and (2) generative operations which create memory stores. The first of these corresponds to the *reading* operation in the new computational theory, and the second corresponds to the writing operation. These reading and *writing* operations are far more sophisticated than the reading and writing operations of conventional computation. In the new foundations, the reading and writing operations are systems of inference rules and generative operations based on new and deep relations between asymmetry and symmetry, which are embodied in new classes of groups I call symmetry-breaking wreath products.

The purpose of this paper is to give an intuitive introduction to some of the basic issues involved.

2 Conventional geometry: Euclid to Einstein.

Radical as Einstein's theory of relativity might seem to be, it in fact goes back to the simple notion of *congruence* that is basic to Euclid. Thus to understand the foundations of modern physics (including quantum mechanics), we should first look at the notion of congruence.

Figure 1 shows two triangles. To test if they are congruent, you translate and rotate the upper one to try to make it coincident with the lower one. If exact coincidence is possible, you say that they are *congruent*. This allows you to regard the triangles as essentially the *same object*.

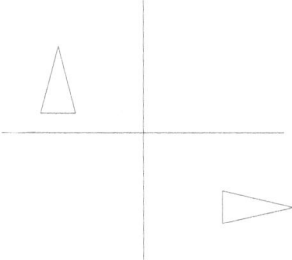

Figure 1. Memory storage or memory erasure?

In contrast, in the theory of geometry which I have developed, the two triangles are different because they must have different histories. For example, to convert the upper one into the lower one, it is necessary to add a history of translation and rotation.

Let us return to the Euclidean view. Simple as the notion of congruence is, it has been the major driving force of geometry for nearly 3,000 years, and was generalized in the late 19th century by Felix Klein in what is probably the most famous single lecture in the entire history of mathematics – his inaugural lecture at Erlangen. In this lecture, Klein defined a program, which became the basis not only of all geometry, but of all mathematics and physics:

> **KLEIN'S ERLANGEN PROGRAM**
>
> *A geometric object is an invariant under a chosen group of transformations.*

Let us illustrate by returning to the two triangles Figure 1. Consider the upper triangle: It has a number of properties:

(1) Three sides.

(2) Points upward.

(3) Two equal angles.

Now apply a movement to make it coincident with the lower triangle. Properties (1) and (3) remain invariant (unchanged); i.e., the lower triangle has three sides and has two equal angles. In contrast, property (2) is not invariant; i.e., the triangle no longer points upwards. Klein said that the *geometric* properties are those that remain invariant; i.e., properties (1) and (3).

Now a crucial part of my argument is this: Because properties (1) and (3) are unchanged (invariant) under the movement, *it is impossible to infer from them that the movement has taken place*. Only the non-invariant property, the direction of pointing, allows us to recover the movement. In the terminology of my books, I therefore say that *invariants* are *those properties that are memoryless*; i.e., they yield no information about the past. That is, I argue:

> **INVARIANTS CANNOT ACT AS MEMORY STORES**
>
> *Because Klein proposes that a geometric object consists of invariants, I say that Klein views geometry as the study of memorylessness.*

Klein's approach became the basis of 20th century mathematics and physics. As an example, let us turn to Special and General Relativity.

3 Special and General Relativity

The significance of Einstein's theory of relativity is that it is the first theory in physics that was founded on Klein's program for geometry. Since then, all the other branches of physics, such as quantum mechanics and quantum field theory, have been based on Klein's program. Let us now show the relationship between Einstein's work and Klein's. Einstein's theory of relativity is founded on the following principle:

> **EINSTEIN'S PRINCIPLE OF RELATIVITY**
>
> *The proper objects of physics are those that are invariant (unchanged) under changes of reference frame.*

To illustrate this principle, let us suppose that a collection of observers, each in frames of different velocities, measure the length of the same spatial interval. They will find that the length is different in their different frames. Therefore according to Einstein's Principle of Relativity, length is not a proper object of physics – because it is not an invariant under changes of reference frame. However, if one takes a *space-time* interval, rather than just a space interval, this turns out to have the same length in the different frames; i.e., it is an invariant under changes of reference frame. Thus, according to Einstein's Principle of Relativity, space-time intervals are proper objects of physics.

Now the changes of reference frame are given by transformations that form a group. Einstein's Principle of Relativity says that the proper objects of physics are those that are invariant under the group of transformations that change reference frames.

It is important to see that this is an example of Klein's theory of geometry. Recall from section 2 that Klein said that a geometric object is an invariant under some chosen group of transformations. In relativity, the chosen transformations are those that go between the reference frames of different observers. Since Klein says that a *geometric object* is an invariant under the chosen transformations, Einstein's Principle of Relativity says that a space-time interval is a *geometric object*; i.e., an invariant under the transformations between reference frames. Generally therefore, Einstein's Principle of Relativity says that the proper objects of physics are the *geometric objects* (invariants) of the transformations between reference frames. Thus Einstein is credited with what is called the *geometrization* of physics.

In Special Relativity, the chosen transformations (between reference frames) are called the Lorentz transformations. In General Relativity, the chosen transformations are more general – they are smooth deformations of local coordinate systems. The invariants are tensors. Therefore we say:

EINSTEIN'S GEOMETRIZATION OF PHYSICS

The proper objects of physics are the geometric objects (invariants) under changes of reference frame.

Special Relativity: The geometric objects are the invariants of the Lorentz transformations.

General Relativity: The geometric objects are the invariants of the local diffeomorphism group.

Now recall (section 2) that I argued that Klein's invariants program concerns memorylessness. This is because past action cannot be inferred from an invariant. Thus, I argue the following:

EINSTEIN'S PROGRAM vs. MEMORY

Since Einstein's Theory of Relativity is aimed at the extraction of invariants, and invariants cannot act as memory stores, Einstein's program concerns memorylessness.

The same situation exists in Quantum Mechanics. For example, the modern classification of particles was invented by Eugene Wigner, and is carried out by the extraction of invariants of measurement operators. We can therefore say that the two cornerstones of modern physics – Relativity and Quantum Mechanics – are founded on memorylessness.

It is important now to observe that Klein's invariants program really originates with Euclid's notion of congruence: Invariants are those properties that allow congruence. Therefore, the congruent properties are those that cannot store memory. In conclusion, I argue the following: The basis of modern physics can be traced back to Euclid's concern with congruence. We can therefore say that the entire history of geometry, from Euclid to modern physics, has been founded on the notion of memorylessness. Furthermore, I argue:

MEMORYLESSNESS

FROM EUCLID TO MODERN PHYSICS

The entire field of geometry, from Euclid to modern physics, including Einstein, has been concerned with the maximization of invariants. Therefore, it has been concerned with the maximization of memorylessness.

---

4 New foundations to geometry

The previous sections have described the conventional foundations of geometry. I have argued for many years that the conventional foundations are completely *stupid* in a rigorous sense: *They cannot embody intelligence.* The reason is that intelligence is founded on the use of memory storage. A human being cannot function without memory. As a consequence, the medical profession fights diseases such as Altzheimer's because these diseases attack memory, and memory not only allows intelligence, but is equated with the person's identity. Furthermore, people require memory not only from themselves but their computers. No one goes into a computer store and asks for a computer with the least amount of memory – their first question is for the computer with the maximal amount of memory. In addition, memory is not just a hardware issue, but a software one: Even the most basic programs, for example word-processing programs, allow the user to retrieve previous steps (via the undo operation) so that alternative steps can be taken. Indeed, in more sophisticated programs, such as mechanical CAD and solid modeling, entire histories with their inheritance structures, can viewed and edited. The point is that for intelligent systems, *memory is the premium*, and one fights for the maximization of memory storage not the minimization, as is done in the conventional foundations for geometry.

As a consequence, I embarked on a 30-year project to build up entirely new foundations for geometry – a system I recently completed and published as the book, *A Generative Theory of Shape* (Springer-Verlag, 550 pages). Rather than basing geometry on the *maximization of memorylessness* (the aim from Euclid to Einstein), I base geometry on the *maximization of memory storage*. The result is a system that is profoundly different, both on a conceptual level and on a detailed mathematical level. Let us consider some of the major differences. First, there is a fundamental difference in the definition of a *geometric object*:

STANDARD FOUNDATIONS FOR GEOMETRY

(Euclid, Klein, Einstein)

A *geometric object is an* invariant; *i.e.,* memoryless.

NEW FOUNDATIONS FOR GEOMETRY

[Leyton, 1992] and [Leyton, 2001]

A *geometric object is a* memory store.

The result of the two different definitions of geometric object is this: The conventional system of geometry is based on statements like this: "Any pair of distinct points lie on exactly one line". This is a descriptive statement. In contrast, my system is founded on *inference rules* for extracting history from objects. These rules are literally rules of deduction, rather like a forensic expert would use. An observer is presented with the current state of some object, e.g., the current shape of a tumor, and the inference rules deduce the past history of the object. Notice that, because *memory is information* about the past, the rules are extracting the memory (information about the past) from the object. In fact, I speak of the rules as *converting the object into a memory store*.

Furthermore, I argue that the part of the object which yields the past information is the shape. In relation to this, one of the fundamental statements of the theory is that *shape is equivalent to memory storage*. Indeed, the statement is more strong than this: *Shape equals memory storage*; i.e., what one means by shape is memory storage and what one means by memory storage is shape. Most importantly, in the new foundations:

Geometry = Theory of Memory Storage.

---

5 The world as memory storage

Let us begin by defining memory in the simplest possible way:

> Memory = Information about the past.

Consequently, we will define a memory store in the following way:

> Memory store = Any object that yields information about the past.

Under the new foundations of geometry, the entire world is viewed as memory storage. Let us consider some examples. A *scar* on a person's face is, in fact, a memory store. It gives us information about the past: It tells us that, in the past, the surface of the skin was cut. A *dent* in a car door is also a memory store; i.e., information about the past can be extracted from this. The dent tells us that, in the past, there was an impact on the car. Any *growth* is a memory store. For example, the shape of a person's face gives us information of the past history of growth that occurred, e.g., the nose and cheekbones grew outward, the wrinkles folded in, etc. Similarly, the shape of a tree gives us information about how it grew. Therefore, from both a face and a tree, we can retrieve information about the past. A *scratch* on a piece of furniture is a memory store, because we can extract from it information that, in the past, the surface had contact with a sharp moving object. A *crack* in a vase is a memory store because it informs us that, in the past, the vase underwent some impact; i.e., this information is retrievable from the crack.

6 The reading operation: Part I

According to the new foundations, memory storage can take many forms – e.g., scars, dents, growths, scratches, etc. In fact, there are probably infinitely many forms that it can take. However I argue that, on a deep level, all memory stores have only one form:

> **FIRST FUNDAMENTAL LAW OF MEMORY STORAGE**
>
> [Leyton, 1992] and [Leyton, 2001]
>
> *Memory is stored only in asymmetries.*

> **SECOND FUNDAMENTAL LAW OF MEMORY STORAGE**
>
> [Leyton, 1992] and [Leyton, 2001]
>
> *Memory is erased by symmetries.*

That is, information about the past can be recovered only from asymmetries. And correspondingly, information about the past is erased by symmetries.

Let us begin with a simple example from [Leyton, 1992]: Consider the sheet of paper shown on the left in Figure 2. Even if one had never seen that sheet before, one would conclude that it had undergone twisting. The reason is that the asymmetry in the sheet yields information about the past. In other words, from the asymmetry, one can *recover* the past history. That is, the *asymmetry acts as a memory store for the past action* – as stated in the First Fundamental Law of Memory Storage (above).

Figure 2. Example from [Leyton, 1992]. A twisted sheet is a source of information about the past. A non-twisted sheet is not.

Now let us un-twist the paper, thus obtaining the straight sheet shown on the right in Figure 2. Suppose we show this straight sheet to any person on the street. Would they be able to infer from it the fact that it had once been twisted? The answer is "No". The reason is that the symmetry of the straight sheet has wiped out the ability to recover the preceding history. This means that the *symmetry erases the memory store* – as stated in the Second Fundamental Law of Memory Storage (above).

This means that symmetry is the absence of information about the past. In fact, from the symmetry, one concludes that the straight sheet had always been like this. For example, when you take a sheet of paper from a box of paper you have just bought, you do not assume that it had once been twisted or crumpled. Its very straightness (symmetry) leads you to conclude that it had always been straight.

The two diagrams in Figure 2 illustrate the two fundamental laws of memory storage given above. These two laws are the very basis of my new foundations for geometry. It will be useful to have these laws in the following form:

THE REALIZATIONS OF THE FIRST AND SECOND FUNDAMENTAL LAWS

[Leyton, 1992] and [Leyton, 2001]

ASYMMETRY PRINCIPLE. *Given a data set D, a program/process for generating D is recoverable from D only if each asymmetry in D goes back to a symmetry in one of the previously generated states.*

SYMMETRY PRINCIPLE. *Given a data set D, a program/process for generating D is recoverable from D only if each symmetry in D is preserved backwards through the generated sequence.*

My books show that the *reading of a memory store* is the application of the two above laws in the following way:

READING OPERATION (PART I)

[Leyton, 1992] and [Leyton, 2001]

(1) *Partition the data set D into its asymmetries and symmetries.*

(2) *Apply the Asymmetry Principle to the asymmetries.*

(3) *Apply the Symmetry Principle to the symmetries.*

According to the new foundations, the act of reading a data set as a memory store *converts* it into a memory store; i.e., it is not a memory store until the inference rules, the Asymmetry Principle and Symmetry Principle, are applied to it.

An extended example will now be considered that will illustrate the power of the reading operation defined above. In a set of psychological experiments that I carried out in the psychology department in Berkeley in 1982, I found that, when subjects are presented with a rotated parallelogram, as shown in Figure 3a, they refer it in their heads to a non-rotated parallelogram, Figure 3b, which they then refer in their heads to a rectangle, Figure 3c, which they then refer in their heads to a square, Figure 3d. It is important to understand that the subjects are presented with only the first shape. That is, the first shape is the data set D. The remaining sequence of shapes is a psychological response to the presented shape.

Close examination reveals that what the subjects are doing is *recovering* the process-history that created the rotated parallelogram (the data set

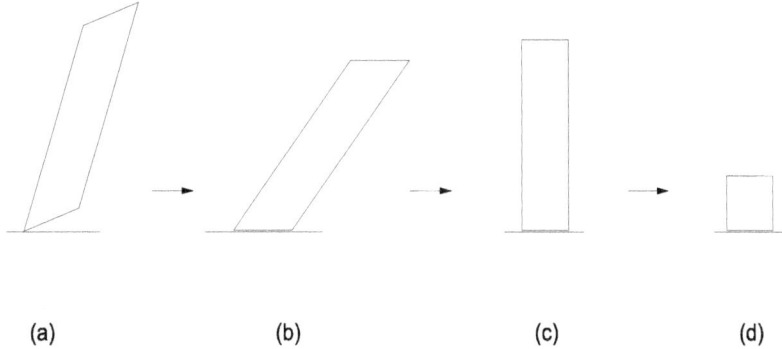

Figure 3. The external history inferred from a rotated parallelogram.

D). That is, they are saying that, prior to being a rotated parallelogram Figure 3a, it was a non-rotated one Figure 3b; and prior to this, it was a rectangle Figure 3c; and prior to this, it was a square Figure 3d. Therefore, it is crucial to note: The sequence, from the rotated parallelogram to the square, represents the *backward* time direction;

In recovering the process-history that produced the rotated parallelogram, they are *converting it into a memory store*; i.e., they are retrieving the history stored in the rotated parallelogram. That is, they are *reading it as a memory store*.

We shall now see that what they are going is using the reading operation defined above. In this reading operation, one first partitions the data set, the rotated parallelogram, into its asymmetries and symmetries. Let us begin by identifying the asymmetries. It is necessary to note that *asymmetries mean distinguishabilities*. In the rotated parallelogram, there are three distinguishabilities:

(D1) The distinguishability between the orientation of the shape and the orientation of the environment – indicated by the difference between the bottom edge of the shape and the horizontal line which it touches.

(D2) The distinguishability between adjacent angles in the shape: they are different sizes.

(D3) The distinguishability between adjacent sides in the shape: they are different lengths.

It is clear that what happens in the sequence, from the rotated parallelogram to the square, is that these three distinguishabilities are removed successively backwards in time. The removal of the first distinguishability, that between the orientation of the shape and the orientation of the environment, results in the transition from the rotated parallelogram to the non-rotated one. The removal of the second distinguishability, that between adjacent angles, results in the transition from the non-rotated parallelogram to the rectangle, where the angles are equalized. The removal of the third distinguishability, that between adjacent sides, results in the transition from the rectangle to the square, where the sides are equalized.

Therefore, each successive step in the sequence is a use of the Asymmetry Principle, which says that an asymmetry must be returned to a symmetry backwards in time.

Having identified the asymmetries in the rotated parallelogram and applied the Asymmetry Principle to each of these, we now identify the symmetries in the rotated parallelogram and apply the Symmetry Principle to each of these. First we need to note that *symmetries mean indistinguishabilities*. In the rotated parallelogram, there are two indistinguishabilities:

(1) The opposite angles are indistinguishable in size.

(2) The opposite sides are indistinguishable in size.

The Symmetry Principle requires that these two symmetries in the rotated parallelogram must be preserved backwards in time. And indeed, this turns out to be the case. That is, the first symmetry, the equality between opposite angles, in the rotated parallelogram, is preserved backwards through the sequence: i.e., each subsequent shape, from left to right, has the property that opposite angles are equal. Similarly, the other symmetry, the equality between opposite sides in the rotated parallelogram, is preserved backwards through the sequence: i.e., each subsequent shape, from left to right, has the property that opposite sides are equal.

In conclusion: We have seen that the subjects are reading the presented figure, Figure 3a, as a memory store, and they are doing so via the reading operation defined above.

I call the successive reading operations in Figure 3 *external* because they produce successively backwards data sets (left to right) that are not subsets of the data sets from which they are inferred. However, I have shown that this sequence of three external reading operations is followed by a sequence of two *internal* reading operations, i.e., operations that produce additional backward data sets that are subsets of the data sets from which they are inferred. These are as follows: When one reaches the square, there are two distinguishabilties in the square itself:

(D4) The distinguishability between the orientations of the sides within the square.

(D5) The distinguishability between the positions of the points within a side.

The removal of these two distinguishabililites recovers the process that traced the square (e.g., by drawing). That is, the removal of D4 recovers the starting side, from which the other sides were obtained by 90-degree rotations. Then, from this starting side, the removal of D5 produces the starting point of the side, from which the other points on that side were obtained by translations. That is, the successive removal of D4 and D5 gives the sequence:

$$\text{Square} \rightarrow \text{Starting Side} \rightarrow \text{Starting Point}.$$

These two transitions are added to the end of the three transitions given in Fig 3, resulting in a total of five successive transitions obtained by the use of the reading operation. The full details, including the ordering rules on transitions, are given in [Leyton, 1992] and [Leyton, 2001].

7 Curvature as memory storage

My books argue that the reading operation defined in section 6 creates far more powerful memory stores than are currently used in computers. As an example, we shall look at the reading operation applied to curvature extrema. Consider any closed smooth curve in the 2D plane. This will be the data set D. In particular, we will convert the curvature extrema into powerful memory stores. The partitioning phase of the reading operation will be given by a theorem that I proved, the Symmetry-Curvature Duality Theorem [Leyton, 1987b], which says that, to each curvature extremum, there is a unique symmetry axis leading to and terminating at the extremum. The curvature extremum is the asymmetry (it violates the rotational symmetry of the shape) and the symmetry axis is the symmetry. I show that, applying the Asymmetry Principle and Symmetry Principle, one recovers a past process that pushed the boundary along the symmetry axis leading to the extremum. Figure 4 shows the recovered process-histories on all shapes with up to eight curvature extrema. The process-histories inferred from the extrema are shown by the arrows. That is, the diagrams show the memory storage extracted from the curvature extrema.

Furthermore, the reading operation extracts even more memory storage from these shapes. Not only does it infer the arrows *within* each shape, but it infers the evolution laws that prescribe the transitions *between* the

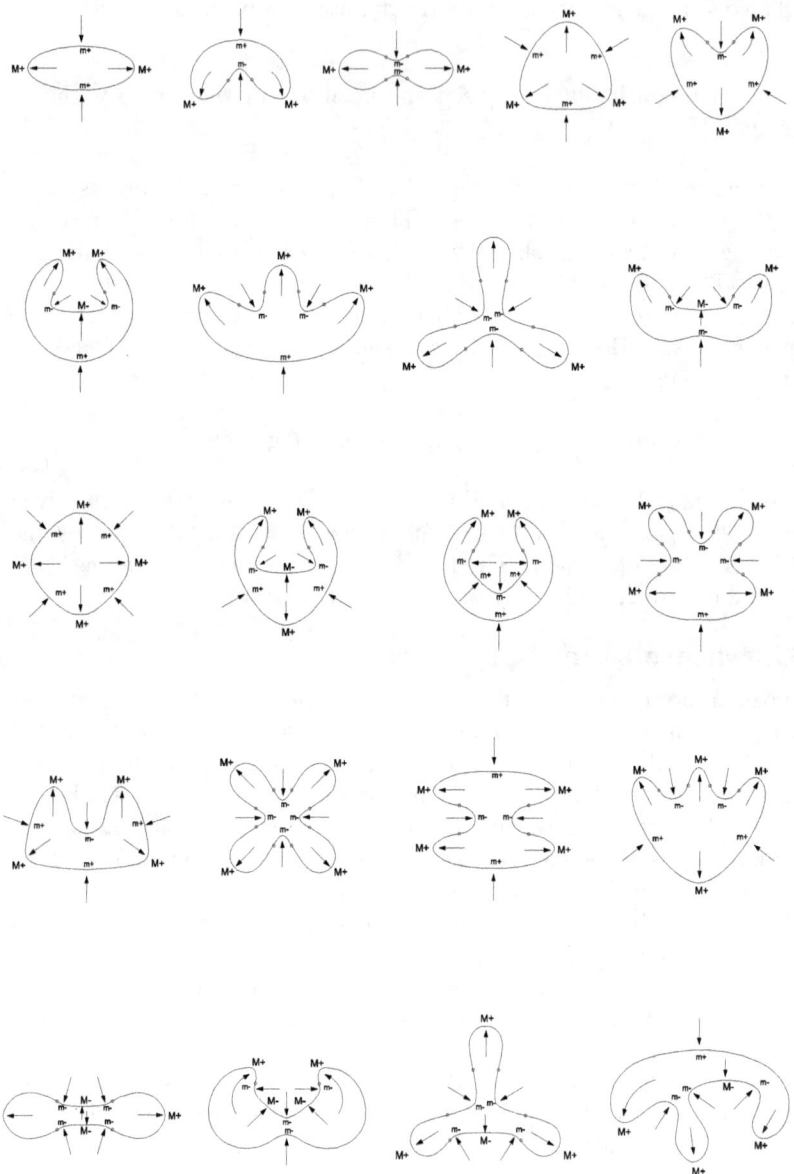

Figure 4. The memory storage extracted from the curvature extrema in all shapes with up to 8 extrema.

shapes. I call these laws, the Process-Grammar, and elaborated them in [Leyton, 1988]. After I published the grammar, it was applied by scientists in many disciplines. For example Steve Shemlon applied it to MRI human brain scans, dental radiographs, transmission electron microscopy (TEM), serial sections of hemocytes (insect blood cells), and neuronal growth models; Evangelos Milios to meteorology [Milios, 1989]; J.P. Lee to chemical engineering [Lee, 1991]; Jean-Philippe Pernot et al to develop nurbs deformation operators for computer-aided design [Pernot *et al.*, 2003].

8 The two fundamental principles

Recall that my new foundations for geometry are based on the maximization of memory storage, whereas the conventional foundations, due to Euclid, Klein and Einstein, are based on the minimization of memory storage. In the new foundations, [Leyton, 2001], I argue that maximization of memory storage requires the fulfillment of two fundamental principles. The first is this:

MAXIMIZATION OF RECOVERABILITY [Leyton, 2001]. *Maximize the recoverability of past states.*

The means by which recoverability is made possible has been shown in the previous sections.

However, my books show that the maximization of memory storage is based on one another principle, *maximization of transfer*, which means this: Situation B is not seen as new, but as the transfer of a previous situation A This represents B as a memory store for A. In fact, in the new foundations, transfer is not described in terms of "situations" but in terms of actions – a set of actions is transferred from one part of the generative history to another part of the generative history.

MAXIMIZATION OF TRANSFER [Leyton, 2001]. *Make one part of the generative history a transfer of another part of the generative history, whenever possible.*

According to this, transfer involves two levels of actions: the set of actions being transferred, and the set of actions doing the transfer. I call the actions to be transferred, the *fiber group*, and the actions doing the transfer, the *control group*. That is, the control group transfers the fiber group. The new foundations contain an extensive mathematical theory of transfer, showing that the transfer relation is best modeled by a group-theoretic product called a wreath product $\circledw$. That is, a structure of transfer is given by a group that has the following structure:

Fiber Group $\circledw$ Control Group.

This group models the transfer component of my theory of memory storage.

Now, in [Leyton, 2001], I show that when one puts together the two principles, the Maximization of Recoverability and the Maximization of Transfer, one obtains (by the Asymmetry Principle) the condition that the control group is symmetry-breaking on the fiber group. This gives a far more powerful theory of symmetry-breaking than is currently used in mathematics, physics, and chemistry. In the conventional theory, symmetry-breaking corresponds to reduction of symmetry group.

CONVENTIONAL VIEW OF SYMMETRY-BREAKING.
Symmetry-breaking is a reduction of symmetry group.

I have argued that this is inherently weak because this causes a reduction in the algebraic structure describing the situation. In my theory, symmetry-breaking causes an increase in the group. It does so in the following way:

NEW VIEW OF SYMMETRY-BREAKING. *The breaking of a symmetry group G_1 is given by its extension by another group G_2 via a wreath product thus: $G_1 \, \textcircled{w} \, G_2$, where G_2 is the symmetry group of the asymmetrizing action.*

$$G_1 \to G_1 \, \textcircled{w} \, G_2.$$

The group $G_1 \, \textcircled{w} \, G_2$ is a new class of group, which I invented in the new foundations, and which I call *symmetry-breaking wreath products*. Notice that the increase in group, on symmetry breaking, means that it is better not to talk of symmetry groups, but *asymmetry groups*. It is part of the way in which my new foundations for geometry invert everything in the conventional foundations of geometry.

An important feature of the above theory is this: The wreath product symbol $\textcircled{w}$ in the above statement implies that the *past symmetry G_1 is transferred onto the present broken symmetry.*

To illustrate further, let us return to the rotated-parallelogram in section 6. According to the above theory, the full generative sequence that produced the rotated parallelogram is given thus:

$$\{e\} \, \textcircled{w} \, R \, \textcircled{w} \, Z_4 \, \textcircled{w} \, A \, \textcircled{w} \, N \, \textcircled{w} \, SO(2).$$

where $\{e\}$ denotes the group of the point; R denotes the group of translations along a line (i.e., corresponding to a side); Z_4 denotes the rotation group around the square; A denotes the group of stretches; N denotes the group of shears; and $SO(2)$ denotes the group of continuous rotations in the 2D plane.

Two crucial points are these: (1) The five successive control levels correspond to the generation of the five successive distinguishabilities identified in section 5. (2) Each level *transfers* the previous level, as given by the wreath product operation; and does so by breaking the symmetry of the previous level.

THEORY OF MEMORY STORAGE

Any memory store is structured as a symmetry-breaking wreath product.

I have shown in [Leyton, 2001] that the ordering of group levels in the symmetry-breaking wreath product is determined by the following rule: Lower control groups are less symmetry-breaking on their fibers than higher control groups. This is determined by a system of rules involving eigenspace dimension, etc. I call the resulting ordering the *algebraic stability ordering*, and have discussed it in considerable detail in [Leyton, 1986a] [Leyton, 1986b], [Leyton, 1986c], [Leyton, 1987b], [Leyton, 1992], and [Leyton, 2001]. Using the above, it is now possible to give the fundamental statement of my new foundations for computation:

NEW FOUNDATIONS FOR COMPUTATION: THE READING OPERATION

The reading operation in a computational process converts a data set into a symmetry-breaking wreath product.

The operation extracts the maximal memory from the data set if the symmetry-breaking wreath product is ordered by algebraic stability.

My books show that the new foundations to geometry are equivalent to new foundations to computation. The reason is as follows. As I said, in my new foundations for geometry, shape is equivalent to memory storage. Furthermore, the new foundations consist of two inter-related components: (1) Inference rules for the extraction of history from shape; and (2) Generative operations which create the shape. In the new foundations, the first of these are defined as the *reading* operations in a computational process, and the second of these are defined as the *writing* operations.

As we can see, these reading and writing operations are far more sophisticated than the reading and writing operations of conventional computers, where reading the state of a memory store means merely registering the state, and writing the state means merely switching it (by interchanging 0's and 1's). In contrast, in my new foundations, the reading and writing operations are, respectively, the extraction and creation of stored history, and this is based on new and very deep relations between asymmetry and symmetry invented in the geometric theory; i.e., symmetry-breaking wreath products and the conceptual and mathematical theory that is developed from them in my books.

BIBLIOGRAPHY

[Lee, 1991] J.P. Lee. Scientific visualization with glyphs and shape grammars. New York, 1991. Msc Thesis, School for Visual Arts.

[Leyton, 1986a] M. Leyton. Principles of information structure common to six levels of the human cognitive system. *Information Sciences*, 38:1–120, 1986. Entire journal issue.

[Leyton, 1986b] M. Leyton. A theory of information structure I: General principles. *Journal of Mathematical Psychology*, 30:103–160, 1986.

[Leyton, 1986c] M. Leyton. A theory of information structure II: A theory of perceptual organization. *Journal of Mathematical Psychology*, 30:257–305, 1986.

[Leyton, 1987] M. Leyton. Symmetry-curvature duality. *Computer Vision, Graphics, and Image Processing*, 38:327–341, 1987.

[Leyton, 1987b] M. Leyton. A limitation theorem for the differential prototypification of shape. *Journal of Mathematical Psychology*, 31:307–320, 1987b.

[Leyton, 1988] M. Leyton. A process-grammar for shape. *Artificial Intelligence*, 34:213–247, 1988.

[Leyton, 1989] M. Leyton. Inferring causal-history from shape. *Cognitive Science*, 13:357–387, 1989.

[Leyton, 1992] M. Leyton. *Symmetry, Causality, Mind*. The MIT Press, Cambridge (Mass.), 1992.

[Leyton, 1999] M. Leyton. New foundations for perception. In *Invitation to Cognitive Science*, pages 121–171. Blackwell, Oxford, 1999.

[Leyton, 2001] M. Leyton. *A Generative Theory of Shape*. Springer-Verlag., Berlin, 2001.

[Leyton, 2004] M. Leyton. Musical works are maximal memory stores. In Mazzola G. and T. Noll, editors, *Mathematical and Computer-Aided Music Theory*. Osnabruck Music Publishi, Osnabruck, 2004.

[Leyton, 2005] M. Leyton. Shape as memory storage. In Y. Cai, editor, *Ambient Intelligence for Scientific Discovery*, pages 81–103. Springer-Verlag, Berlin, 2005.

[Milios, 1989] E.E. Milios. Shape matching using curvature processes. *Computer Vision, Graphics, and Image Processing*, 47:203–226, 1989.

[Pernot et al., 2003] J-P. Pernot, S. Guillet, J-C. Leon, B. Falcidieno, and F. Giannini. Interactive operators for free form features manipulation. In *SIAM conference on CADG*, Seattle, 2003.

[Shemlon, 1994] S. Shemlon. *The Elastic String Model of Non-Rigid Evolving Contours and its Applications in Computer Vision*. PhD thesis, Rutgers University, 1994.

Michael Leyton
DIMACS Center for Discrete Mathematics
and Theoretical Computer Science
Rutgers University
Email: mleyton@dimacs.rutgers.edu

Theories Looking for Domains Fact or Fiction?
Reversing Structuralist Truth Approximation

THEO KUIPERS

ABSTRACT. The structuralist theory of truth approximation essentially deals with truth approximation by theory revision for a fixed domain. However, variable domains can also be taken into account, where the main changes concern domain extensions and restrictions. In this paper I will present a coherent set of definitions of "more truthlikeness", "empirical progress" and "truth approximation" due to a revision of the domain of intended applications. This set of definitions seems to be the natural counterpart of the basic definitions of similar notions as far as theory revision is concerned. The formal aspects of theory revision strongly suggest an analogy between truth approximation and design research, for example, drug research. Whereas a new drug may be better for a certain disease than an old one, a certain drug may be better for another disease than for the original target disease, a phenomenon which was nicely captured by the title of a study by Rein Vos [1991]: *Drugs Looking for Diseases*. Similarly, truth approximation may not only take the shape of theory revision but also of domain revision, naturally suggesting the phenomenon of "Theories looking for domains". However, whereas Vos documented his title with a number of examples, so far, apart from plausible cases of "truth accumulation by domain extension", I did not find clear-cut empirical instantiations of the analogy, only, as such, very interesting, non-empirical examples.

1 Introduction

This paper starts by recapitulating the structuralist theory of truth approximation, as developed in [Kuipers, 2000], and then elaborates a point that was already made explicit by Sjoerd Zwart [1998]. In [Kuipers, 2000, p. 207] I wrote, in concluding the second chapter on truth approximation (chapter eight):

> Finally, variable domains can also be taken into account, where the main changes concern extensions and restrictions. We will

not study this issue, but see [Zwart, 1998, chapters two to four], for some illuminating elaborations in this connection, among other things, the way in which strengthening/weakening of a theory and extending/reducing its domain interact.

More specifically, in this paper I will present a coherent set of definitions of "more truthlikeness", "empirical progress" and "truth approximation" due to a revision of the domain of intended applications. This set of definitions seems to be the natural counterpart of the basic definitions of similar notions in [Kuipers, 2000] as far as theory revision is concerned.

The formal aspects of theory revision strongly suggest an analogy between truth approximation and design research, for example drug research. Whereas a new drug may be better for a certain disease than an old one, a certain drug may be better for another disease than the original target disease, a phenomenon which was nicely captured by the title of a study by Rein Vos [1991]: *Drugs Looking for Diseases*. Similarly, truth approximation may not only take the shape of theory revision but also of domain revision, naturally suggesting the phenomenon of "Theories looking for domains". However, whereas Vos documented his title with a number of examples, so far, apart from plausible cases of "truth accumulation by domain extension", I did not find clear-cut empirical instantiations of the analogy, only, as such, very interesting, non-empirical examples.

Section two briefly states the structuralist theory of truth approximation by theory revision for a fixed domain and concludes with the formal analogy with design research. Suggested by this analogy, the corresponding theory of "truth approximation by domain revision for a fixed theory" is presented in section three, which ends with a synthesis of truth approximation by theory and/or domain revision. As a kind of meta- and self-application of the "Theories looking for domains" theme, section four is devoted to the search for historical cases exemplifying this theory. I conclude with arguing that it is plausible to expect that truth approximation by domain revision in the empirical sciences frequently occur, surprisingly enough, in particular as far as the instrumentalist method is practiced.

2 Truth approximation by theory revision

My favorite, qualitative theory of truth approximation is best represented within the structuralist theory of theory representation. Surprisingly enough, and if I am not mistaken, the latter theory of representation has not been presented in earlier MBR-conferences as a pre-eminent case of model-based reasoning in the sense that not the sentences but the models of a theory form the primary domain of discourse[1].

Starting from a fixed vocabulary, and a suitable similarity type of structures, usually provided by a research program, let Mp indicate the set of structures of that type, also called the *potential models* of the theory. Let the subset M of Mp indicate the set of *models* of the theory. Finally, assuming that our target is a fixed domain of physically or, more broadly, nomically possible phenomena, let I indicate this domain "as seen through Mp", hence a subset of Mp, and be called the set of (intended) nomic possibilities or the *domain of intended applications*. Note that we do not yet suppose to dispose of a general characterization of I as a subset of Mp. What we only know is that each intended application can be represented as a potential model. A general characterization of I is "the great unknown" where theories, as represented by their models, are looking for. More formally, a theory is a triple of the form <Mp,M,I>, together with the *weak claim* that I is a subset of M (I ⊆ M) or the *strong claim* that I = M. A theory is said to be true (false) in the weak sense when the weak claim is true (false). It is easy to check that (a general characterization of) I represents the strongest true weak claim, and may be called "the truth" in this context, that is, the truth about the given domain in the given vocabulary.

Now it is plausible to define what it means, for fixed <Mp,I>, and hence theories of the same research program, that one theory M2 is closer (or more similar) to the truth than another M1. Intuitively, when M2 is moving from M1 in the direction of I. Formally, when M2-I ⊆ M1-I (∅2-area empty in Figure 1) and I-M2 ⊆ I-M1 (∅1-area empty) and at least once it should be a proper subset relation (at least one *-area non-empty). In terms of symmetric differences (A∆B=$_{df}$ (A-B)∪(B-A)), M2∆I should be a proper subset of M1∆I. In [Kuipers, 2000] I have argued that, among other formulations, the two clauses of this definition amount to: (relative to I) M2 has more true consequences than M1 and M2 has more correct models than M1, respectively.

Knowing what it means that one theory is closer to the truth than another is one thing, in view of the fact that we don't know the truth, judging that this is in fact the case, in the light of our evidence is another. Although there is no theorem guaranteeing that more-successfulness entails

---

[1]With the kind help of Lorenzo Magnani, I only found a quasi-relevant reference by Ronald Giere [1999] in his MBR98-contribution, entitled "Using models to represent reality". Giere pays some attention to the pioneering work of Patrick Suppes in this respect. However, Suppes' equally pioneering work with respect to "Using models to represent *theories*" is not touched upon. This work of Suppes was the starting point of Joseph Sneed and later followed by Stegmüller, Balzer, Moulines, and others, and called the structuralist representation of theories, i.e. theories as sets of structures, forming their models.

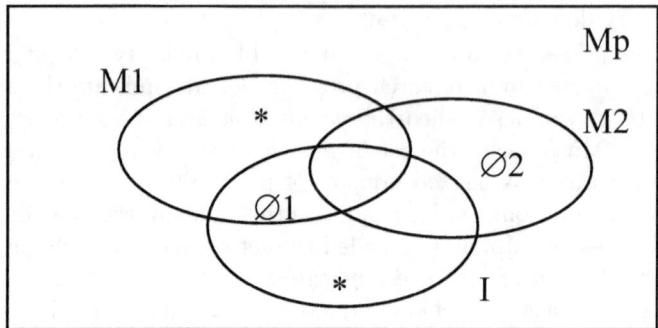

Figure 1. Theory M2 is closer to the truth (about domain I) than theory M1.

closer-to-the-truth, there is a theorem, the *Success Theorem*, guaranteeing almost the reverse entailment, viz., closer-to-the-truth entails at-least-as-successfulness.

Let R represent the set of (documented) realized applications at a certain time, that is, the experimentally or otherwise realized nomic possibilities at that time. Let S indicate the strongest law induced on the basis of R. Of course, R will partly be the result of testing hypothetical laws. If no mistakes have been made in representing the realized possibilities, R is not only a subset of Mp, but even of I, for nomic *im*possibilities, by definition, can't be realized. Moreover, in the structuralist representation S is also a subset of Mp, such that it has been concluded, at least provisionally, that conceptual possibilities outside S, that is, in Mp-S, are nomically impossible. Finally, if our inductive jump from R to S is correct, S has to be a superset of I. To sum up: $R \subseteq I \subseteq S \subseteq Mp$.

Let us call R/S the data set. We define what it means that M2 is, relative to the data set R/S, more successful than M1 as follows: $M1 \cap R \subseteq M2 \cap R$ and $S \cup M2 \subseteq S \cup M1$. The first condition amounts to "no loss of established examples" and the second to "no loss of explained established laws".

Now the *Success Theorem* states, assuming that R/S is correct, that is, $R \subseteq I \subseteq S$, then M2 is at least as successful relative to R/S as M1 if M2 is closer to the truth. This theorem is such that *persistently* being-more-successful is functional or instrumental for truth approximation, though not guaranteeing it, in the sense that it is very difficult/special for M1 to be at least as successful as, let alone more successful than, M2 when M2 is not closer to the truth than M1. More in detail we may argue as follows.

Let M2 be more successful than M1 relative to R/S. This suggests the *Comparative Success Hypothesis*, according to which it is hypothesized that this will remain the case, whatever experiments we design and perform. Testing of this hypothesis may result in the conclusion, at least for the time being, that this is in fact the case: M2 persistently remains more successful than M1. This is the paradigm situation for speaking of *empirical progress*, and for applying the "instrumentalist rule of success", viz. replace, for the time being, M1 in favor of M2. Under this condition it is very hard to imagine that M2 is, despite appearances, not closer to the truth than M1. To be sure, if this would nevertheless be the case, we have not been creative enough in designing new experiments. In other words, the situation even allows the tentative conclusion that M2 is in fact closer to the truth.

So far I have presented the naïve or basic structuralist theory of truth approximation. It is basic in the sense that several idealizations have been made, requiring concretization. In Kuipers [Forthcoming] I have extensively illustrated the (philosophical) method of concept explication by idealization and concretization by the example of truth approximation. Let us review the main concretizations as they have been elaborated or indicated in [Kuipers, 2000].

Above we did not make a distinction between (relatively) observational and theoretical terms. In chapter nine I have spelled out how this distinction works out, with the main conclusion that empirical progress, of course, in observational terms, remains functional for truth approximation on the theoretical level, though with some greater risk of being wrong in this.

Another idealization was that we implicitly assumed that any established counterexample of a theory, that is, a realized possibility not being a model of the theory, is as bad for one theory as for any other. However, in chapter ten I have introduced the idea that one structure may be more similar to another than a third. In this way, the possibility arises that a counterexample is less dramatic for one theory than for another, because the former has a model that is more similar to the "countermodel" (i.e., the potential model representing the counterexample), than any model of the other theory. Adapted definitions of more-successfulness and closer-to-the-truth lead again to the conclusion that empirical progress is functional for truth approximation. As a matter of fact we only know of real life scientific examples of (potential) truth approximation, e.g. the Law of Van der Waals as successor of the ideal gas law, when this concretization is introduced. Without this, we only know of toy examples.

Finally, in [Kuipers, 2000, chapter thirteen] I have indicated that it is possible to introduce a domain vocabulary, as a subvocabulary of the (observational) vocabulary, such that it is possible to define the domain explicitly in that vocabulary, leaving the further behavior of the applications as the unknown to be specified. The advantage of this is that domains become comparable sets, that is, it enables us to say that an intended application of one theory is or is not an intended application of another. This will be an important, though implicit, assumption in dealing with the main topic of this paper. But first I will introduce the analogy with design research in order to make that topic a plausible subject of theoretical and (meta-) empirical research.

Analogy with design research

The formal pattern of (the development of) design research, as presented in [Kuipers, 2001, chapter 10], is to some extent analogous to that of (descriptive or explanatory) "nomological" research. Starting from a set of relevant characteristics, let us call the subset of characteristics of the intended product (or process) its *intended profile*. As soon as we have a prototype we can delineate the subset of its actual characteristics, the *prototype profile*. A second prototype amounts to genuine progress when the symmetric difference between its profile and the intended profile is a proper subset of that between the profile of the first and the intended profile. In this way, for example, improving a (medical) drug for a given disease becomes formally analogous with improving a theory for a specific domain. See [Forbus and Gentner, 1997] for another illustration of progress in design research, viz., designing "qualitative simulators" to be evaluated in terms of useful and problematic properties derived from qualitative reasoning.

Rein Vos [1991], who elaborated the formal description of progress in design research for the case of drugs, entitled his book *Drugs Looking for Diseases*, because by historical research he came to the conclusion that pharmaceutical research often goes in the other direction. That is, there may be an interesting chemical substance that does not do what it is supposed to do for a given disease, even after some manipulation. However, instead of discarding the substance for this reason, one starts to look for other diseases, whether or not related to the first one, for which the substance may have curing effects. In this way, for example, bèta blockers have been discovered as a drug for certain heart diseases.

Hence the plausible question arises whether in the context of nomological research something similar, viz., "theories looking for domains" or, equivalently, "models looking for applications", does also exist. One type of example is well known, viz., the case that a theory which is or is not successful for a certain domain turns out to be (also) successful for a totally unrelated domain. In this case we have a purely formal analogy. A famous example is Keynes' transfer of hydraulic modeling from fluid mechanics to macroeconomics. Our interest in this paper concerns however cases in which the new and old domains are to some extent related, that is, genuine domain revision.

3 Truth approximation by domain revision

It is plausible to define what it means that, for fixed <Mp,M>, (the truth about) domain I2 is closer (or more similar) to M than (the truth about) domain I1. Intuitively, when I2 is moving from I1 in the direction of M. Formally, when I2-M $\subseteq$ I1-M ($\emptyset$2-area empty in Figure 2) and M-I2 $\subseteq$ M-I1 ($\emptyset$1-area empty) and at least once it should be a proper subset relation (at least one *-area non-empty). Note that the *-areas differ from those in the case of theory revision, since I2 is now moving from I1 in the direction of M, whereas M2 was moving from M1 in the direction of I. In terms of symmetric differences, MΔI2 should be a proper subset of MΔI1. Now it is not difficult to argue that this definition amounts to: M has more true consequences and more correct models relative to I2 than relative to I1.

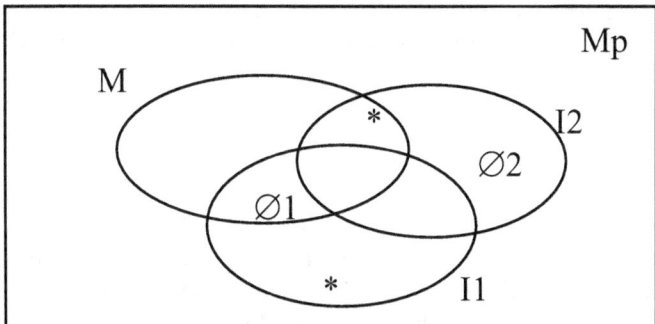

Figure 2. (The truth about) domain I2 is closer to theory M than (the truth about) domain I1.

Empirical progress by domain revision becomes a bit more complicated because we have to compare two data sets, relative to I1 (R1/S1) and to I2 (R2/S2). M is at least as successful relative to R2/S2 as to R1/S1 if and only if M $\cap$ R1 $\subseteq$ M $\cap$ R2 & S2 $\cup$ M $\subseteq$ S1 $\cup$ M. The first condition now amounts to the claim that M does not loose established examples of I1 by going from R1 to R2; and the second condition amounts to the claim that M does not loose explained established laws with respect to I1 by going from S1 to S2.

Assuming again correct data, that is, R1 $\subseteq$ I1 $\subseteq$ S1 and R2 $\subseteq$ I2 $\subseteq$ S2, it is now easy to derive the *Success Theorem for Domain Revision*: "M is closer to I2 than to I1" entails "M is at least as successful relative to R2/S2 as to R1/S1".

Starting from one set of realized possibilities R and one strongest established empirical law S, that is, when they are not yet specified relative to the two domains, we may also define these specifications for i = 1, 2 as follows: Ri $=_{df}$ R ∩ Ii and Si $=_{df}$ S ∪ Ii, which even guarantees the condition Ri ⊆ Ii ⊆ Si.

The present *Success Theorem* is again such that *persistently* being-more-successful relative to one domain than to another is functional or instrumental for truth approximation. Let M be more successful relative to R2/S2 than to R1/S1. This suggests the *Comparative Success Hypothesis*, according to which it is hypothesized that this will remain the case, whatever experiments we design and perform. Testing of this hypothesis may result in the conclusion, at least for the time being, that this in fact is the case: M persistently remains more successful relative to R2/S2 than to R1/S1. Whether we should call the relevant domain revision a case of making empirical progress may be disputed. However, in the indicated situation it is very hard to imagine that I2 is, despite appearances, not closer to M than I1, and this certainly is progress in knowledge of a kind, again guided by the rule of success.

Synthesis

Again we have restricted ourselves to the naïve or basic theory of truth approximation by domain revision, containing several idealizations. Similar concretizations as in the case of theory revision can be made. However, we will turn to a plausible formal synthesis of basic truth approximation by theory and domain revision, being a generalization of both.

As mentioned before, the basic definitions of truth approximation by theory revision and domain revision, respectively, amount to comparisons of symmetric differences. Both can be generalized qualitatively, leading to the same result:

Definition: <M2, I2> *fits better than* <M1, I1> if and only if M2ΔI2 ⊂ M1ΔI1.

This leads to four special cases, viz., when M*⊂M and I*⊂I (as depicted in Figure 3) then there is "better fit" if and only if, for some i, j = 1, 2, 3, 4, the i- and j-area are empty (i&jE) and the k- and/or l-area are non-empty (kvlN), as follows:

I <M*,I*> is a double specialization of <M,I> iff 3&4E/1v2N
II <M,I> is a double generalization of <M*,I*> iff 1&2E/3v4N

III <M*,I> is a double strengthening of <M,I*> iff 1&3E/2v4N

IV <M,I*> is a double weakening of <M*,I> iff 2&4E/1v3N

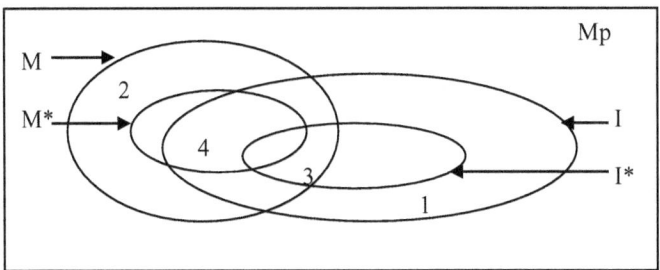

Figure 3. (Four special cases of "<M2,I2> fits better than <M1,I1>".

We obtain some interesting "special-special cases" of "better fit" if we assume in addition that $I^* = I$ or $M^* = M$. Let us start with the first possibility, depicted in Figure 4. When $M^* \subset M$ and $I^* = I$ there is "better fit" if and only if the i-area is empty (iE) and the j-area is non-empty (jN), as follows:

I/III ⇒ <M*,I> is a theory (specialization or) strengthening of <M,I> iff 3E/2N

II/IV ⇒ <M,I> is a theory (generalization or) weakening of <M*,I> iff 2E/3N

where the first condition amounts to: "not fewer correct models and fewer incorrect models" and the second to "no new incorrect models and some new correct models".

The second possibility, viz., $M^* = M$ and $I^* \subset I$, leads to the following "special special cases", depicted in Figure 5. In this case there is "better fit" if and only if the i-area is empty (iE) and the j-area is non-empty (jN), as follows:

I/IV ⇒ <M,I*> is a domain *restriction* of <M,I> iff 4E/1N

II/III ⇒ <M,I> is a domain *extension* of <M,I*> iff 1E/4N

where the first condition amounts to "not fewer examples and fewer counterexamples" and the second to "no new counterexamples and some new examples". In the first case we might also speak of "domain specialization" or

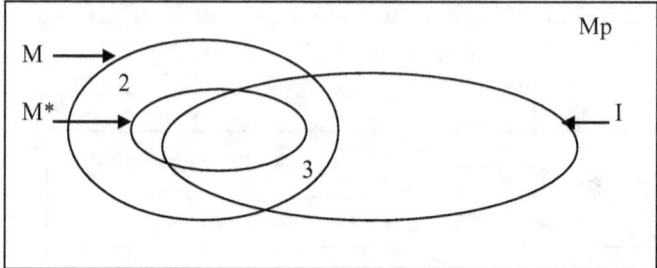

Figure 4. (Truth approximation by theory strengthening or weakening.

"domain weakening" but I prefer to speak of "domain restriction". Similarly for the second case. Although one might speak of "domain generalization" or "domain strengthening", I prefer to speak of "domain extension".

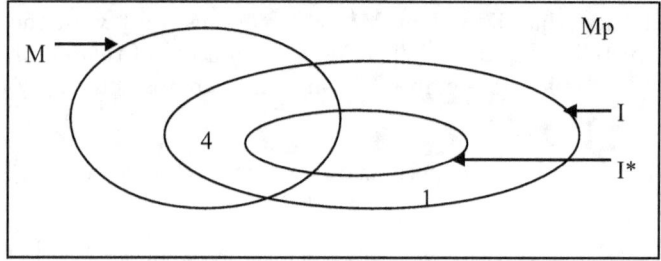

Figure 5. Truth approximation by domain restriction or extension.

4 In search of applications of "Truth approximation by domain revision"

So far we have shown that it is perfectly possible to explicate the idea of "Truth approximation by domain revision", and hence of the possibility of "Theories looking for domains". Now the question arises, as a kind of meta- and self-application of the "Theories looking for domains" theme, whether there are historical cases exemplifying this possibility.

Apart from a plausible type of cases the "predicted" phenomenon may seem rare in the empirical sciences. Of course, there are well-known examples of theories of which the formal core, i.e., <Mp,M>, finds a totally new area of applications. It may be that the new type of applications behave in relevant respects analogously to the originally intended ones, in which case we have a kind of domain extension, say, domain extension by analogy.

However, and here I depart from the *Drugs Looking for Diseases* perspective, in which it doesn't really matter whether the diseases are or are not related, my main interest is in truth approximation by genuine domain revision in the sense of domain manipulation within an area of, in some sense, related empirical phenomena, that is, the vocabulary constituting Mp should have the same kind of interpretation. Given this assumption, we may distinguish between basic (or set-theoretic) kinds of domain revision and refined kinds.

Of the basic kinds we can distinguish domain restriction, domain extension, and their combinations: domain restriction followed by domain extension or vice versa. Note that speaking of "truth approximation by domain extension from I* to I", in accordance with our definition, viz., I* is proper subset of I, such that I-I* is a (proper) subset M (the 1-area empty and the 4-area non-empty in Figure 5), may sound somewhat strange in the case that the *weak claim* that I* is a subset of M (I* $\subseteq$ M) is true and hence the weak (but stronger) claim that I is a subset of M is true as well. For this amounts to the case that we correctly extend the domain for obtaining a true weak claim without having to correct the original domain. Let us call this "truth accumulation by domain extension" (with the counterpart of "truth accumulation by theory strengthening"). Historical cases of this type are also well known: successful extensions of the domain with respect to which the theory was already (assumed to be) successfully applied. The successful search for new domains of application of Newton's theory, by Newton and his followers, belongs to this type[2] [3].

Of course, later findings may relativize the correctness of the original and/or later claims. In this case one option is theory revision, as in the case of the discovered deviations from Newton's predictions, leading to Einstein's special and general theory of relativity. The special theory is, relative to Newton's theory, a typical case of "truth approximation by (idealization followed by a, to be sure, fundamental) concretization of the theory". In

---

[2] It is not always unambiguous whether we have a case of pure domain extension or a case of domain extension by analogy. For example, from the point of view of Newton's theory of gravitation the Rutherford-Bohr planetary model of the atom is a kind of domain extension by analogy with the planetary system, since the Coulomb forces behave analogously to the gravitational forces. However, looked from Newton's general theory of motion we have a proper case of pure domain extension.

[3] Of course, starting from an *un*interesting (purported) truth, domain extension will be equally uninteresting. In other words, it is perfectly possible to let the search for domain revision be guided by "interesting (purported) truths", which reflects, apart from the terminology, a typically instrumentalist perspective. It may well lead to domain restriction, rather than domain extension.

order to define this type of truth approximation by theory revision I have developed in part three of [Kuipers, 2000] a general theory of "refined truth approximation by *theory* revision", which was indicated in section two. As a side remark, in contrast to Rivadulla's general claim [Rivadulla, 2006], for *refined* truth approximation it is not at all relevant that the theories concerned are incompatible, e.g., those of Newton and Einstein, of Copernicus and Kepler, and of Boyle/Gay-Lussac (ideal gas law) and Van der Waals. Unfortunately, it needs some preparation to explain this possibility of refined truth approximation in detail. The reader is referred to [Kuipers, 2000, chapter ten, pp. 268–271], where the gas example is presented in detail.

Besides concretization of the theory, when later findings relativize the correctness of the original and/or later claims related to "truth accumulation by domain extension", another option is to concretize the (specification of the) domain. More generally, it is plausible that also a general form of "refined truth approximation by *domain* revision" can be defined with "truth approximation by (idealization followed by) concretization" as a special subtype.

Hence, the remaining question is whether there are historical examples of truth approximation by basic or refined kinds of domain revision, apart from cases of (putative) "truth accumulation by domain extension". As far as the basic kind is concerned it may be not surprising if there are none. Also in the case of "truth approximation by theory revision" basic kinds are either toy examples or examples of "truth approximation by theory strengthening", that is, the counterpart of "truth accumulation by domain extension". However, so far, I also did not find convincing refined cases, but I think this is a matter of insufficient knowledge of the history of science. Moreover, from the structuralist perspective it is tempting to focus on empirical theories of a mathematical nature, while it may be easier to find cases in areas where this is less typically the case, e.g., biology and psychology, because the target phenomenon of domain revision may occur in these areas at least as frequently.

It is easier to find related types of non-empirical cases. Let us begin with cases of "truth accumulation by domain extension". Mathematics provides evident cases. For example, every extension of the set of natural numbers for which Goldbach's conjecture has been proved belongs to it. Similarly, even before the full proof of the four-color problem, every new type of map for which the four or, for that matter, the five color problem was positively solved, provides another illustration. Mathematical economics also provides beautiful illustrations. As Bert Hamminga [1983] has documented, in the development of the theory of international trade new results frequently amounted to an extension of the proven domain of validity of a certain "interesting theorem", such as the "factor prize equalization theorem"; for a brief further indication see [Kuipers, 2001, section 1.2.7]. By the way, this type of research, though not empirical as such, can be motivated in terms of truth approximation with respect to the *actual* world, because any extension of the proven domain of validity increases the chance that the theorem also holds in the actual world.

I do not yet know of basic cases of "truth approximation by domain revision" in mathematical economics other than of the cumulative type. However, (mixed) refined cases, based on distances between structures in Mp, have been studied. In particular, "truth approximation by (idealization followed by) concretization of the domain *and* the theory", where the theory is an interesting theorem in the sense indicated above. In Cools *et al.* [1994], see also [Kuipers, 2000, chapter eleven, section two], important developments in the theory of the capital structure of firms have been reconstructed along these lines, more specifically, the transition from the theory of Modigliani and Miller to that of Kraus and Litzenberger.

At the moment I only know of one, beautiful, case study of (presumably[4]) basic truth approximation by domain revision in mathematics, viz., Lakatos' [1976] reconstruction of the history of Euler's theorem about polyhedra. As is well-known, Lakatos shows convincingly, using a lively class room setting, with historical notes, that the history of Euler's claim that the polyhedra satisfy $V - E + F = 2$ (V the number of vertices, E that of edges, and F that of faces) is a history of domain revision. More specifically, it is domain revision by all kinds of revision of the definition of a polyhedron. Hence, the case also illustrates that basic truth approximation by domain revision may well take the form of "truth approximation by *definition* revision".

David Atkinson provided me with a clear case in physics of "truth approximation by domain extension" in the form of definition revision, more precisely, in the form of successive definition extension: (putative) truth accumulation of the law of conservation of energy by extending the definition of "energy". With his kind permission, I quote his description of the case in full.

> In the elastic collision of balls, there is a conserved scalar quantity, the kinetic energy: if we know the mass of each ball, and measure its speed at a given time, we can calculate from these numbers the total kinetic energy of the system, and this does

---

[4]To be honest, I did not (yet) check whether there is an episode that satisfies the basic definition perfectly, without being merely a case of domain extension, but this seems very likely.

not change in time, at least if the effects of inelasticity and friction are negligible. If the balls are not free, however, but are attached to one another by an array of springs, the total kinetic energy of the balls does not remain constant in time, i.e. the law of conservation of energy breaks down. Or does it? By extension of the definition of energy to include the potential energy that is stored in an extended or a compressed spring – which can be calculated if one knows the (Hooke) spring constant –, one finds that the law of conservation of energy holds, on condition that one defines the energy as the sum of the total kinetic energy of all the balls, and the sum of the potential energies stored in all the springs. For the motion of cannon balls in the gravitational field of the earth, one finds that the same law is valid, where now the potential energy is stored in the gravitational field, and is proportional to the height of the ball at any instant.

A proviso in the above is that friction should be negligible. If it is not – or if one makes measurements with sufficiently great accuracy, or waits long enough – one finds that energy, as defined above, does not seem after all to be conserved, but appears to diminish slowly with time. The law is reinstated by a further extension of the notion of energy to include heat. A precise equivalence was found in the XIX century (Joule) between heat and work, which is a form of energy. If one adds the heat produced by friction, and that produced by any inelasticity in collisions, to the kinetic and the potential energy to produce the total energy, one finds that this is indeed conserved.

The conservation of kinetic energy + potential energy + heat was thought to be exact until, at the end of the XIX century, it was found that, in the decay of radioactive substances, there is a small mismatch in energy before and after decay. The law was reinstated by Einstein, who showed in his theory of special relativity that mass and energy are equivalent. In radioactive decay the sum of the masses of the constituents before and after decay was not the same, and if one extends the notion of energy once more to include mass, m, and adds mc^2 to the kinetic energy + potential energy + heat, one finds that the law of conservation of energy is reinstated.

Since it is a case of successive domain extension by successive definition extension, it is, however, not the type of whimsical domain revision as in the case of polyhedra.

Gordana Dodig-Crnkovic provided me with an example that may be less straightforwardly cumulative with respect to the (definition of the) domain, viz., the study of magnetism. With her kind permission, I quote her indication of this case as well.

> First, a number of observations (starting in ancient Greece) are made about what magnets are and how they behave. Later on the concept of magnetic field is introduced. Then the fact that the moving electric charge induces magnetic field was established. Earlier concepts about magnetism were transplanted to the new environment. Maxwell predicted the existence of electromagnetic waves and identified light as an electromagnetic phenomenon. One has noticed that even subatomic particles possess magnetic properties, so the concept and the theory have shifted the physical domain again. Evidently, the domains of the first proto-theories of magnetism were very different from the domains of the theory of electromagnetism of today, that is, electromagnetism and quantum electrodynamics.

Only a full structuralist reconstruction could show whether this is a case of not purely cumulative domain revision. However, without that we have to face that there is no fixed theory, that is, there is no fixed theory of magnetism originating in ancient Greece.

5 Concluding remarks

The question remains whether there are convincing cases in the empirical sciences of basic or refined truth approximation by domain revision, other than pure domain extension, whether or not in the form of definition revision, other than successive definition extension. One reason to expect many explicit or hidden cases of this kind is the frequently confessed instrumentalism; for domain revision in practice evidently is a plausible form of instrumentalist practice. Happily enough for epistemological realists, it is possible to show [Kuipers, 2000] in general that the instrumentalist methodology with respect to theory revision is in theory, and hence in practice, functional for truth approximation and, even more surprisingly, more efficient than the falsificationist methodology. It is rather easy to see that a variant of the argumentation for this claim can be used for domain revision. The first claim is argued by realizing that, starting from a falsified theory, it may well be that the shortest route to the truth will go via intermediate theories that are also false. The falsificationist method instead forces one to start over again with a new theory as soon as the next theory again gets falsified. Similarly, given a theory, starting from a "non-fitting" domain, that is, a domain with respect to which the (weak claim of the) theory is false, it may well be that the shortest route to a maximally fitting domain, that is, a domain with respect to which the strong claim of the theory is true, hence the truth for that domain, will go via intermediate nonfitting domains. Again, the plausibly adapted version of the falsificationist method to domain revision, if that is acceptable from the falsificationist perspective at all, forces one instead to start over again with a new domain as soon as the next domain again turns out not to fit the given theory.

Of course, in practice, neither the domain nor the theory will be fixed stubbornly, while the other is revised. As I wrote at the end of [Kuipers, 2000, p. 332], "science in the making", that is, truth approximation in practice, is a matter of dialectical interaction between vocabulary, domain and theory. The above mentioned example of magnetism illustrates this process nicely.

Let us consider from this point of view the various attitudes that are possible with regard to a counterexample of the theory, assuming that we at least fix the basic vocabulary. Hence, I leave it to the reader to think about the way in which revision of the vocabulary can be brought into the picture [Kuipers, 2000, pp. 129–130]. So far we have suppressed the plausible distinction between the "core theory" and auxiliary hypotheses. Hence, in the spirit of the Duhem-Quine thesis, we may *first* look for auxiliary hypotheses that might be blamed for the counterexample, which may lead to revision of one or more of the auxiliary hypotheses, leaving the "core theory" intact. However, as soon as we accept the counterexample as a counterexample of the core theory for the domain, we may, *secondly*, look for an appropriate theory revision, or, *thirdly*, for an appropriate domain revision.

A similar story can be told about dealing with explanatory problems of a theory. An explanatory problem amounts to an established empirical law that cannot be derived from the theory; hence it is too weak for the domain. Again we may revise an auxiliary hypothesis, the core theory, the domain, the vocabulary, or a combination of these.

Of course, in all cases, we want to avoid *ad hoc* revision; we prefer revisions that lead to successful new predictions. However, we have not paid attention to "*ad hoc*-ness", because from the truth approximation analysis it is clear that *ad hoc* changes are unproblematic if and only if they survive the severe testing of the relevant "comparative success hypothesis" [Kuipers, 2000, pp. 168–169], and this we have assumed throughout.

In sum, truth approximation by revision of theories, domains and vocabularies is perfectly possible, in particular when choices are guided by the instrumentalist rule of success. This is in sharp contrast to one of the main claims of Rivadulla [2006, end of section two]: *empirical success is no indicator of truth, verisimilar or truth approximation*. As a matter of fact, his illustrations of "historical failures of the Newtonian gravitational model in astrophysics and cosmology" nicely provides an historical case (section 4.1) I have been looking for. According to the Kelvin-Helmholtz gravitational collapse hypothesis the luminosity of the stars originates from the fact that a star is the result of a collision of smaller bodies, due to mutual gravitation. This collision produces, according to the theory of Joule, an amount of heat equivalent to the kinetic energy lost in the collision. As Eddington noted, this explanation leads to bizarre conclusions about the age of the sun, reason for which "the source of stellar energy", was no longer considered to belong to the domain of the theory of gravitation: hence, here we have a case of domain restriction.

It is important to note that the conclusion was not that the theory should be revised. As is well known, and reported by Rivadulla in section 4.2, this was essentially done in a rather drastic way by Einstein's general theory of relativity, in particular, in light of the classical failure of Newton's theory, viz., to explain the advance of Mercury's perihelion. This was soon followed by the successful prediction of light deflection by the Sun and, much later, by the successful explanation of the phenomenon of *red shift*. Of course, these phenomena disappeared out of the domain of Newton's theory, if they were ever considered as such at all, which is only the case for Mercury's behavior. Moreover, one might say that the domain of Newton's theory became essentially restricted to phenomena with small velocities and weak gravitational fields, that is, velocities and fields for which the theory is approximately true.

For completeness, I add that Rivadulla also shows that Newton's theory can be successfully used to calculate the (relative) masses of binary stars (section 3.1), illustrating that the orbiting of binary stars, as might be expected, belongs to its domain. Moreover, he reports (section 3.2) the way in which Newton's theory (or model, as he prefers) can explain the stability of stars, that is, that they do not collapse. A clear case of domain extension.

The illustration of domain revision by restriction, i.e., the exclusion of the problem of the source of stellar energy from the domain of Newton's theory, suggests that several examples of so-called Kuhn-loss might be reconstructed as a case of domain restriction. Hence, we may conclude that truth approximation by domain revision is, at least in the form of domain restriction, fact rather than fiction in the empirical sciences.

Acknowledgements

This paper was first presented on the Workshop "Structures and Dynamics of Science", in Paris in 2004, organized by Ulises Moulines. I am grateful for all comments that I received in Groningen, in Paris, and, last but not least, in Pavia. Special thanks go to David Atkinson, Gordana Dodig-Crnkovic, Nico Krijn, Andrés Rivadulla, Jan-Willem Romeyn and Rein Vos for their useful comments and suggestions on an earlier draft.

BIBLIOGRAPHY

[Cools et al., 1994] K. Cools, B. Hamminga, and T. Kuipers. Truth approximation by concretization in capital structure theory. In B. Hamminga and N. B. de Marchi, editors, *Idealization VI: Idealization in Economics*, pages 205–228, Amsterdam, 1994. Rodopi.

[Forbus and Gentner, 1997] K. Forbus and D. Gentner. Qualitative mental models: Simulations or memories? In *Proceedings of the Eleventh International Workshop on Qualitative Reasoning*, pages pp. 1–8, Cortona, Italy, 1997.
http://www.qrg.northwestern.edu/papers/Files/QMM_QR97.pdf.

[Giere, 1999] R. Giere. Using models to represent reality. In L. Magnani, N. J. Nersessian, and P. Thagard, editors, *Model-Based Reasoning in Scientific Discovery*, pages 41–57, New York, 1999. Kluwer Academic/Plenum Publishers.

[Hamminga, 1983] B. Hamminga. *Neoclassical Theory Structure and Theory Development*. Springer, Berlin, 1983.

[Kuipers, 2000] T. Kuipers. *From Instrumentalism to Constructive Realism. On some Relations between Confirmation, Empirical Progress and Truth Approximation.* Kluwer Academic Publishers, Dordrecht, 2000. Synthese Library, vol 287.

[Kuipers, 2001] T. Kuipers. *Structures in Science*. Kluwer Academic Publishers, Dordrecht, 2001. Synthese Library, vol 301.

[Kuipers, Forthcoming] T. Kuipers. Empirical and conceptual idealization and concretization. The case of truth approximation. In *Liber Amicorum for Leszek Nowak*, Amsterdam, Forthcoming. Rodopi.

[Lakatos, 1976] I. Lakatos. *Proofs and Refutations: the Logic of Mathematical Discovery*. Cambridge University Press, Cambridge, 1976.

[Rivadulla, 2006] A. Rivadulla. The role of theoretical models in the methodology of physics. In L. Magnani, editor, *Model-Based Reasoning in Science and Engineering*. This volume, 2006.

[Vos, 1991] R. Vos. *Drugs Looking for Diseases*. Kluwer Academic Publishers, Dordrecht, 1991.

[Zwart, 1998] S. Zwart. *Approach to The Truth. Verisimilitude and Truthlikeness*. PhD thesis, ILLC, 1998. Revised version: *Refined Verisimilitude* (Synthese Library, vol. 307), Kluwer Academic Publishers, Dordrecht, 2001.

Theo Kuipers
Department of Philosophy
University of Groningen
Groningen, The Netherland
Email: T.A.F.Kuipers@philos.rug.nl
URL: http://www.rug.nl/filosofie/Kuipers

Scientific Cognition as Model-Based Reasoning

PING LI AND DACHAO LI

ABSTRACT. Recent work on model-based reasoning (MBR) in science has focused on scientific discoveries and conceptual change. This paper argues that model-based reasoning may provide a framework to explain the reasoning in every scientific context at the level of cognitive mechanisms, and attempts to account for normal science and scientific explanation within a model-based framework (model-based reasoning thesis).

1 Introduction

The view of Scientific Cognition in terms of *model-based reasoning* (MBR) has increasingly occupied the literature of the last two decades, and the accounts of mental modeling have provided a crucial understanding of the cognitive basis of scientific reasoning. This kind of approach to scientific cognition is mainly a direct reaction to the received view, especially related to the problems of the nature and structure of theories and to the syntactic account of scientific reasoning, as well as a reaction to Kuhn's theory of scientific revolution as *Gestalt* shift. Consequently, it is natural for this approach to focus on the problems of the nature of theories and of their generation and on the changes of conceptual structures in the contexts of discovery and development. Given the general claim that scientific cognition can be studied in terms of model-based reasoning, it seems to us that it is also particularly interesting to use the conceptual framework of MBR to capture details of scientific practices as considered in other traditions, including those involved in the context of justification developed by logical positivists, Kuhn's normal science, and even Lakatos' degenerative research programs.

In this paper, we will consider scientific cognition as a kind of model-based reasoning claiming that this can provide a new framework for the explanations of reasoning practices of every scientific context at the level of cognitive mechanisms. We will make a distinction between the singular problem of mental models and the plural one of model-based framework of science: the singular problem is not concerned with specific forms (including structures and formats) of models on which reasoning operates; on the contrary, the plural problem is related to specific forms of mental models. Based on this distinction, we claim that model-based reasoning is a semantic process implemented by cognitive operations on instantiated models in working memory[1]. After having provided an analysis of the case of Maxwell's electromagnetic theory with the aim to illustrate that scientific discoveries and revolutionary changes share the same kind of model-based processes with other reasoning practices (say, in normal science), this paper will provide a tentative account for model-based practices in normal science and in explanations, to favor the construction of a whole model-based framework for science.

2 Scientific cognition as model-based reasoning thesis

... [T]he cognitive sciences might come to play the sort of role that formal logic played for logical empiricism or that history of science played for the historical school within the philosophy of science. This development might permit the philosophy of science as a whole finally to move *beyond* [Italic emphasis by the authors of this paper] the division between "logical" and "historical" approaches that has characterized the field since the 1960s. [Giere, 1992, p. xv]

It is this *"beyond"* that suggests that scientific cognition as MBR Thesis can be taken as a general claim, independent of any traditional demarcations of scientific contexts. In addition, the continuum hypothesis held between science and ordinary cognition also provides an underlying support to the thesis as a universal claim. "While there is currently little direct evidence on the issue of continuity" [Brewer, 1999, p. 492], a lot of important research work and theories developed recently in the fields of science studies (e.g., that of Dunbar, Gentner, Giere, Gooding, Magnani, Nersessian, Thagard, Tweney), developmental psychology (e.g., that of Carey, Gopkin, Keil, Perner, and Wellman) and science education (e.g., that of Chi, McClosky, and Clement) shows that the continuum assumption has been a fruitful working hypothesis. In this case, the former itself can be considered as an indirectly empirical support to the hypothesis of continuity. On the other hand, according to Johnson-Laird's mental model theory (MMT) of ordinary human reasoning, endowed with an extensive experimental con-

---

[1]Nersessian's definition of mental modeling is: "Mental modeling, a semantic process thought to utilize perceptual mechanisms in inference, is hypothesized by many cognitive scientists to be a fundamental form of human reasoning" [Nersessian, 2003, p. 197].

firmation[2], ordinary reasoning too is the kind of model-based reasoning. Following the continuum hypothesis, it is natural to claim that scientific reasoning is merely an extension of ordinary human reasoning; namely, the former is a refinement and complication of ordinary cognitive strategies. Hence, there are no reasons for us to believe that model-based reasoning is a salient feature *only* of some scientific contexts (say, discovery and conceptual change). In other words, the continuum hypothesis suggests that the processes of model-based reasoning can constitute a unified cognitive basis underlying both ordinary and scientific reasoning-practices, looking for reasonable explanations of scientific practices at the level of cognitive mechanisms.

As Giere writes, "[...] adopting a model-based framework makes it possible to employ resources in cognitive psychology to understand the structure of scientific theories in ways that may illuminate the role of theories in the ongoing pursuit of scientific knowledge" [Giere, 1999, p. 99].If we consider the flourishing current development of cognitive studies in the area of model-based reasoning in science, we should have to take this suggestion seriously. In the last two decades, indeed, many authors contributed a lot to the development of the thesis, and much of their work appeared in several important books (e.g., [Magnani *et al.*, 1999; Magnani and Nersessian, 2002; Giere, 1992; Carruthers *et al.*, 2002; Gorman *et al.*, 2005] and some special journal issues (e.g., *Foundation of Science* 5(2), 9(3); *Philosophica* 61(1), 62(2); and *Mind and Society* 2(4), 3(5)). However, much of research on the thesis is restricted to scientific practices *only within* the context of discovery. This over-concentration of this kind of research would bring potential problems. For example, overlooking and undervaluing the legacies of the traditional philosophy of science instead of reconsidering its central topics in terms of cognitive analyses, and misunderstanding the role of model-based reasoning in science thus creating the illusion that scientific creativity and the generation of novel conceptual structures own their peculiar cognitive processes or mechanisms fundamentally different from those of other scientific practices. This would reproduce the distinction between justification and discovery, instead of developing a unified model-based framework for all kinds of reasoning practices in science.

It seems there are some confusion and inconsistencies within the current model-based framework. Among them, the most common one in the current literature is the use of the concept "mental model", which is so popular that

---

[2]The theory has been tested by a lot of experimentation involved in various reasoning tasks, including reasoning with logical connectives, reasoning with quantifiers, modal or probabilistic reasoning, relational reasoning, everyday inferences and arguments, counterfactual reasoning, reasoning by psychotic individuals, etc.

we are not surprised that there are few responses to Brewer's challenge and clarification[3] [Brewer, 1999; Brewer, 2003]. A salient case of this confusion is the mix of Johnson-Laird's concept [Johnson-Laird, 1980; Johnson-Laird, 1983] with Gentner et al.'s term [Gentner and Stevens, 1983] under the same label of "mental model"[4].

To avoid potential conceptual confusion in the current literature, the first question we have to answer is about the real status of the models involved in the processes of reasoning. It is obvious that both external physical and internal mental models often play an important role in the processes of reasoning. While the manipulations of external physical models would cause the changes of corresponding mental models (see [Dogan and Nersessian, 2005]), it is the latter that is responsible for the mental operations of reasoning. Hence we are sure that the models employed to account for cognitive operations of reasoning are the knowledge structures constructed with mental representations. Moreover, we are also sure that certain kinds of scientific reasoning operate on many kinds of knowledge structures other than Johnson-Laird's mental models. Hence, it is not appropriate to appeal for some specific form of mental representations in defining the term of "model-based reasoning", since mental models in question are constructed with different contents (kinds and levels of information), structures, and formats of internal representations and thus may take many forms[5]. For example, we learn of some forms such as images, prototypes, frames, schema, mental models, perceptual symbols, and theories from the field of psychology.

Based on the above considerations, we suggest a distinction between the singular problem of mental models and the plural one regarding the model-based framework of science. On the one hand, it is a singular problem just because this problem of mental models is not concerned with their specific forms (including structures and formats); that is, the singular problem is a question of what is the nature of mental models in general – in this sense, we call them "generalized mental models", in order not to be confused with both of Johnson-Laird's and Gentner's concepts. Mental models in the

---

[3]We agree with Brewer's clarification. But it seems that there never is a linguistic reform in cognitive science though we have had Brewer's proposals [Brewer, 1999].

[4]It is important to note that Johnson-Laird's mental models are kinds of knowledge structures constructed temporarily in working memory at the organizational level of schema, and that Gentner et al.'s mental models refer to a class of knowledge structures in long-term memory. These are, according to Brewer, "the subclass of theories which use causal/mechanical explanatory frameworks" [Brewer, 1999, p. 500]. To distinguish between them, Brewer suggests the term "mental model" still for Gentner's concept and "episodic model" or "constructed schemata" for Johnson-Laird's concept [Brewer, 1999]; also see [Brewer, 2003] for an overview on mental models.

[5][Brewer, 1999; Brewer, 2003] and [Nersessian, 2002a] discussed alternative forms of mental models.

singular problem only refer to internal mental representations functioning in reasoning and in the generation of new mental representations; thus, they are characterized not by their particular forms, but by their internal status and their general cognitive functions in representing reality and in reasoning processes. Mental models in the generalized sense cover all forms of knowledge structures involved in processes of reasoning, and should be used in arguments for model-based reasoning as a semantic process. Mental models in the "plural" sense, on the other hand, related to the kinds of specific forms of mental models, deal with a variety of classes both of mental models and of reasoning.

In addition, it seems that model-based reasoning in specific cognitive tasks operates on some specific and integrated structures in working memory (WM) – we call them "instantiated models" – instead of operating directly on knowledge structures in long-term memory (LTM). Working memory plays a central role in integrating both retrieved information from long-term memory and/or perceptual information into instantiated models and in generating new mental representations based on instantiated models. In this way, working memory acts as a kind of *assembling* apparatus, considering that more common and more realistic kinds of reasoning are those based on multi-models with multi-forms of mental representations. It is in working memory that many forms of mental representations, including perceptual and imaginary ones, are combined together in an integrated model functioning in a process of reasoning. In her studies on Maxwell's vortex-idle wheel model, Nersessian writes: "The practices of analogical modeling, visual modeling, and thought experimenting (simulative modeling) are frequently used together in a problem-solving episode" [Nersessian, 2002a, p. 137]. That is, several forms of mental representations involved in those modeling processes come together and generate an integrated model for a reasoning task. Constituent forms or models are not just supervening things or epiphenomena: a mental image of the wheel in Maxwell's case is helpful to generating an external representation, such as drawings [Nersessian, 2002a, p. 138].

In the working memory assembling (WMA) model of mental representations, we divide mental models generated in WM by information in long-term memory into three kinds: theoretical models (including those of Gentner's mental models), schematic models (such as Johnson-Laird's mental models), and imagery models (e.g., images in Kosslyn's account). Following the generations of these three kinds of models, an integration of kinds of mental models with different formats happens in working memory for a variety of cognitive tasks.

The picture characterized above might be helpful to reduce the confusion in the current literature and to make the model-based framework a more reliable tool able to account for the reasoning practices of other contexts of science and not only for discovery and conceptual development. In the following section, a case analysis will show that revolutionary changes share the same model-based processes with reasoning practices of normal science.

3 Maxwell's electromagnetic theory: the continuity between normal science and scientific revolution and the continuity of scientific revolution

Nersessian [2002a] says that model-based reasoning is a central characteristic of scientific reasoning during scientific revolution and theory innovations. She uses the cognitive-historical approach to study the development of Maxwell's electromagnetic theory, and points out that model-based reasoning causes conceptual changes in science. Indeed some processes of model-based reasoning, such as analogical modeling and visual modeling, can draw and check the constraints of the current conceptual system in terms of the constraints derived from the target, and generate new constraints which thus are integrated in a new revised conceptual system. It is the process of "generic abstraction" that completes the tasks of drawing and integrating constraints derived from many sources, and thus that causes the genuine creativity in science. It seems clear that Nersessian's analysis of model-based reasoning focuses on the contributions of these processes of reasoning to conceptual and theoretical innovations, and provides a new perspective on the understanding of scientific revolutions, considered as continuous and non-accumulative process.

Maxwell and even other physical scientists before Einstein's rejection of the concept "ether", as [Nersessian, 2002b] points out, still believed that electromagnetic phenomena are mechanical phenomena of ether by nature, and that a complete explanation of electromagnetic field has to involve a mechanical theory of ether. Also, Maxwell insisted that the electromagnetic theory was not yet a complete theory, but he believed that there would be a mechanical theory of ether in the future that would meet with his electromagnetic theory. It is obvious that Maxwell's theory still belongs to Newton's conceptual framework at a large extent; thus his "conceptual innovation" is not a conceptual change and innovation in Kuhn's sense (revolution). The significance of Maxwell's work, however, is that his concept of "electromagnetic field" brings new constraints that conflict with those implied in the conceptual system of Newton' mechanics. It is this conflict, as we know, that was resolved by Einstein's radical revision of the conceptual foundations of Newton's mechanics, which eventually establishes electromagnetic phenomena as a realistic domain different from that of mechanical phenomena. It is this last shift of the concept of "electromagnetic field" that is a genuine conceptual innovation in the sense of scientific revolution.

The interesting facts related to the shift is that the result of Michelson-Morley experiment had no impact on Einstein in the generative process of his theory of relativity, and that instead the experimental result of the constancy of the velocity of light – a predication derived from Maxwell's theory – and the elegancy of Maxwell's theory impressed Einstein. Therefore, when facing the conflict between Maxwell's theory and the mechanical theory, Einstein chose the former and renovated Newton's mechanics. Informed by this historical case model-based reasoning provides undoubtedly the possibility of genuine scientific creations: we maintain that it is not only characteristic of scientific reasoning in the processes of revolutionary changes. The case of Maxwell's theory does show that model-based reasoning is a basic cognitive process of scientific practices of reasoning in the stage of normal science too: this that becomes one of indicators to the continuity between normal science and scientific revolution.

The above analysis suggests that the general claim maintaining that scientific cognition consists in model-based reasoning seems more important from a perspective of cognition than only emphasizing on model-based reasoning as the basic cognitive process or mechanism of scientific discovery and/or conceptual change in science. The next section will instead characterize scientific practices of normal science within the conceptual framework of scientific cognition as model-based reasoning, a perspective which has received little attention in the current literature on model-based reasoning in science. In section five, we will discuss a central topic of the standard philosophy of science, scientific explanation, based on our perspective on model-based reasoning.

4 Reasoning practices in normal science

Following Nersessian's account of mental modeling, we consider modeling constraints as an aspect important for understanding scientific cognition. Thus we set out the characterization of the core of normal science, namely Kuhn's concept of "paradigm", in light of modeling constraints. Here our working hypothesis suggests that knowledge background and paradigms have to be considered as modeling constraints in scientific reasoning. This suggestion is consistent with Nersessian's idea of concepts as modeling constraints and Giere's conception of scientific theories as incomplete models.

Nersessian points out that it is necessary to understand concepts of scientific theories on the basis of modeling constraints. In addition, [Giere, 1999] claims that scientific laws are not really statements about the world but *part* of characterization of theoretical models, such that only combining them with a specific problem context one has a model that can be compared with a real system. It is obvious that one has to understand scientific laws and theories in terms of theoretical models but not in terms of modeling constraints if laws and theories are merely linguistic descriptions of incomplete theoretical models. Thus, we use modeling constraints to understand a theoretical framework or paradigm that is presupposed for scientific practices in normal science: a paradigm is a set of modeling constraints. The constraints provided by a paradigm can be divided into two main kinds: explicit constraints which constitute a disciplinary matrix, and implicit/tacit constraints which consist in exemplars. According to this viewpoint, the practices of puzzle-solving in a normal science are the processes of model-based reasoning within one fixed set of constraints. Consequently, there are three important aspects in our analysis of the concept of paradigm as a set of modeling constraints: a disciplinary matrix as explicit constraints; exemplars as implicit/tacit constraints; and puzzle-solving as model-based reasoning.

The claims mentioned above merit investigation in two directions. On the one hand, the empirical support for them derived from cognitive psychology shows that the processes of model-based reasoning in ordinary tasks can be used to understand the remarkable features of scientific practices in normal science. This is a kind of indirect but substantive evidence for the continuum hypothesis as an indispensable component of the conceptual framework of scientific cognition as model-based reasoning. In other words, this makes it clear that (at least) the two kinds of cognition, ordinary and scientific, are continuous. On the other hand, that empirical evidence in turn provides a support for the model-based view of theories developed by [Giere, 1999], along the line of semantic approaches to the nature of scientific theories. The latter would become (Nersessian [1992; 2002a; 2002b]) a conceptual framework employed in an analysis of cognitive features of scientific practices and even common human knowledge.

The continuum hypothesis, if taken for granted, should be also taken as a reason why we expect that (at least) some of basic features of ordinary reasoning would be present in scientific reasoning. According to the mental model theory of ordinary human reasoning put forward by [Johnson-Laird, 1980] and developed by him and his followers and proponents, human reasoning is a process of model-based reasoning in nature. The findings of psychological experiments indicate that mental models play an indispensable role in ordinary human reasoning, and that there are several common characteristics of the use of mental models, such as

1. A mental model represents one possibility satisfying constraint of mod-

eling, but captures what is common to all the possibilities.

2. With regards to modeling constraints, mental models do not represent what is false, but what is true.

3. Procedures of model-based reasoning rely on counter-examples (alternative models) to refute invalid inferences.

4. The greater the number of models that a task needs, the poorer the performance is[6].

Let's interpret the above features taking advantage of the problem of relational reasoning.

According to Johnson-Laird and Byrne, human beings' reasoning relies on the construction and manipulation of mental models and can be characterized as a three-step procedure:

- They imagine a state of affairs in which the premises are true; in other words, they construct a mental model of the premises.

- They come up with a putative conclusion compatible with this model.

- They try to falsify this conclusion by constructing alternative models of the premises. If there are no such models, then the conclusion is a valid inference from the premises.

In the following problem of relation reasoning (Problem 1), for example, the premises are:

1. A is to the right of B

2. C is to the left of B

3. D is in front of C

4. E is in front of A

The subject is asked to answer the spatial relation between D and E. Problem 1 is compatible with the model:

C B A

D   E

---

[6]Among the characteristics of the use of mental models discussed by Johnson-Laird [1983; 1999; 2001] Byrne and Johnson Laird [1989], these four features are very useful for our characterization of scientific reasoning.

Based on this model, the subject would draw an initial conclusion: D is to the left of E. No other models are compatible with the premises. Thus, Problem 1 is called as the "problem of one-model".

Problem 2:

1. B is to the right of A

2. C is to the left of B

3. D is in front of C

4. E is in front of B

The problem is compatible with one model:

C A B

D   E

Based on the above model, the subject would draw an initial conclusion: D is to the left of E. After further searches for models, the subject would find out that the below model

A C B

D E

is also compatible with the premises and supports the same conclusion.

According to the theory of mental model, Problem 2 should be more difficult than Problem 1 because it is harder to deal with two models than with one model. This claim is confirmed by experiments.

The model-based reasoning in the ordinary inferential tasks such as the one described above permits to draw and to integrate the constraints derived from the premises in virtue of mental models – if the premises in question are taken as the linguistic descriptions of modeling constraints. If the claim that model-based reasoning is a fundamental process underlying both ordinary and scientific reasoning is true, it is possible to account for scientific practices of normal science in light of the key features of model-based reasoning revealed by cognitive psychology. First, as mentioned above, a mental model in the sense of the mental model theory captures what is common to all the possibilities even though it represents only one possibility satisfying modeling constraints. In the practices of normal science, the central function of exemplars is to represent the modeling constraints involved in a paradigm in a cognitively manageable way, such that scientists develop other models through the processes of inference by similarities.

In particular, the tacit constraints derived from exemplars are necessary to the practices of problem-solving in normal science and make them processes of similarity-grouping since the similarity to an exemplar ensures that the necessary tacit constraints are satisfied by the processes of model-based reasoning.

Second, mental models represent what is true relative to modeling constraints, and thus may lead to systematic errors. Similar situations would happen in reasoning practices of normal science (for example, the historical cases of the particle theory, when it was used to explain phenomena of light such as reflection, refraction, and Newton ring). Of course, it is has to be noted that, due to the function of generic abstraction, new constraints can emerge in the processes of model-based reasoning, even those that suggest us to reject the old entrenched constraints, as shown in Nersessian's analysis of the Maxwell's case [Nersessian, 2002b]. Moreover, this second feature of mental models is responsible for the conventionality of normal science and accounts for the fact that the practices of puzzle-solving would (at least usually) not challenge the fundamental theoretical hypotheses or principles of a paradigm.

Third, the procedures of model-based reasoning in ordinary inferential tasks rely entirely on alternative models to refute invalid inferences. The subject would confirm the inference from the initial models in the case that there are no incompatible alternative models or that he/she could not discover alternative models. This may be the reason why a genuine valid refutation against an existent hypothesis/model in reasoning practices of normal science is that of putting forward incompatible alternative hypotheses/models. That is, the presence or construction of competing theories is a prerequisite or an essential way to frustrate an existing theory; on the contrary, the fact that there are no alternative hypotheses/models which could be developed is a strong argument for an existent hypothesis/model.

Finally, the greater the number of models that a task needs, the poorer the performance is. In fact, the number of mental models involved in a task is inversely proportional to the amount of modeling constraints. Kuhn expounds how the "would-be" researchers acquire capacities to do research work in a specific discipline through the inferential training similar to that of exemplar-exercises in textbooks. The cognitive process similar to learning based on exercises in textbooks is required for the training of researchers even though the two processes are fundamentally different in the sense that a question involved in exercises usually has a single definite resolution (model), but researchers in actual scientific practices have to find out a variety of unknown constraints of modeling.

5 Model-based explanations in science

Explanation is one of the most typical and important functions of scientific theories, and the covering-law model of scientific explanation is one of the most valuable historical legacies of logical positivist philosophy of science. From the recent literature, a novel idea emerges: many forms of mental representations can be used to produce explanations and thus lead to the feelings of understanding (see [Brewer, 1999] for a short review; and [Brewer et al., 1998] for a psychological account of explanation). Accordingly, it is possible to develop a psychological account that expounds the cognitive basis of scientific explanation within the conceptual framework of model-based reasoning.

Scientific explanation as model-based reasoning is a kind of goal-guided cognitive process. Goals, derived from specific cognitive tasks, play an important role in determining what kinds of modeling-constraints need to be abstracted from an explanandum and which levels of explanations should be reached. We divide the complex forms of mental representations of scientific knowledge into three suitable levels: instance, schema, and theory, which can produce three basic levels of explanations respectively. In fact, most discussions on the representational forms of conceptual structures in the current cognitive-historical analyses of science focus on schematic models and theoretical models (e.g., mental models discussed in [Gentner and Stevens, 1983]. Schematic models provide explanations at the law-like level, as Brewer says: "schemata are the forms of mental representation that are appropriate to account for laws in the psychology of science and for the large class of empirical generalizations in nonscientists" (p. 496).

According to the model of explanation as model-based reasoning, there are four basic steps of cognitive operations in a process of model-based explanation: (1) generic abstraction; (2) ascription of feature constraints; (3) generation of instantiated models in working memory; and (4) the feelings of understanding. Usually, an explanation begins at the stage of generic abstraction, in which two kinds of constraints on modeling a phenomenon or conceptual construct (e.g., Boyle's law) are temporarily fixed. One is a set of feature constraints that characterize the phenomenon or construct; another is a set of variables (and/or a set of constants) that describe the initial and boundary conditions of the explanandum. In contrast to the mental modeling way of generating a new conceptual structure (see [Nersessian, 2002a, p. 152]), the following step is not to construct an initial model for target, but to search for a suitable relation of ascription under some representational forms of knowledge stored in long-term memory. It should be noted that this process is often involved in a selection of one ascription in light of specific cognitive goals and background knowledge. At the third stage of explanation by model-based reasoning, an instantiated model is generated through the information stored in long-term memory and thanks to the constraints that describe the initial and boundary conditions of the explanandum. Finally, the agent who undertakes the explanation undergoes the experience of understanding the explanandum through an internal process of mapping.

Therefore, explanation is a semantic process of understanding based on mental models, in which tacit or implicit constraints are often used to construct instantiated models. This is the fundamental reason why it is impossible in principle to construct a logical structure which links the explanandum with the explanans. In other words, instantiated models are not the explanans in Hempel's models of scientific explanation even though such kinds of explicit knowledge contained in the explanans are necessary constraints which are used to construct instantiated models. Thus, information stored in long-term memory that covers the explanandum cannot produce an explanation if it does not support the construction of an instantiated model.

6 Conclusion

Cognitive approaches are not a kind of panacea that can save the philosophy of science and make it be perfectly recovered from the illness of logicism and historicism. While they are able to overcome some shortcomings of the traditional philosophy of science and to open new areas and research, cognitive studies of science have their own limitations. The proposal of this paper was an extension of Giere's model-based view of theories and of Nersessian's mental modeling account, which defend a model-based framework for the explanations of scientific practices at the level of cognitive mechanisms.

Acknowledgement

The research work of this paper is supported by the Human and Social Sciences Foundation of the Education Ministry (98JAQ720010) and by the Social Sciences Foundation of Guangdong Province (02B68), P. R. China. The authors are grateful to Nancy Nersessian' comments on the draft of the paper and greatly appreciate Lorenzo Magnani's editorial help.

BIBLIOGRAPHY

[Brewer et al., 1998] W.F. Brewer, C.A. Chinn, and A. Samarapungavan. Explanation in scientists and children. *Minds and Machines*, 8:119–136, 1998.

[Brewer, 1999] W.F. Brewer. Scientific theories and naive theories as forms of mental representation: Psychologism revived. *Science and Education*, 8:489–505, 1999.

[Brewer, 2003] W.F. Brewer. Mental models. In L. Nadel, editor, *Encyclopedia of Cognitive Science*, pages 1–6, London, 2003. Nature Publishing Group.

[Byrne and Johnson-Laird, 1989] R.M.J. Byrne and P. N. Johnson-Laird. Spatial reasoning. *Journal of Memory and Language*, 28:564–575, 1989.

[Carruthers et al., 2002] P. Carruthers, S. Stich, and M. Siegal, editors. *The Cognitive Basis of Science*, Cambridge, 2002. Cambridge University Press.

[Dogan and Nersessian, 2005] F. Dogan and N.J. Nersessian. Design problem solving with conceptual diagrams. In B.G. Bara, L. Barsalou, and M. Bucciarelli, editors, *Proceedings of the XXVII Annual Conference of the Cognitive Science Society*, pages 600–605, Mahwah, NJ, 2005. Erlbaum. CD-Rom.

[Gentner and Stevens, 1983] D. Gentner and A.L. Stevens, editors. *Mental Models*, Hillsdale, NJ, 1983. Erlbaum.

[Giere, 1992] R.N. Giere, editor. *Cognitive Models of Science*, Minneapolis, MN, 1992. University of Minnesota Press.

[Giere, 1999] R.N. Giere. *Science without Laws*. University of Chicago Press, Chicago, IL, 1999.

[Gorman et al., 2005] M.E. Gorman, R.D. Tweney, D.C. Gooding, and A. Kincannon, editors. *Scientific and Technological Thinking*, Mahwah, NJ, 2005. Erlbaum.

[Hempel, 1966] C.G. Hempel. *Philosophy of Natural Science*. Prentice-Hall, Englewood Cliffs, NJ, 1966.

[Johnson-Laird and Byrne, 1991] P.N. Johnson-Laird and R.M.J. Byrne. *Deduction*. Erlbaum, Mahwah, NJ, 1991.

[Johnson-Laird, 1980] P.N. Johnson-Laird. Mental models in cognitive science. *Cognitive Science*, 4:71–115, 1980.

[Johnson-Laird, 1983] P.N. Johnson-Laird. *Mental Models*. Harvard University Press, Cambridge, MA, 1983.

[Johnson-Laird, 1999] P.N. Johnson-Laird. Deductive reasoning. *Annual Review of Psychology*, 50:109–135, 1999.

[Johnson-Laird, 2001] P.N. Johnson-Laird. Reasoning and rationality. In *The International Symposium on the Ontological Question*, Rome, 2001. Pontificial Lateran University.

[Kuhn, 1962] T.S. Kuhn. *The Structures of Scientific Revolutions*. University of Chicago Press, Chicago, 1962.

[Magnani and Nersessian, 2002] L. Magnani and N.J. Nersessian, editors. *Model-Based Reasoning. Science, Technology, Values*. Kluwer Academic/Plenum Publishers, New York, 2002.

[Magnani et al., 1999] L. Magnani, N.J. Nersessian, and P. Thagard, editors. *Model-Based Reasoning in Scientific Discovery*. Kluwer Academic/Plenum Publishers, New York, 1999.

[Nersessian, 1992] N.J. Nersessian. How do scientists think? Capturing the dynamics of conceptual change in science. In R. Giere, editor, *Cognitive Models of Science*, Minnesota Studies in the Philosophy of Science, pages 3–44, Minneapolis, 1992. University of Minnesota Press.

[Nersessian, 2002a] N.J. Nersessian. The cognitive basis of model-based reasoning in science. In P. Carruthers, S. Stich, and M. Siegal, editors, *The Cognitive Basis of Science*, pages 133–153, Cambridge, 2002. Cambridge University Press.

[Nersessian, 2002b] N.J. Nersessian. Maxwell and "the method of physical analogy": model-based reasoning, generic abstraction, and conceptual change. In D. Malament, editor, *Essay in the History and Philosophy of Science and Mathematics*, pages 129–166, Lasalle, IL, 2002. Open Court.

[Nersessian, 2003] N.J. Nersessian. Kuhn, conceptual change, and cognitive science. In T. Nickles, editor, *Thomas Kuhn*, pages 178–211, Cambridge, 2003. Cambridge University Press.

[Salmon, 1998] W.C. Salmon. *Causality and Explanation*. Oxford University Press, Oxford, 1998.

Ping Li and Dachao Li
Department of Philosophy
Sun Yat-sen University
Guangzhou, P.R.China
Email: `hsslip@sysu.edu.cn, dachao_sandy@hotmail.com`

Simulation for the Shift of Paradigm
STANISLAVA MILDEOVÁ

ABSTRACT. In the last decade all fields of science experienced a revolution in the use of information technology. Its use depends mainly on people and their ways of thinking. The first question that comes to mind is how we usually think. To answer this question, we have to analyze our mental models. It is obvious that mental models as a representation of the world around are quite limited in comparison to the world's complexity. The objective of this paper will discuss the possibility of overcoming the drawbacks given by the characteristics of people's mental models; the possibility of how we can use the knowledge gained from computer simulation for thinking development. The paper communicates practical experiences that arose in the stage of creation of the user interface; it deals with the process of visualization during the interface design and its two aspects – content and form. Simulators are shown as a tool that allows the enhancing of our mental models and can help us with the desired shift in reasoning; as "learning labs", allowing description of the complex system.

1 Reasoning

We do not perceive surrounding reality directly, but attach to it an image and meaning through our mental models. During a human life these mental images constantly change form and they influence our way of seeing the surrounding world. We transform the perceived reality in our mind using mental models into signs and pictures, which then influence our decisions [Chermack, 2003]. Man is constantly formed during his life through decisions he makes, the decisions form him back and make him human, as stated in concepts of the existentialist Jean Paul Sartre.

Let us continue in this spirit and start from the definitions of mental models that highlight the role of mental models in decision-making and the action of the subject on the basis of getting to know the object rather than based on the concept common in cognitive psychology. To analyze mental models means to answer the question of how each individual is capable of creating thought processes that he uses afterwards. Moreover, thinking is a very complex process that has not yet been explained completely. However, there are several guides for "correct thinking".

In relation to complex systems we should see that mental models which represent the external world in our mind (corresponding term "worldview" is presented by P. Checkland in his soft system methodology that we have to mention in conjunction with complex systems) and are based on unclear and unsupported assumptions and are often incomplete and fuzzy [Checkland, 1981]. The topic that we examine in conjunction with the important characteristics of the complex systems is that system thinking [Richmond, 1993] and an understanding of dynamic relations are not sufficiently evolved. The traditional ways (described as "bounded rationality") do not fulfill the requirements set by the complexity of systems defined within the new paradigm and this new socially formed and culturally shared mental model.

The paradigm, which is currently changing, dominated our culture for centuries, during which it formed modern western society and significantly influenced the rest of the world as well. This paradigm includes a number of ideas and values, which include belief in the scientific method as the only valid path to knowledge. Using systems thinking we can understand systems as wholes. Mastery of systems thinking brings not only certain skills and competencies, but a significant shift in understanding the world. The author feels this necessary shift in thinking, so called "metanoia" – thinking about things in the world surrounding us, not as separated unlinked events, but as a flow of events mutually interconnected and influencing each other in time and space. Metanoia – mental shift – represents in this situation something deeper than just learning something. It also means some change of oneself. It does not mean just receiving information, but rather this self-transformation – for example, the transformation of our mental models. The mental models then are essentially filters, by which we interpret our experiences, change plans and choose options. Philosophical systems thus can be seen as mental models of their authors.

2 Systems thinking and structuralism

It is very difficult to comprehend "things" as they really are. Understanding "things" is related to our paradigm of thinking. In the currently dominant paradigm of thinking we probably create a long list of problem causes in answer to the question "What is the cause of something?". The pre-requisite of this thinking process is that every factor contributes to the effect as a cause, causal effect runs only one-way and every factor is independent of the other factors. The systems thinking paradigm presents an alternative to each of these assumptions. According to this paradigm each cause is connected to its effect, but also to the next cause in the feedback loop. The direction in which systems thinking is moving is from the linear to the dynamic one, considering the changes in the environment in mental models.

Virtually all of the important characteristics of the systems approach can be found in the way of thinking called "structuralism" [Hawkes, 1999]. The author values the effort for reconciliation between philosophy, social and natural science and that although philosophical structuralism can be applied in sciences and represents the connection of theory and empiricism. Structuralism highlights the knowledge that facts can only be analyzed in their relationships, in their function and in the context of dynamically changing entity. Its central term "structure" corresponds to the term "system" and the way of its application is close to systems thinking.

3 Simulation for mental models

How to solve the problem? The contribution to thinking process improvement is the learning [König, 1996], which represents a change in mental model "vdouble-loop learning", a term first introduced by Chris Argyris. This concept is very useful contributing from the prospect of utilization in problem solution. Other authors are working on another issue: there are a lot of materials published dealing with the topic that a human is only capable of performing one conscious process at a time. Another frequently mentioned limit is the capability to understand only quite simple issues. As a result, there is a need to divide problems by a deductive method into groups comprising simpler and simpler ones until we are able to resolve them, as described by Descartes.

Will deduction be sufficient? Can we learn anything new in such a complex world without using a simulation? If we had a sufficiently developed intuition, if systems thinking were practiced from pre-school age and if we were able to recognize systems structures and their consequences, would we be able to improve our awareness of the systems dynamics and deem simulation useless? The Latin word "simulare" means "to imitate". A simulation model imitates the behavior of a real system. Its basis is an understanding of the relationships and links of the object being modeled. This has a close connection to mental models. As was mentioned earlier, our decisions and actions are not based directly on the real world, but on our mental images of the surrounding world, relationships between individual items and ideas about the consequences of our actions. We have conducted verification, where we compared how each practical problem was solved: 1) using only mental models and 2) when enhancing the mental models by a computer simulation.

Participants in the verification were students of the Prague University of Economics. The outcomes were very different. It has to be said that the student's decision making in the original strategies without using the simulation was largely defined by their fear of risk and implicit image of the company that was the subject of the problem solving exercise. The results of our study confirmed that when using mental models, decisions are not based on rational evaluation of goals, or possibilities and consequences, as we will see further on. What is important, however, are traditional procedures and stereotypes of thinking and acting. In the follow-up, problem-solving, use of the simulation, it was obvious to the students that their perception of reality and their mental models were somewhat absurd. Using iterative modeling and simulation approaches students were able to learn from their mistakes and improve their current mental models of reality.

4 Visualization and design in simulator development

The fact that problem-solving using simulation approximated a probable optimal solution was also given by the quality of the simulation used. Special modeling tools are used for simulations. Simulation models usually have the form of a computer program, which emulates the key features of the system being modeled. In our study, the simulation model was transformed to a simulator. The simulator is a model enhanced by a user-interface, supporting an interactive user approach to the model itself. This allows the user to change the conditions and create alternative situations. Simulations using the simulator allow elegant experimentations with the models.

The main reason for building our simulators [Mildeová, 2004] is the fact that experimenting directly with economic systems is unacceptable. We perceive the simulation as an experiment with the model that reflects the important characteristics of the system being monitored. The computer simulation was also applied as the methodology allowing also solving tasks, which are analytically solvable with difficulties or not at all. At an acceptable level of simplification, these models offer the possibility to simulate the strategy choice and other section of decision-making with an emphasis on dynamic relations within the first years of a company life cycle. An user-friendly interface allows understanding and simulating of various strategies and their consequences.

We used Powersim Constructor for our simulator development. Powersim is a modern piece of simulation software, which allows its users to develop complex simulators. The name Powersim arises from the words Powerful Simulation, efficient, effective simulation, which Powersim definitely is. The design of the front-end varies widely from the first computer simulators that were in a plain text mode to today's multimedia interface. Recently, computer simulator makers have tried to approximate standard graphics with web presentation. The use of the graphical environment and control elements arrangement, that most users are familiar with because of other program applications, makes for quicker insight of the user into the particular simulated problem. Similar rules to those for making presentations are valid here: suppress unsubstantial information, stress essential information.

Our effort during the interface design was the "modal representation" format not the "amodal representation", in which case the form is not related to the content. The user-interface was made as the element of user and simulator interaction, whose design can be very diverse. It points out that the simulator creators have to respect the manner of using the simulator and that for teaching purposes, it is important to have a clear and understandable interface. One working window system for the graphical interface was graphically designed. Graphical items are arranged in sections according to displayed blocks of information in this window. There is an information section (it displays important simulation output), a control-button section (it displays further windows), a system regulators section (it influences simulation behavior), and a time-development control section (simulation, initialization, sequencing). Graphs are used to facilitate understanding input and output of the simulation. Achieving high-quality graphical interface creation was not a simple business from the point of view of time consuming work, in particular. Graphical adjustments of the interface and testing took up most work time.

5 Getting over mental barriers

How did decision-making influenced only by mental models look like in our verification? We have seen that mental models used as a basis for making a decision, were very simple, even primitive. Often they were not correct; the consequences of the decision were not evaluated correctly. The verification confirmed that mental models are significantly limited by the "5 ± 2 rule", which has been proven by psychologists: at one moment man can grasp maximum of 5 ± 2 variables in their mutual interaction. They were solving problems with certain limitations; moreover they were not able to imagine dynamic causalities.

The clear implication is that understanding complexity, dynamics and behavior of complex production, technical or biological systems are impossible for an individual. The simulation modeling presented in our verification provided an opportunity to overcome this restriction. The simulation served for manipulation of mirrored models of our mental models. It helped us to understand causalities and connections between cause and effect in our mental models. Modeling and simulation helped us to understand how things work. It helped us to avoid mistakes and expanded our mental models, so we could understand causalities, causes and their effects.

Reality changes in real-time. Dynamic models, in contrast, change in compressed time, which allows us to see future possible developments. The simulation allowed us to look into the dynamic complexity (based on interaction between components), because it is hidden in time, in contrast to complexity based on the model structure. Through the simulation we were able to control time. So if we want to learn the basic principles of dynamic behavior in complex systems, we need to use the modeling technique of a simulation. The sense of this work is that students verifying that mental models alone is not enough. Enhancing them by a possibility of modeling appeared to have contributed to the company's success. From their part, the biggest benefit was in realizing the sensitivity of the parameters and the impact of unsupported assumptions on decision-making; the unclear influence of some parameters on the final output. The students valued that the simulation improved their perspective as to the quantity of possible solutions. We believe that this ,,getting over mental barriers" is the biggest asset for students (see Figure 1).

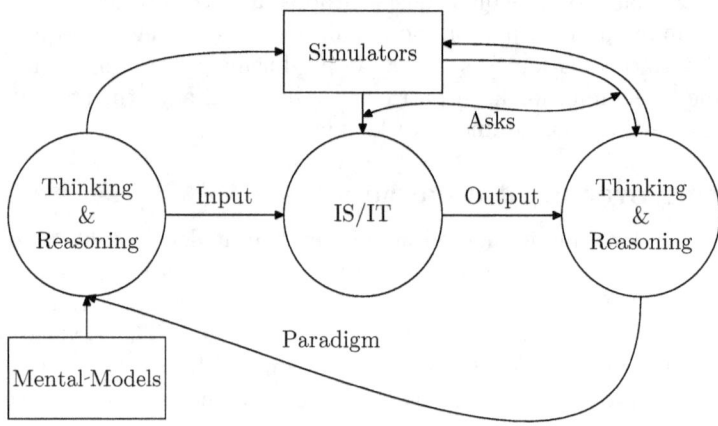

Figure 1. Getting over mental barriers

6 Conclusion

Various physical, social and ecological subsystems, shaping our world, are inter-related in a very complex manner. We can not keep up with this level of mutual interdependence when developing our mental capabilities. There still exists certain influence of conventionality and the current paradigm of perception of reality – mainly the linearization and tendency to omit feedback and delays. Our verification confirmed that we are significantly constrained when deducting logical consequences from mental models. We are especially limited in estimating dynamic relations. As opposed to the current paradigm of world perception (in the sense of Thomas Kuhn [1997]), it is important to view the processes not from the linear, independent point of view, but to perceive them as a group of interacting actions. It must be mentioned that these isolated causal loops are responsible for a system's behavior and that external factors can only influence it slightly. There is a significant mind shift, in this sense, in the way of problem-cause perception that will certainly change our overall attitude to the responsibility for deciding in social, economic, environmental [Mildeová, 1999] and other problems.

On the basis of our experiences from the teaching process in our courses in the Faculty of Informatics and Statistics, Prague University of Economics, we can state that the use of simulators improve systems thinking and understanding of dynamic relations. Students may see such simulators as a "micro world" or a "virtual laboratory". Simulators can help with the desired shift in reasoning (so called "metanoia" – perception of reality and things around us not as individual and separate but rather interacting and interdependent). Simulating and modeling cannot be taught from books or by communication with other people. The only way of improving oneself is by creating models and simulations – based on practical applications. We need to constantly develop our educational system, i.e. change the standard method of teaching and to further use computer-based teaching environments. One of the main barriers to these changes being implemented in an educational environment is the low ability of teachers, and their students, to change their paradigm of thinking. If we overcome these barriers and change our thinking, we will be able to better discern problems and their influence on the behavior of people.

Acknowledgements

The paper is supported by the project GA402/05/0502 from the Czech Science Foundation.

BIBLIOGRAPHY

[Checkland, 1981] P. B. Checkland. *Systems Thinking, Systems Practice*. John Wiley & Sons, New York, 1981.

[Chermack, 2003] T. J. Chermack. Mental models in decision making and implications for human resources development. *Advances in Developing Human Resources*, 5(4):408–422, 2003.

[Hawkes, 1999] T. Hawkes. *Structuralism and Semiotic*. Host, Brno, 1999.

[König, 1996] U. H. König. *Use of Simulation in Mangement and Management Education – Speeding up the Wheel of Learning?* Universität Mannheim, Mannheim, 1996.

[Kuhn, 1997] T. S. Kuhn. *The Structure of Scientific Revolution*. OIKOYMENH, Prague, 1997.

[Mildeová, 1999] S. Mildeová. *Structural economy – environment simulation model (SEESM), Final Report, PHARE Project OSS no. 85.2200.10.* VŠE, Prague, 1999.

[Mildeová, 2004] S. Mildeová. Creating business flight simulators for education. In Ch. Hofer and G. Chroust, editors, *IDIMT-2004*, pages 253–265, Linz, 2004. Trauner Verlag Universität.

[Richmond, 1993] B. Richmond. Systems thinking: critical thinking skills for the 1990s and beyond. *Systems Dynamics Review*, 9(2):113–133, 1993.

Stanislava Mildeová
Department of Systems Analysis
Faculty of Informatics and Statistics
University of Economics
Prague, Czech Republic
E-mail: mildeova@vse.cz

The Role of Theoretical Models in the Methodology of Physics

ANDRÉS RIVADULLA

ABSTRACT. The fruitfulness of the use of theoretical models is evident in nowadays physics. Indeed there is no branch in theoretical physics where models are not used. Models are hypotheses restricted to single phenomena or to limited number of phenomena, and they are particularly required in domains where there are not yet any theories available, or there are no theories at all. Any case models make use of extant theories in order to do their job: to save observable phenomena or to provide empirically testable predictions in their domains of intended applications. Nonetheless they are not true or verisimilar representations of reality, but merely instruments in order to deal predictively with Nature. The use of theoretical models offers an excellent argument against realism in physics. In order to support my view I will present in this paper some examples both of successful applications and failures of the Newtonian gravitational model in modern astrophysics and cosmology. These examples will show firstly that a priori we have no reason to be suspicious about what does not count as an intended application of a model. Since, moreover, revolutionary theory displacement embodies theoretical incompatibility, it becomes senseless to claim that models represent verisimilarly or approximately the empirical reality they are concerned with.

1 Introduction

Since the development of mathematical physics theoretical models have played an increasing role in the methodology of physics, and they have become indispensable for a better understanding of how physics deals with Nature. The awareness of the role of theoretical models in physics is one of the most important discoveries in the philosophy of science, as it sheds new light on the epistemological debate between realism and instrumentalism in the epistemology of physics.

Models are not theories -we might claim they are at most *hypotheses concerned* with a single phenomenon, or with a limited number of phenomena. But basically there is no fundamental difference between models and theories. Popper himself maintained that models might be called theories, although not all theories are models. Models, affirmed Popper too, incorporate theories. Thus Kepler's astronomy supplied an idealized geometrical model of planetary motion, and Newtonian celestial mechanics provided a very successful model for gravitational phenomena, which remained accepted until it was substituted by the more successful one provided by relativity theory: the pseudo-Euclidean four-dimensional spacetime. Theoretical models in atomic and nuclear physics would be unthinkable without quantum mechanics. And so would theoretical models in astrophysics without classical and quantum mechanics, and thermodynamics. Statistical mechanics suggested gas models for quantum particles like bosons and fermions. Hydrodynamics offered a model of the Universe as a fluid of galaxies in relativistic cosmology. Etc.

The common feature shared by all theoretical models is that they are intended to save the observed phenomena and to provide empirically testable predictions in their domains. But they are not committed to be true or verisimilar representations of reality. This is precisely the point where I disagree with most philosophers of science. Neither verisimilitude (Karl Popper), nor isomorphism (Max Black, Bas Van Fraassen), analogy (Mary Hesse, Giovanni Boniolo), similarity (Ronald Giere), or likeness (Sklar) are the alleged relationships between models and the world. The reason is that the fundamental demand of any model is empirical success, and since we do not know the phenomena under investigation *as they really are*, the inference from success to truth or to verisimilitude is not legitimate. Empirical success is no indicator of truth or verisimilitude. Weaker demands like analogy or similarity are even less justified. Following we are not entitled to claim that models *represent* the phenomena as they really are. Cartwright's approach to models as those objects of which theories are true only, is arguable as well, since it is not always the case that theories are available in all scientific domains [Cartwright, 1983].

In absence of theories, like in nuclear physics or in astrophysics, physicists postulate theoretical models in order to deal with the phenomena they are concerned with. As already suggested, they must be consistent with the theories they incorporate. Actually models are intended to fulfil experimental dialogues with Nature. But the fact that models allegedly refer to something *out there in the world* does not constitute any cogent reason for claiming that they represent it, in any realist sense of representation, since the access to Nature is possible only throughout the models themselves.

2 Truth and the progress of science

Despite the logical impossibility to achieve true theories about the world (logical problem of induction!), the view that truth is the aim of science, is widely accepted in modern theory of science. For instance, *scientific realist* Karl Popper [Popper, 1979, p. XXII] maintains a theory of truth as *correspondence*: a sentence is true when it agrees with, or corresponds to, facts or reality, and that "science aims at truth, or at getting nearer to the truth, however difficult it may be to approach truth, even with very moderate success" [Popper, 1994, pp. 173-174].

Thus there is a *tension* in Popper's picture of science. He maintains since his *Logik der Forschung*,1935, that absolute truth is the main regulative idea of science, and that we can compare theories in order to judge which one comes nearer to the truth. But, on the other hand, by logical reasons truth remains *unknown*. In spite of this crude reality, Popper [Popper, 1994, p. 161] insists on that "Though we can never justify the claim to have reached the truth, we often give some very good reasons, or justification, why one theory should be regarded to be nearer to it than another". Thus, in [Popper, 1994, p. 176], he claims that "There are many examples in physics of competing theories which form a sequence of theories such that the older ones appear to be better and better approximations to the (unknown) truth. For example, Copernicus' model appears to be a better approximation to the truth than Ptolemy's, Kepler's a better approximation than Copernicus', Newton's theory a better approximation still, and Einstein's better again". However, the different types of cases in which, according to [Popper, 1963, p. 232], we would say of a theory that supersedes another one, that it corresponds better to the facts, are merely different ways of claiming that the first one is more successful than the other. Thus the question remains, how can it be justified that success is an indicator of truth approximation.

Among contemporary physicists scientific realism enjoys wide acceptance. Einstein's realist view is well known. [Penrose, 1989, chapter 6] assumes a realist position too, and so does [Weinberg, 2001] and [Weinberg, 1998, p. 48]: "the task of science is to bring us closer and closer to objective truth". Many philosophers of science are concerned with the elucidation of the idea of *truth approximation*. From this it has resulted a family of different versions of realism, all with an indubitable *Popperian flavor*, like Philip Kitcher's *optimistic-induction* realism, Lawrence Sklar's *on-the-road-to-the-truth* realism, and Theo Kuipers' *evaluation methodology* realism, among others. It is obviously impossible to discuss here meticulously these views, which I will simply sketch in the following.

Every one acquainted with Popper's epistemology immediately sees that Philip Kitcher shares many ideas with him, to wit: that the most obvious epistemic aim is truth, that the kind of truth we are interested in is *significant* truth, i.e. true answers to significant questions, that scientists make discoveries about a world that exists independently of human cognition, that there exists a correspondence between conceptual/linguistic elements, like words, sentences, schemes, and the independent reality elements, what is "really there", etc. These philosophical theses allow [Kitcher, 1993, p. 122 and 137] to suggest that the wave theory proposed by Fresnel was nearer to the truth than Brewster's corpuscular theory, that Lavoisier improved Priestley, Dalton improved Lavoisier, and Avogadro improved Dalton. This approach to the idea of truth approximation bases on Kitcher's *optimistic induction*, as radically opposed to Laudan's pessimistic induction. Optimistic induction allows, according to [Kitcher, 1993, p. 137], to affirm that, from the perspective of a theory T_j, another theory T_{i+1} ($j > i+1$) "appears closer to the truth than T_i". The applicability of this rule depends on the use of more primitive terms like "more correct", "more complete", "more adequate", etc., and the whole enterprise aims at evaluating theories in terms of our expectations "that our theories will appear to our successors to be closer to the truth than those of our predecessors". Again the question remains open, whether empirical success is an indicator of truth approximation.

Also Lawrence Sklar defends the idea that aiming at the truth is the correct idea in scientific method. [Sklar, 2000, pp. 87–88] expresses in a very lucid way his view as follows: "We don't believe, now, that in the future we will believe our current best available theories to be true. But we may very well believe now, and with good reasons, that we are entitled, now, to believe of our current theories that they are 'on the road to the truth' and that in the future they will be looked upon as having 'been ahead in the right direction' ". Sklar's position is very daring, since he must convince us of the goodness of the "good reasons" he is assuming scientists have for believing that their present theories are approximately true. In spite of what "the general induction from past scientific revolutions" teaches – here sticks out again Laudan's pessimistic induction- [Sklar, 2000, p. 91] maintains firmly the idea that we may very well be willing to believe that our present theories are "heading in the right direction", "heading toward the truth", "pointing to the truth". Sklar's ultimate argument on behalf of the thesis that we can rationally believe that our current best theories are "stages along the road to the truth" is that despite the scientific revolutions "the theories inevitably bear deep conceptual relationships to one another at the level of their most abstract and most theoretical concepts" [Sklar, 2000, p. 94]. I myself believe that Newtonian mechanics constitutes a limiting case of relativity theory [Rivadulla, 2004a] [Rivadulla, 2004b, chapter IV], and that this fact is a paradigmatic case of the rationality of scientific change. But since both theories are incompatible with each other, as it will be emphasized in the Conclusion of this paper, it is impossible to claim that the latter comes objectively nearer to the truth than the other one. Only the *predictive balance* allows scientists to make a rational choice among incompatible theories, not the subjective belief that a theory is "pointing to the truth".

Among the three mentioned philosophers of science, I sympathize at most with Theo Kuipers' *evaluation-methodology* realism, because of its *instrumentalist* flavor. In the syncretism of his proposal lies the attractiveness of Kuipers' evaluationist methodology. Indeed, although [Kuipers, 2001, p. 62] opposes to Popper's falsificationism an instrumentalist methodology, according to which "the selection of theories should exclusively be guided by empirical success, even if the better theory has already been falsified", he shares Sir Karl's view that truth approximation can be warranted in the methodology of science. In this sense Kuipers is a realist, a *constructive realist* [Kuipers, 2000, chapter XIII], truth being for him a product of language and reality. Kuipers' *instrumentalist* methodology of theory selection, which is governed *à la Laudan* by the *rule of success*, is claimed to be more efficient for the purpose of truth approximation than Popper's falsificationist methodology. I agree completely with Kuipers' recommendation to aim at the most successful theory, because this fits in with my own view on the rationality of theory choice. Kuipers' strong thesis is that "if one applies the evaluation methodology, one comes, as a rule, closer to the truth, whether one likes it or not" [Kuipers, 2000, p. 241].

In order to present my own view I will concentrate in the following on the Newtonian gravitational model, as a case study with general philosophical implications. My aim will be to show that, despite its great success in many cases, it fails in at least as many other intended applications, which it should account of. My conclusion is going to be that *empirical success is no indicator of truth, verisimilitude or truth approximation.*

3 Some examples of the successful application of the Newtonian gravitational model in modern astrophysics

Newtonian mechanics, and in particular the Newtonian gravitational model, has been celebrated as one of the most successful theoretical tools in the history of physics. The discoveries of the Halley comet and the planet Neptune were flattered as a proof of its incomparable predictive power. Moreover, in *Newtonian approximation* it has been possible to calculate the masses of celestial bodies, like the Sun and the Earth [Rivadulla, 2004a, p.

141/144]. Etc. In order to reinforce the applicability of the Newtonian gravitational model, I will dwell briefly in the following with two more examples taken from modern astrophysics.

3.1 The computation of the masses of binary stars

Let us assume a system of visual binaries of masses M_1 and M_2, orbiting with period P around the system's gravity center and with the orbital plane perpendicular to the observer's vision line. The distance between both stars being $r_1 + r_2$, according to the *Newtonian gravitational model* both stars attract mutually with a force whose value is given by the *gravitation law*. By elementary arithmetic we obtain [Böhm-Vitense, 1989, p. 68] *Kepler's Third Law*:

$$P^2 = \frac{1}{M_1 + M_2} (r_1 + r_2)^3,$$

which, together with equation

$$\frac{M_1}{M_2} = \frac{r_2}{r_1},$$

due to the physical situation of the system, enable for the calculation of the stars masses.

3.2 Non-collapsing stars. How the Newtonian model *saves* the stability *spherically symmetric and static* stars.

Despite its own gravitation stars do not usually collapse. In order to account for this stability, the Newtonian approach starts with the simplifying assumption that stars are spherically symmetric, and that their interior exerts a pressure outwards, $F(p)$, which compensates the gravitational contraction force $F(g)$. Dimensional analysis requires for the situation to be described by equation [Ostlie and Carroll, 1996, pp. 317–318]

$$dm \frac{d^2r}{dt^2} = F(g) - dF(p).$$

Since by definition $dF(p)=AdP$, substituting for $F(g)$ the value given by *gravitational law* and applying the definition of density ρ, elementary arithmetic calculations allow us to conclude:

$$\rho \frac{d^2r}{dt^2} = -G_N \frac{M_r \rho}{r^2} - \frac{dP}{dr}.$$

In the ideal case of *static* stars, for which it applies that

$$\frac{d^2r}{dt^2} = 0,$$

we get

$$\frac{dP}{dr} = -\rho g,$$

where g denotes the gravity intensity at distance r from the star's center. In stellar astrophysics this expression is known as the *hydrostatic equilibrium equation*. It is one of the fundamental equations of the theoretical model of spherically symmetric and static stars interiors.

4 Some historical failures of the Newtonian gravitational model in astrophysics and cosmology

The extraordinary success of the Newtonian celestial model is no guarantee of its approximation to the truth. Indeed, as it will be shown in the following sections, not every intended application of this model is actually an application of it, and, what is much more serious, many intended applications do fail loudly, because the real systems do not fit in with theoretical predictions of the model.

4.1 The *Kelvin-Helmholtz* gravitational collapse hypothesis

The first hypothesis about the source of the energy of stars was put forward by Hermann von Helmholtz in a lecture given in Königsberg on February the 7th, 1854, on occasion of the 50th anniversary of Immanuel Kant's death. According to [Kelvin, 1867, pp. 493–494], Helmholtz's *meteoric theory*, later on known as the contraction hypothesis, "consists in supposing the sun and his heat to have originated in a coalition of smaller bodies, falling together by mutual gravitation, and generating, as they must do according to the great law demonstrated by Joule, an exact equivalent of heat for the motion lost in collision". For Kelvin it could scarcely be doubted, "That some form of the meteoric theory is certainly the true and complete explanation of solar heat".

Assuming the truth of the gravitational contraction hypothesis, stars luminosity was conceived of by Helmholtz and Kelvin as an intended application of the Newtonian gravitational model. In the hypothetical case of a star with constant density $\rho \approx \bar{\rho}$ and mass $M \approx \frac{4}{3}\pi R^3 \bar{\rho}$, the *total potential gravitational energy* of the star can be obtained [Ostlie and Carroll, 1996, p. 330] integrating over all star shells of thickness dr, to wit:

$$V_g = -4\pi G_N \frac{4}{3}\pi \bar{\rho}^2 \int_0^R r^4 dr \approx -\frac{3}{5} G_N \frac{M^2}{R}.$$

Now, according to the *virial theorem*, the total energy of a system in equilibrium is the half of its potential energy. Thus the *total mechanical energy of a star* should be

$$E \approx -\frac{3}{10} G_N \frac{M^2}{R}.$$

Applying this formula to the Sun, assuming a constant *luminosity* during the Sun's whole life, and given that luminosity is power, i. e. energy per time unity, our Sun's present age should be circa 10^7 years. This age is bizarrely short, since it contradicts all paleontological, geological, and astronomical available data, as [Eddington, 1930, p. 290] already recognized.

What can we learn from this episode of early astrophysics? The most immediate reaction could be to doubt whether it was right to consider the source of stellar energy as an intended application of the Newtonian gravitational model, i. e. if Helmholtz and Kelvin themselves should be blamed, not the Newtonian model itself, as they did "exceed" its application limits. Maybe. But, as a matter of fact, we can only recognize the uselessness of a model after it has proven its inefficacy to deal with some aspects of the reality. A priori, we have no reason to be suspicious about what does not count as one of its intended applications. This is a good reason to consider theoretical models merely as provisional tools for dealing with Nature.

4.2 *Classical failures* of the Newtonian gravitational model

It is widely known that the Newtonian model fails in many intended applications, like Mercury's perihelion, the light deflection by the Sun, and the gravitational red shift, which it should account for, *if it were true of gravitational phenomena*. Indeed, Einstein affirms, that the Newtonian account of the light deflection by the Sun and of the advance of Mercury's perihelion show the failure of classical mechanics [Einstein, 1996].

That the light deflection by the Sun must be considered an intended application of the Newtonian gravitational model is for me indubitable. Newton himself was interested in this question [Rivadulla, 2004b, note 18], and as Clifford Will [Will, 1986, chapter IV] relates, during the eighteenth and nineteenth centuries John Michell, Pierre Simon Laplace and Johann Georg von Söldner were looking for a solution of this problem. It is relatively easy to find out that in *Newtonian approach* the light deviation by the Sun amounts to 0.87 seconds [Kittel, 1965, Chapter 14]. Unfortunately, this quantity amounts to the half of the value predicted by general relativity. On the contrary, according to Clifford Will [Will, 1993, section 14.3], recent measurements of light deflection agree with general relativity theory to 0.1%.

The history of Mercury's perihelion is generally better known. In order to save the observations, Le Verrier assumed the existence of Vulcan, a planet whose gravitational interaction with Mercury would be responsible for the observed *anomalous perihelion*. Indeed, this *ad hoc* hypothesis increased the falsifiability of the Newtonian celestial mechanics. But since the hypothesized planet was not found, this assumption made the situation of Newtonianism still more hopeless. According to [Kittel, 1965, chapter XIV] Kittel's calculation, in *Newtonian approach* Mercury's perihelion precession amounts to 13 seconds per century, very distant from the observed 43 seconds. On the contrary, general relativity predictions fit in very well with observations.

In modern astrophysics and in cosmology theoreticians deal with a quantity z, which is used both to measure the expansion of the Universe and the effect of the gravitational field of a star on a photon leaving it. This quantity is called *red shift* and is given by $z = \nu_e/\nu_o - 1$, where ν_e and ν_o denote respectively the frequencies emitted and observed of light. If the star at issue is a black hole, it determines an *event horizon* beyond which photons cannot escape. Relativistic cosmology accounts precisely for this fact, as it allows to deduce that the red shift of a photon trying to escape from a black hole is infinite. Curiously the gravitational red shift can be considered too an intended application of the Newtonian model. Indeed, as early as 1783 John Michell adventured the hypothesis of the existence of black holes, as he called *dark stars* a kind of possible celestial objects, whose escape velocity should be, because of their enormous masses, larger than the velocity of light. Unfortunately the *Newtonian approach* to this problem fails completely [Rivadulla, 2004b, note 20] since it allows black holes to confer on photons positive red shift values, thus not distinguishing between usual stars and black holes.

5 Conclusion

It might arise the question, whether or not there is still some room left for dealing with truth approximation in the philosophy of physics. The answer must be: No, there is not. Indeed the discrepancy between Newtonian mechanics and relativity theory is deeper, than merely conceiving of the latter as some kind of *extension* of the former. The reason is that both are mathematically incompatible with each other: Newtonian mechanics bases on a three-dimensional Euclidean world, whereas relativity theory bases on a four-dimensional pseudo-Euclidean world. Thus, since the incompatibility is total, it cannot be claimed that the Newtonian model – the most successful tool for dealing with gravitational phenomena until its displacement by relativity theory – represents gravitational phenomena verisimilarly or approximately, or that it bears a certain analogy or similarity to reality.

But the situation is still more dramatic, if we resort to nuclear physics.

Here there is no theory available embracing nuclear phenomena. Theoretical models of the nucleus are only here at work. The theoretical *drama* lies in the fact that incompatible nuclear models are at work for different nuclear phenomena; i.e., different and incompatible nuclear "hypotheses" can be assumed, according to which kind of empirical phenomena they are supposed to save.

All this has serious consequences for the epistemology of physics. There is no sense in the claim that models represent aspects of the world in a realist sense of the term *representation*. I insist that since the inference from empirical success to verisimilitude, analogy or similarity cannot be justified, we are not allowed to maintain that models represent the reality. Moreover, theoretical models do not provide any explanation either of the phenomena they are dealing with. Thus, they have to be conceived of merely as conceptual tools in order to deal scientifically with Nature.

Acknowledgements

This paper is part of a research on *Theoretical Models in Physics* supported by Spanish Ministry of Education and Science.

BIBLIOGRAPHY

[Böhm-Vitense, 1989] E. Böhm-Vitense. *Introduction to Stellar Astrophysics*. Cambridge University Press, Cambridge, 1989.
[Boniolo et al., 2002] G. Boniolo, C. Petrovich, and G. Pinsent. Notes on the philosophical status of nuclear physics. *Foundations of Science*, 7:425–452, 2002.
[Cartwright, 1983] N. Cartwright. *How the Laws of Physics Lie*. Clarendon Press, Oxford, 1983.
[Eddington, 1930] A. S. Eddington. *The Internal Constitution of the Stars*. Cambridge University Press, Cambridge, 1930.
[Einstein, 1996] A. Einstein. Über die spezielle und allgemeine relativitätstheorie. In A. J. Fox et al., editor, *The Collected Papers of Albert Einstein, Vol. 6*. Princeton Univ. Press, 1996.
[Kelvin, 1867] L. Kelvin. On the age of the sun's heat. *Macmillan's Magazine*, 1867. Reprinted as Appendix E of his Treatise on Natural Philosophy, University Press, 1867, 1903,Cambridge.
[Kitcher, 1993] P. Kitcher. *The Advancement of Science: Science without legend, objectivity without illusions*. University Press, Oxford, 1993.
[Kittel, 1965] Ch. et al. Kittel. *Mechanics*. McGraw-Hill Book Company, New York, 1965. 2nd Edition.
[Kuipers, 2000] Th. Kuipers. *From Instrumentalism to Constructive Realism. On some relations between confirmation, empirical progress, and truth approximation*. Kluwer Academic Publishers, Dordrecht, 2000.
[Kuipers, 2001] Th. Kuipers. *Structures in Science. Heuristic Patterns based on cognitive structures*. Kluwer Academic Publishers, Dordrecht, 2001.
[Ostlie and Carroll, 1996] A. Ostlie and D. Carroll. *An Introduction to Modern Astrophysics*. Addison-Wesley, Reading, Massachusetts, 1996.
[Penrose, 1989] R. Penrose. *The Emperor's New Mind*. University Press, Oxford, 1989.
[Popper, 1963] K. Popper. *Conjectures and Refutations. The Growth of Scientific Knowledge*. Routledge and Kegan Paul, London, 1963.

[Popper, 1979] K. Popper. *Die beiden Grundprobleme der Erkenntnistheorie*. J. C. B. Mohr (Paul Siebeck), Tbingen, 1979.
[Popper, 1983] K. Popper. *Realism and the Aim of Science*. Hutchinson, London, 1983.
[Popper, 1994] K. Popper. Models, instruments and truth. In K.R. Popper, editor, *The Myth of the Framework. In defence of science and rationality*. Routledge, 1994.
[Rivadulla, 2004a] A. Rivadulla. *Éxito, Razón y Cambio en Física. Un enfoque instrumental en teoría de la ciencia*. Editorial Trotta, Madrid, 2004.
[Rivadulla, 2004b] A. Rivadulla. The newtonian limit of relativity theory and the rationality of theory change. *Synthese*, 141:417–429, 2004.
[Sklar, 2000] L. Sklar. *Theory and Truth. Philosophical Critique within Foundational Science*. University Press, Oxford, 2000.
[Weinberg, 1998] S. Weinberg. The revolution that didn't happen. *The New York Review of Books*, XLV/15:48–52, 1998.
[Weinberg, 2001] S. Weinberg. *Facing up*. Harvard Univ. Press, Cambridge, MA, 2001.
[Will, 1986] C. Will. *Was Einstein Right?* Basic Books, New York, 1986.
[Will, 1993] C. Will. *Theory and Experiment in Gravitational Physics*. University Press, Cambridge, 1993. Revised edition.

Andrés Rivadulla
Universidad Complutense
Facultad de Filosofía, Madrid
E-mail: `arivadulla@filos.ucm.es`

The *Evolutionary History* of Models as Representational Agents

DEMETRIS PORTIDES

ABSTRACT. The Nuclear model research program is characterized by two distinct stages of development, each of which gave birth to models that represent the nuclear structure. Despite the fact that aspects of the models of the first stage are present in those of the second, the latter models are the result of a conceptual framework which is distinct from its predecessors. In moving from the first stage of development to the second quantum theory plays a minimal role. In fact, the evolutionary history of nuclear models demonstrates that scientific models possess a partial independence from theory, which is primarily based on the causal reasoning involved in their construction and the explanatory power they acquire because of the former. Through an analysis of the causal reasoning involved in the construction of the various nuclear models I explore the ways by which the two stages of the nuclear research program relate. In order to articulate an adequate logical analysis of the relation that accounts for the evolutionary history of the nuclear models, I argue, we must treat the models as partially independent from quantum theory. The approach suggested in this paper conflicts with the suggestion of the Semantic View that models of the theory represent their respective physical systems. I argue that scientific models acquire representational power because the assertions of the theory used in their construction are supplemented or enriched with theory-independent causal reasoning. It is the causal reasoning involved in the construction that gives scientific models their explanatory power, without which they lack the capacity to represent actual physical systems.

1 Introduction

The question of how the assertions of a scientific theory relate to the phenomena has been explored in philosophy of science over time long and has been traditionally addressed by the use of formal methods through either of two approaches: (1) that the deductive consequences of the theory stretch all the way to the phenomena, or (2) that the theory represents the phenomena structurally via its semantic models.

The first approach, which was held by the logical positivists and came to be known as the Received View of scientific theories [Putnam, 1962], is the conception of scientific theories as Hilbert-style formal axiomatic calculi [Suppe, 1977] whose logical interpretation is partially supplied by meta-mathematical models, i.e. structures in the Tarskian sense. Models in this sense are not vehicles of representation of physical systems, but the means by which subsets of sentences of a theory are interpreted, i.e. they are structures that make a set of sentences all true. In this view, the media of scientific representation are sentences (i.e. linguistic entities); models are a secondary form of scientific theorizing that facilitates the understanding of the formal calculus and hence, in a sense, are representationally-redundant. The Received View was criticized – and to a large extent abandoned – primarily on grounds that are by–products of its focus on syntax. Namely, it implies a theoretical/observational distinction and an analytic/synthetic distinction in the vocabulary and sentences of a theory both of which are not tenable distinctions in any clear sense. Also for the reason that, it relies on the notion of correspondence–rules for providing partial interpretation of the formal calculus, which is not a clearly defined notion. Lastly, because it withholds from models their representational role which is a conspicuous feature of scientific practice.

The second approach, which is the successor of the received view, is known as the Semantic or Model-theoretic conception of scientific theories. In the Semantic View (SV), a theory is identified with a class of mathematical models which would have been interpretations (i.e. the intended models) of a formal calculus were the theory formalized [Van Frassen, 1990; Giere, 1988]. The proponents of the SV hold, however, that this class of models can be directly defined without recourse to a formal language, in other words models are given the status of being constitutive parts of theories. In this sense models are mathematical structures that represent their target systems through isomorphism or other forms of mapping [Van Frassen, 1990; Suppe, 1989; Giere, 1988; Da Costa and French, 2003]. Scientific representation consists merely in mapping the elements and relations of a theoretical model onto the elements and relations of a data structure of the target system. Furthermore once the theory structure, i.e. the class of models, is defined an indefinite number of models becomes available for the representation of physical systems.

The SV, just like its predecessor, is meant to be a rational reconstruction of scientific theories and not as a precise description of how actual scientific modeling is carried out. Nevertheless, it entails that scientific models are characterized by specific features, attributes and functions that, it could be argued, are not met in actual scientific theorizing and modeling. For instance, the semantic view entails that theoretical models are the vehicles of representation of physical systems, and furthermore that every scientific model that represents its target is a constituent part of theory or could be subsumed under the theory structure. If it could be argued that this is not what is observed in scientific practice but that actual scientific models that are proposed for the representation of physical systems differ significantly from theoretical models [Morrison, 1999; Cartwright, 1999], then we could justifiably infer that the representation relation that the SV dictates is not sufficient to understand scientific representation and the nature of scientific models. And if scientific models do not represent physical systems merely by virtue of being mathematical structures (subsumed or not under a theory structure), then what the SV advocates about models and representation and about the relation between theories, models and phenomena is not necessary for explicating the role of models in scientific representation [Portides, forthcoming]. Although these considerations give rise to important questions in the context of a discussion on the representational role of models, they will not be addressed directly in the sequel. Instead, in this paper I shall focus on the fact that rational reconstructions that attempt to formulate a logical framework into which theories can be essentially reformulated, in effect fail to capture the function of models that renders them the primary media of the historical evolution of our theoretical representations of phenomena.

My argument is twofold. The positive aspect of the argument is to try to establish that the function of models as representational devices could be fully understood only if their evolutionary history is assembled and analyzed. The negative aspect is to show that because rational reconstructions, like the SV, overlook the evolutionary history of models, they obscure the latter's representational function.

Within the diverse number of roles that have been attributed to scientific models that of being representational devices stands out. It is central to the quest for clarifying and understanding the notion of "model", as used in science, that we analyze scientific models in connection to their role in theoretical representation of phenomena. In this paper I try to see how antecedently available information (whether experimental, semi–empirical, or theoretical), which at earlier stages of development of a particular scientific domain is used fragmentally in isolation from each other within different models to represent different aspects of the target physical system, is integrated into new representational models. The latter amalgamate all the accumulated fragments of the relevant antecedent information in order to give a representation of all known aspects of the target physical system. This is what I call the evolutionary history of models. In looking at the evolutionary history of models the focus is not on mathematical structure, as dictated by rational reconstructions of the above kind, but on the known aspects of the causal structure of the target system and the reasoning involved in developing the causal explanatory mechanisms that feature in the model's representation of the target system.

One of the cases from the history of physics that exemplifies the thesis I want to motivate, that the evolutionary history of models is a crucial factor in understanding the models' representational role, is the development of models of the nuclear structure. After Heisenberg proposed the proton-neutron hypothesis, quantum mechanics could not be applied directly to the nuclear domain because it gives rise to the nuclear many-body problem. The problem could be understood as follows. Consider the internal energy E of a nucleus of A nucleons, given as an eigenvalue of the Schrödinger equation $\left(i\hbar \frac{\partial \psi}{\partial t} = -H\psi\right)$, where H is the Hamiltonian of the nucleus, the eigenfunction $\psi(r_1, ..., r_A)$ is the wavefunction, and r_i denotes the position of the ith nucleon:

(1) $H\psi(r_1, ..., r_A) = E\psi(r_1, ..., r_A)$

If we were to apply the Schrödinger equation to compute the eigenvalues for E, we would express the Hamiltonian as the sum of the kinetic energy operator for nucleonic motion and the potential energy operator for interaction between the nucleons, $H = T + V$, where for nucleon mass m:

(2) $T = \left(-\hbar^2/2m\right) \sum_{i=1}^{A} \nabla_i^2$, and, $V = \sum_{i>j}^{A} \sum_{j=1}^{A} V_{ij}$

The V_{ij} corresponds to the interaction potential between nucleons i and j. It is known that expressing the potential energy as a sum of pair-wise terms is an idealization, since influences on the pair-wise interactions are exerted by the presence of other nucleons, but this fact would merely add redundant complexities to the problem. The nucleus can exist in different bound states, characterized by different wavefunctions and different values of E, as well as of other observable quantities. The different eigensolutions to equation (1) would correspond to the different states of the nucleus. Consequently, if we could solve the Schrödinger equation, the eigenvalues E would give us the energy of the different states and from the corresponding eigensolutions ψ we would be able to extract all possible information concerning all other conceivable properties of these states. All this is what quantum theory instructs us; yet solving the Schrödinger equation for the nucleus poses enormous problems. Firstly, the nature of the pair-wise nucleon-nucleon interaction is not completely known, nor is the influence on it from the presence of other nucleons. Secondly, even if this interaction is specified, we encounter insurmountable computational difficulties for the cases of more than two nucleons that force us to resort to variational techniques for solving the nuclear many-body problem at the expense of acquiring any significant insight into the nuclear structure and properties.

These difficulties are indicative of the fact that none of the available quantum mechanical theoretical models (e.g. the harmonic oscillator, the hydrogen atom, etc.) fit the world of nuclear physics. In order to apply quantum mechanics to the nuclear domain we need to construct the Hamiltonian operator for the nucleus, more or less, from scratch. It is therefore not surprising that the nuclear structure research program developed through two temporally and methodologically distinct stages.

In the first stage, two conflicting hypotheses coexisted, each of which gave rise to a different category of nuclear models. In accord with the first hypothesis the nucleus is assumed as a collection of closely coupled particles and the models (e.g. the *liquid drop model*) developed in this category accounted only for collective modes of nuclear motion, and ignored any plausible relative motion between the nucleons. In accord with the second hypothesis the nucleons are assumed to move in rather independent ways in an average nuclear field and the models (e.g. the *single particle shell model*) developed in this category accounted for nuclear motion only as an aggregate of independent nucleon motion. Whichever hypothesis is used as the starting point of the conception of the nucleus, quantum theory plays a minimal role in model construction. The primary criterion for model acceptance is its explanatory power and one of the characteristics that distinguish the two categories of models is that each explains well the phenomena related to the nucleus for which the other category fails to provide adequate explanations. Models of the first category, for instance, explain well nuclear fission and the electric quadrupole moments of the nuclei, whereas those of the second category explain well the "magic numbers" of the nuclei.

The second stage consists in the attempt to integrate all the preceding results and the acquired knowledge about the nucleus, into one, more or less, "all-inclusive" model. This model, which came to be known as the *unified model* of nuclear structure, is based on a rather complex hypothesis about the nucleus, which is the product of amalgamating the results of the two categories of models of the first stage. Despite the fact that aspects of both kinds of earlier models are present in the unified model, the unified model itself is constructed within a conceptual framework which is distinct from those of its predecessors. In moving from the first to the second stage of development, quantum theory also plays a minimal role. In fact, the evolutionary history of nuclear models demonstrates that scientific models possess a certain degree of partial independence from theory, as argued by [Morrison, 1999], which is primarily based on the causal reasoning involved in their construction and the explanatory power they possess. The unified model exemplifies this thesis.

2 The first stage of development

The *liquid drop model* of nuclear structure [Moszkowski, 1957] is based on the analogy that the mean free path of nucleons must be significantly small compared to the nuclear radius, just as the mean free path of molecules in a liquid drop is small compared to the radius of the drop. According to the model, because any energy acquired by a nucleon is quickly shared, nuclear excitations involve collective displacements of many nucleons. Thus the motions of individual nucleons are ignored and the nuclear wavefunction is entirely described in terms of the position of the nuclear surface. To set up the energy equation a series of idealizing classical assumptions are employed, such as: (1) The nucleus in its stable state has spherical shape, and for small deviations from sphericity, where the surface undergoes deformation oscillations at constant density, the surface tension of the nucleus acts as a restoring force. (2) The energy of the nucleus is the sum of the volume energy, surface energy and Coulomb energy and that, on the assumption of incompressibility, the volume energy is independent of the nuclear shape, the surface energy is least for spherical shape and increases with deviation from sphericity, and the Coulomb energy decreases with deviation from spherical symmetry. The result is a classical energy function for the collective motion, where $E(0)$ is the energy for spherically symmetric shape, C_λ are nuclear–deformation–resistance (classical) coefficients, the quantities B_λ are mass parameters, and $\alpha_{\lambda\mu}$ are deformation classical time–dependent spherical tensors:

$$(3) \quad H = E(0) + \sum_\lambda \sum_\mu \left(\tfrac{1}{2} B_\lambda \left| \dot{\alpha}_{\lambda\mu} \right|^2 + \tfrac{1}{2} C_\lambda \left| \alpha_{\lambda\mu} \right|^2 \right)$$

The classical functions are then quantized by introducing momenta $\pi_{\lambda\mu}$, canonically conjugated to the $\alpha_{\lambda\mu}$, so that the Hamiltonian operator takes the form:

$$(4) \quad H = E(0) + \sum_\lambda \sum_\mu \left(\tfrac{1}{2} \frac{\left| \pi_{\lambda\mu} \right|^2}{B_\lambda} + \tfrac{1}{2} C_\lambda \left| \alpha_{\lambda\mu} \right|^2 \right)$$

Though the liquid drop model is what we would call a semi-classical model, it could be claimed that it represents the nucleus, albeit partially, because it explains certain properties of nuclei, namely nuclear fission and the electric quadrupole moments; it could also be claimed that the SV cannot accommodate such a model because it cannot be related to any theoretical model of quantum mechanics [Portides, forthcoming]. The representational function of the liquid drop can be discerned in the reasoning involved in its construction. The development of mass-spectroscopy along with other experimental results, led in 1935 to von Weizsäcker's semi-empirical result about the binding energy of the nucleus. His semi-empirical mass formula consisted of five components, where Z is the proton number and A is the total nucleon number:

$$(5) \quad B = C_{vol}A - C_{surf}A^{2/3} - C_{coul}Z^2 A^{-1/3} - C_{sym}(A-2Z)^2 A^{-1} - C_{pair}A^{3/4}\delta$$

These five terms are the causal factors that ostensibly affect the nuclear binding energy. The first three terms are just of the form suggested by the classical analogy with the charged liquid drop. If we consider an infinitely extendible liquid (of constant density) then the energy would be proportional to the number of particles. In the nuclear analogy this volume energy is the average energy due to saturated bonds between the nucleons, which increases B. But since the nucleus is finite, the nucleons near the surface should interact with fewer nucleons (i.e. there should be unsaturated bonds). Thus B should decrease by an amount proportional to the surface area, i.e. to $A^{2/3}$. Furthermore, the binding energy reduces more on account of the Coulomb repulsion between any two protons. This is inversely proportional to the distance between two protons, which turns out to be inversely proportional to $A^{1/3}$. At this point the classical analogy ceases to help, and the following two considerations suggest the addition of the last two terms in equation (5). The tendency of nuclei to have equal numbers of protons and neutrons N gives rise to the symmetry term which for $Z = N$ diminishes. Also a pairing term must be added in order to reproduce the special stability of even-even nuclei and the almost complete absence of stable odd-odd nuclei. Thus in the Weizsäcker formula, $\delta = +1$ for odd-odd nuclei, $\delta = 0$ for odd-nucleon nuclei, and $\delta = -1$ for even-even nuclei.

The liquid drop model is a valuable guide for *explaining* the Weizsäcker formula, despite the fact that more detailed models are required to relate the magnitudes of the various terms to the basic interactions between nucleons. The success of the formula in yielding relatively accurate values and in reproducing all important nuclear trends, except for the lightest nuclei, can therefore be regarded as an indicator of the relative success of the model. One such success is the explanation of the phenomenon of nuclear fission of heavy elements. Nuclear matter is assumed to be incompressible, just as a liquid almost is, but deformation is possible. If a spherical nucleus is deformed into an elongated shape the following things would happen. First the

Coulomb repulsion is diminished because the average distance between protons increases. Second the surface energy increases because the surface area increases. These two changes, that have opposing effects on the magnitude of the binding energy, mean that heavy nuclei will demonstrate instability against deformation. This is so because the Coulomb energy increases with Z^2, whereas the surface energy increases with $A^{2/3}$, hence for large Z the Coulomb energy prevails. For light nuclei, on the other hand, the surface tension is more significant hence the spherical shape is the stable configuration. Therefore, a deformation of a large nucleus, whether spontaneous or initiated by the capture of a particle, may lead to a large deformation and subsequently to a split-up into two or more parts of comparable mass. The liquid drop model does not just offer this qualitative explanation but it also provides, to a first approximation, good quantitative results for fission. Even though some important properties of the nucleus are not adequately accounted for by the model (e.g. the special stability demonstrated by some nuclei with particular numbers of protons or neutrons, known as the 'magic-number' nuclei), the conspicuous element in the discussion of the liquid drop model is that in its construction there is a constant interplay between theory, model and semi-empirical results. The Weizsäcker formula is the primary factor in the construction of the model, because, in addition to being a guiding heuristic in constructing the model, it is the explanandum of the model and what the model must tie to the abstract terms of quantum mechanics. The liquid drop provides an explanation for three of the causal terms of the Weizsäcker formula by providing a description of the physical mechanism that represents the nucleus and gives rise to the terms of the formula, thus it represents some of the factors that are responsible for such effects as nuclear fission.

The *single particle shell model* of nuclear structure is based on an analogy with the physics of the atom, and in particular the orbital structure of electrons in complex atoms. In the nuclear analogy, the nucleons are assumed to move approximately independently, in spite of the strong interactions known to exist between free nucleons. The principal presupposition behind this model could be made more precise if it is assumed that for an odd-A nucleus the nucleons are regarded as filling the shells (stationary orbits) in such a way that all of them except the last odd nucleon pair-off to form an inert core. This core is further assumed not to contribute at all to the angular momentum or the electromagnetic moments of the nucleus. Thus, the nuclear picture we are faced with is that of the remaining odd nucleon acted upon by the rest of the nucleus via a prescribed potential. Notice however that unlike the atomic case, in the nuclear case there is no central field produced by an external source (as the nucleus of an atom presents a Coulomb field for the orbiting electrons); one has only the strong attractions between the nucleons. So a corollary to the above assumption is that we consider the motion of the odd nucleon in the nucleus, under the influence of all other nucleons of mass M, as motion in a spherically symmetric *fictitious* central field of force. This idealization, or theoretical distortion, leads to the problem of how to describe the effective mean potential within the nucleus such that it represents the physical system and its mechanisms. The potential energy part of the Hamiltonian operators could be chosen from the list of stock potentials of quantum mechanics. One such postulated potential is the *infinite square well*:

(6) $\quad V(r) = \begin{cases} -V_o, \text{ for, } r < R. \\ \infty, \text{ for, } r > R. \end{cases}$

Another is that of the *infinite harmonic oscillator well*:

(7) $\quad V(r) = -V_o + \frac{1}{2} M \omega^2 r^2$

Both of these well-known potentials bear a very distant resemblance to a real nucleus because, among quite a few other things, they do not provide the possibility of barrier penetration through tunneling[1]. Stock potentials such as a *finite square well* or a *finite square well with rounded edges*, which bear a closer resemblance to the realities of the nucleus, can only be solved numerically. Therefore, although they do indicate significant improvement in predictive success on several fronts, they offer no help in gaining qualitative insight from their solutions. To avoid unnecessary dwelling on the problems faced by such models I will point out just one. All these models predict only some (and not all) of the magic numbers, and since this is one of the primary nuclear features that an independent particle model –in particular, one that assumes spherical symmetry– should account for, questions in regard to the reasons for this discrepancy are sound. It is experimentally known that nuclei with either a proton or a neutron number that coincides with the magic numbers do not possess electric quadrupole moments in the ground state. Quadrupole moments are defined to measure deviation of the nuclear density from spherical symmetry; zero quadrupole moment indicates a spherical nuclear shape and large quadrupole moments indicate large spherical asymmetry. Since shell models assume spherical symmetry it is expected that they should demonstrate zero quadrupole moments and hence predict the magic numbers. But in addition to such predictive discrepancies, no one expects the form of field generated by the nucleons to be

---

[1] The transmission of energy even though the energy lies below the top of the barrier. This is a wave phenomenon and in quantum mechanics it is also exhibited by particles.

that of a harmonic oscillator or an infinite square well. The stock models of quantum mechanics offered no help on the issue, and the Hamiltonian operator was pursued phenomenologically in order to reproduce *inter alia* the magic numbers.

It was suggested [Mayer, 1948] that a non-central force term be added to the central force potential of the shell model. This term was based on an *ad hoc* hypothesis, which postulated that there is an interaction between the orbital and spin angular momenta of the unpaired nucleon, and which implied that the potential energy operator should take the following form:

(8) $\quad V(\mathbf{r},\mathbf{l},\mathbf{s}) = V(\mathbf{r}) + \frac{2\alpha}{\hbar^2}(\mathbf{l}\cdot\mathbf{s})$

Where the central potential $V(\mathrm{r})$ can take any desired form, and for the spin-orbit coupling term the value of the constant α could be considered to have no radial dependence. The mechanism of the interaction was unknown. It derived its usage from the physics of the atom, but in the latter case the spin-orbit coupling of an electron bound in an atom arises from the interaction of the magnetic moments associated with the angular momenta. If an electron moves in an electric field with certain velocity, a magnetic field is induced at its location. This induced magnetic field has an effect on the magnetic moment associated with the electron spin. From elementary electromagnetism it can then be shown that the magnetic moment is measured through its interaction energy with an external homogeneous magnetic field. That is to say, the operator representing the interaction energy is given by the scalar product of the magnetic moment with the induced magnetic field. The magnetic moment is a function of the spin of the electron, and the magnetic field can be shown to be a function of the orbital angular momentum. Consequently, the interaction energy of the electron can be expressed as a spin-orbit coupling (i.e. a scalar product of the orbital and spin angular momenta). For an electron bound in an atom we thus have a theoretically justified way of showing that there is an energy part of the Hamiltonian operator that is associated with the spin-orbit coupling. For the nucleons, however, the spin-orbit coupling cannot arise from an induced magnetic field, because neutrons are uncharged particles and thus their motion does not give rise to magnetic fields. The mechanism of the spin-orbit interaction of nucleons was therefore largely unknown. A dimensionless constant (represented here as α) was introduced and a change was made in the algebraic sign on the spin-orbit coupling operator of the atomic case, which implied that the mechanism which gives rise to the nucleon spin-orbit coupling is different from the electromagnetic mechanism that operates for electrons in an atom.

The spin-orbit coupling was introduced into the potential energy operator of the shell model in an *ad hoc* manner, simply to adjust the model's predictions and thus account for empirical observations. The addition of the spin-orbit coupling makes the empirically known magic numbers occur as closed sub-shells, and with certain adjustments they could be made to appear as clearly separate closed major shells without moving beyond this model by interpolating a potential $V(r)$ between the two extreme cases of the harmonic oscillator and the infinite square well, e.g. the following suggestion due to [Woods and Saxon, 1954]:

$$(9) \quad V(\mathbf{r}) = -V_o \left(1 + e^{(r-r_n)/t}\right)^{-1}$$

In this operator, r_n is the nuclear radius, which can be taken to be equal to the radial distance at which the absolute value of the potential drops to one half of its central value; and t is the surface thickness (or diffuseness of the nuclear surface, if one is considering nuclear scattering) of the potential. Both of these quantities contribute to the magnitude of the central potential.

Although there is no theoretically systematic way to justify the introduction of the spin-orbit force, as is the case for the atom, it does not mean that experimentally motivated arguments are disqualified from providing the justification. Indeed this is what takes place. Because of the piecemeal fashion by which the model is constructed, the concern with substantiating the spin-orbit force shifts to the experimental facts. Numerous scattering experiments were conducted some of which succeeded in showing that the spin-orbit force exists in nuclear matter (e.g. [Adair, 1952]; [Signell and Marshak, 1958]). Empirical evidence supporting a postulated physical mechanism is usually acquired by the use of a mode of scientific inquiry which does not involve the theory directly but is strongly associated with representational models. This mode of inquiry is peculiar to representational models because when used together with experiment it provides a way to render causal mechanisms that are constitutive parts of models explanatory. It involves holding constant the important components of the model-defining hypothesis and altering some of its subsidiary parts, in an attempt to discover empirical evidence in support of a postulated causal mechanism. This was the case for the spin-orbit hypothesis, which can serve as an example to clarify this point. One of the conjuncts in the defining hypothesis of the model (other than the spin-orbit coupling hypothesis) can be modified in ways that enable inferences to be made about the postulated spin-orbit mechanism. One of the hypotheses that underlie the single particle shell model involves the idealizing assumption that the nucleons pair-off to form an inert core. When this hypothesis is modified and all other defining parts of the model remain equal, a new model emerges. The physical intuition behind this particular modification is that those nucleons that fill up shells form an inert core, and the loose nucleons that remain in the unfilled shells contribute to the nucleus's properties. Although, the modified model (being less idealized) generally makes different predictions from the single particle shell model, its predictions about spin are the same as the latter's[2], thus corroborating the postulate that the spin-orbit interaction makes a contribution to the nuclear behavior. The initially *ad hoc* hypothesis of spin-orbit coupling is thus elevated to the status of a property of nuclear matter, despite still being in need of theoretical justification. Hence, the introduction of the spin-orbit hypothesis into the shell model is what supplies the model with explanatory power and gives it its representational capacity.

Both the liquid drop and the single particle shell models are only partial representations of the nuclear structure, but this fact is not sufficient to imply their falsity. The two research programs are rivals in the first stage of development and each is used with the ambition that it will lead to a full explanation of nuclear phenomena. A full explanation is understood as providing the description of all the causal factors responsible for nuclear phenomena. The liquid drop is clearly a semi-classical model that relies very little on quantum mechanics. The shell model is closely linked to quantum mechanics but digresses from the theory when physical mechanisms have to be postulated to achieve the necessary explanatory power. So far this argument merely shows that actual scientific models are constructed to serve as heuristic devices in the attempt to apply quantum theory to the nuclear domain. If this were the case then it could be claimed that actual scientific models are the pragmatic counterparts of theoretical models or that they approximate respective theoretical models, hence the SV would be safely guarded. But if scientific models were merely heuristic devices then this could be achieved simply by having a class of "local" models each of which could be used to explain and predict parts of the target system. However, the accumulating experimental knowledge and the continuous development of our understanding of the causal structure of the nucleus, alters our conception of the nucleus and leads to the conjecture that the two hypotheses that underlie each model may not be incompatible. This results in the emergence of a new research direction, what I call the second stage of development.

---

[2]For the relative experimental results and accompanying explications the reader is referred to any of the following nuclear physics literature, [Preston and Bhanduri, 1975; Segré, 1977; Burcham, 1973].

3 The Second Stage of Development

This stage is characterized by the attempt to assimilate all previous results into one all-inclusive model. This direction is motivated by the experimental evidence, but also from the explanatory and predictive successes and failures of the predecessor models, e.g. it became apparent that the shell model could not furnish acceptable explanations for several nuclear phenomena, and that its fictitious assumption of a central potential could not be adequately corrected by reliance solely on the independent particle hypothesis. It was also motivated by the fact that the liquid drop model was developed to a more generalized and sophisticated version [Bohr and Mottelson, 1953] and its explanatory power for the collective modes of nuclear motion reached acceptable levels. The latter, which is often referred to as the Collective Model of the nucleus, relies heavily on the liquid drop model as described above. It considers the nucleus as a collection of closely coupled particles and by use of the hydrodynamic analogy, the Hamiltonian for the collective motion of the nucleons is developed. The collective Hamiltonian consists of four terms each accounting for the vibration, rotation, giant resonance, and a mixture of vibration-rotation modes of collective nuclear motion. These reasons led to the view that an explanatorily satisfactory model should incorporate both collective and individual modes of motion of the nucleons, and that this could not be achieved by either of the two earlier approaches.

This research direction led to the unified model, which incorporates many of the peculiarities of the collective model in a way that seems to be assimilating them into shell structure. The unification of the single particle shell model with spin-orbit coupling with the collective model could be understood by the proponents of the SV as structurally subsuming the latter into an extended structure of the former. Frequently in the nuclear physics literature, we encounter the view that this unification is a form of an extension of the shell model, in the sense of adding correction factors to the initial shell-structure. I think that both of these views are wrong and that the right way to understand this historical development is to view it as an improvement in explanatory power and hence a better representation of the nucleus. The development of nuclear models began with the two underlying conflicting hypotheses explained earlier, which were conflicting not by virtue of being logically incompatible, but because of the underlying physical intuitions and the different quests for explanation of experimental knowledge. The first hypothesis was motivated by the growing experimental knowledge of the large-strength and short-range nuclear interactions, by the fact that its resulting models provided good explanations for phenomena such as nuclear fission and giant resonance, and for providing a justification for the

Weizsäcker formula. The second hypothesis was motivated primarily by the fact that it makes use of quantum mechanical principles from the outset, and gives rise to models that seem to be closely linked to quantum theory. The development of the unified model exemplifies the case in which the amalgamation of existing physical intuitions takes place at the level of the model defining hypothesis. Indeed, what seems to be a structural subsumption of one model into another is in fact just the use of the existing mathematical representations from the different model Hamiltonians in the context of a new model. This context, which is dictated by the unified model's underlying hypothesis, implies an alteration in the physics of the predecessor models. If we view the unified model as a case of two distinct models structurally-welded together then we fail to see the changes that take place in the physics. In what follows I try to substantiate this claim and clarify its importance.

The unified model is the offspring of a hypothesis that combines all previous physical intuitions on nuclear structure. The goal is to overcome the shortcomings faced by earlier models. On the one hand, we know that strong interaction models are extremely schematic, because despite the large-strength nuclear forces the nucleons are expected to demonstrate some form of independent motion. On the other hand, the basic assumption of the shell models is that the nuclear potential is spherically symmetric. There is enough experimental evidence, however, that slow nuclear surface vibrations and static deformed nuclei occur, depending on the structure of the potential energy surface. Therefore, the single-particle orbits should depend on the form of the nuclear surface, that is, on the spatial distribution of all nucleons. *Inter alia* this thinking led to the conclusion that there could be an interaction between collective and single-particle degrees of freedom, in short, to the unified model. Both of these modes of nuclear motion are combined in the unified model Hamiltonian, which also involves an interaction term for the two modes:

(10) $H_{TOT} = H_{SP} + H_{COL} + H_{INT}$

The first Hamiltonian term is that of the single-particle modes of motion. The general formulation of the unified model does not require a particular specification for this term. For the sake of keeping along the lines of the exposition in the preceding section, let me assume shell structure with spin-orbit coupling by letting H_{SP} be as dictated by equation (8). The H_{COL} term is that of the collective modes of motion. In fact empirical knowledge led to the conclusion that there are four kinds of nuclear collective behavior; the mathematical representations of these are borrowed from the collective model:

(11) $H_{COL} = H_{ROT} + H_{VIB} + H_{ROT-VIB} + H_{GR}$

Nuclear motion due to fission is not an explicit part of the above Hamiltonian, but the phenomenon derives its explanation indirectly by reference to the second and third terms in equation (10). The last term H_{GR} is that of giant resonance that describes density fluctuations in the nucleus that may be caused by the electric field of a coincident photon.

If we were to focus our attention only on the Hamiltonian structure of equation (10), then it is conceivable that we perceive the model as an extension of the single particle shell model. In other words, we could consider the second and third terms of the Hamiltonian as correction factors to the idealized single particle shell model, along the lines of McMullin's [McMullin, 1985] cumulative account of de-idealization. We could possibly even go as far as to claim that the correction factors are introduced on purely pragmatic grounds. These considerations would establish the desired structural link between the model and quantum theory. However, focusing only on the structural aspects of the model allows for the possibility that we invent numerous hypotheses that give rise to the same structure, each one giving the model a different physical character, of course. But the physics of the model are dictated by just one underlying hypothesis, without which the model cannot be understood. The ideas about the physics of the target system are not explicit features of the model, but are tacit components of its defining hypothesis. These physical ideas are not the result of systematic theoretical considerations. They are the development of the physical intuitions and the causal understanding of the nuclear domain, molded by the successes and failures of earlier models. In view of the fact that they are not validated systematically by theory, they are *ad hoc* considerations to be tried out and ultimately judged by their success. This is why considerations about the defining hypothesis are important.

The hypothesis that underlies this model and gives rise to its structure is this: *Nucleons move nearly independently in a common slowly changing non-spherical potential.* When analyzed this complex hypothesis about the nucleus asserts that: (a) the nucleus is a complex system of a collection of particles that exhibit some form of independent nucleon motion, (b) which is constrained by a slow collective motion of a core of nucleons, that is constituted by three distinct kinds of motion (vibration, rotation and giant resonance), two of which demonstrate an interaction mode, and (c) that the two modes of motion interact with each other. This hypothesis is the product of amalgamating the physical understanding of the target system that derives from each of the models of the first stage and it is the starting point for understanding the mathematical terms of the Hamiltonian. We notice that the shell model hypothesis is a specific manifestation of the first clause in the above hypothesis. In fact, any conceivable hypothesis that allows the nucleons to have single-particle degrees of freedom could be substituted for this clause of the unified model hypothesis. The fact that we impose shell structure to the unified model hypothesis is based on other additional hypotheses, which are motivated by reasons independent from quantum theory. In particular, by the fact that shell structure has been so far successful in accounting for single-particle degrees of freedom, for which the spin-orbit interaction hypothesis played an operative role.

The second clause of the above hypothesis is meant as a constraint on the independent motions of the nucleons. This term is borrowed from the collective model and all the terms of equation (11) are justified along the same lines as the liquid drop model: the classical form of the Hamiltonian is established by the use of a classical description of the nuclear collective motion. The modes of motion of rotation, vibration, the interactive mode of rotation-vibration and giant resonance are described in terms of classical parameters and classical collective coordinates and then converted into their quantum mechanical analogues. However when imported into the unified model its purpose is not only to describe nuclear motion but to represent a collective potential that the nucleus as a whole exerts that is non-spherical and affects, or restrains, the motion of its individual nucleons. This clause, therefore, contains three implicit sub-clauses: (i) that there is a collective mode of nuclear motion which constrains the motion of individual nucleons, (ii) that if the nuclear potential is to constrain the motions of individual nucleons then there must be an interaction between the single-particle modes with the collective modes of motion, (iii) that if we assume the nuclear potential to change sufficiently slowly then we can make an approximation that sanctions the separation of the nuclear motion into single-particle and collective motions. The last two implicit sub-clauses indicate that although the mathematical representation of this term does not undergo any significant modification from that of the collective model its role in the description of nuclear motion departs from that of the latter.

The demand of sub-clause (iii), that if the nuclear potential is to constrain the motions of individual nucleons then there must be an interaction between the two modes of motion, explicitly accounted for by the third term of equation (10) and implicitly by various components in the collective term of the Hamiltonian. The interaction term of the Hamiltonian is, in other words, chosen so that the assumption of non-sphericity is introduced into the total Hamiltonian because a correction factor to the single particle shell potential that can bring about such an effect is not feasible. This is a departure from the shell model. The non-sphericity of the potential is introduced into shell structure as a consequence of the fact that single-particle and collective mode coupling is assumed. The H_{SP} term, which represents shell structure, expresses only the individual nucleon motion. It does so, in an idealized way, i.e. by retaining the assumption of spherical symmetry within it. But spherical symmetry is not part of the assumptions underlying the total unified model Hamiltonian. Often physicists talk of the deformed single particle shell model, but I take it that this is just convenient phraseology. The information contained in the total Hamiltonian becomes clear in a closer inspection of its terms: If the interaction between single-particle and collective motions is weak, we talk of *weak coupling*. Such is the case for spherical nuclei, whose physics are represented in terms of quadrupole vibrators. That is, the H_{COL} term of the Hamiltonian reduces to that of an oscillator similar to the liquid drop equation, because the potential is spherically symmetric hence rotary motion can be totally ignored. In addition, only quadrupole deformations need be considered in the collective and interaction terms of the Hamiltonian; as we may recall quadrupole moments are used to measure the deviation of the nuclear density from spherical symmetry. Thus for weak coupling the Hamiltonian in equation (10) reduces to that of a quadrupole vibrator. If this were a good representational model for all nuclei then we could speak of the collective and interaction terms as added correction factors to the single-particle term of the Hamiltonian, because collective and single-particle motions in this model appear as if they function independently of each other. However, this is only a very special case of the unified model Hamiltonian given in equation (10). The latter comprises the general picture of the deformed nucleus, to which the spherical nucleus is a limiting case. Hence in the more general case it is assumed that the interaction between the collective and single-particle motions is strong, i.e. there is *strong coupling*, and this shows that spherical symmetry is not a part of the assumptions underlying the total unified model Hamiltonian but only appears to be so.

Undoubtedly, the unified model is essentially a hybrid of the shell model and the collective model. My question is, "what sort of hybrid is it?" In regard to the physical picture it describes, the unified model is closer to the shell model, that is, the nucleons move approximately independently rather than being strongly coupled. But the crucial connection between the unified model and the shell model is not suggested by the use of the same mathematical representations, but by the fact that while collective motions in nuclei involve all the nucleons, the most loosely bound ones have proportionately the most effect. This connection is suggestive of how the unified model is constructed. By assuming the nuclear potential to change sufficiently slowly, we are making an idealization that sanctions the separation of the nuclear motion into intrinsic and collective motion. The first of these represents the motions of the nucleons in a fixed potential while the second is associated with variations in the shape and orientation of the nuclear field. This separation is in many respects analogous to the separation into electronic and nuclear motion in molecules. I think what motivates this idealization is primarily the surviving success of the shell and collective models. If we consider the intrinsic and collective modes of motion as if they are separated we can represent the separate Hamiltonian terms by borrowing from the two predecessor models. So the unified model merely borrows the mathematical representations of its separate terms and it is not an extension of the shell model. This understanding also shows what sort of correction the interaction term in the Hamiltonian (of equation (10)) is. It is meant as a de-idealization of the hypothesis that the nuclear potential changes slowly.

This understanding could be achieved only by looking at the underlying hypothesis of the unified model. All the physical assumptions involved are contained in the hypothesis, not in the defined structure. If we focus solely on the structure of the model then any story would do. We could, for instance, look at the unified model and see an extension of the shell model that includes several correcting factors, which account for the collective modes of motion. Earlier I explained that according to the SV we define the intended models of the theory directly without recourse to formal syntax. But representational models do not belong to the class of intended models of the theory. They are distinct entities: their function is to represent their target systems. Thus if we do not try to interpret them with reference to their underlying hypotheses we run the risk of making the epistemic mistake of overlooking the causal reasoning involved in their construction.

4 Conclusion

I have given an analysis of the hypotheses that underlie the unified model. Through such analyzes, I argued, we can discern the evolutionary history of representation models. Since a model viewed as a mathematical structure cannot itself exhibit its own evolutionary history, this feature of model construction is obscured by the SV. The above story enables us to discern one important element of scientific inquiry into the world: that progress in the construction of representational models in a particular physical domain relies on the explanatory and predictive successes of predecessor models. This use of representational models indicates how physical knowledge about a particular physical domain is accumulated, which, in its turn, shapes the defining hypotheses of successor models. The models describe causal mechanisms which could initially be postulated in *ad hoc* ways. These, *inter alia*, give the unified model its partial independence from theory and provide its explanatory power. Moreover, it is clear that the unified model is not developed by the use of the theoretical models of quantum mechanics, but by a complex variety of methods that are manifested when the evolutionary history of the model is assembled and which demonstrate that in the construction of such complex models there is a constant interplay between theory, predecessor models, semi-empirical results and experimental evidence whose synthesis is a manifestation of the development of physical intuition about the target system.

BIBLIOGRAPHY

[Adair, 1952] R.K. Adair. Angular distribution of neutrons scattered by helium. *Physical Review*, 86(2):155–162, 1952.

[Bohr and Mottelson, 1953] A. Bohr and B.R. Mottelson. Collective and individual particle aspects of nuclear structure. *Danske Matematiske-Physike Medd.*, 27(16), 1953.

[Burcham, 1973] W.E. Burcham. *Nuclear Physics*. Longman, London, 1973.

[Cartwright, 1999] N.D. Cartwright. Models and the limits of theory: Quantum hamiltonians and the bcs models of superconductivity. In Morgan M. and Morrison M., editors, *Models as Mediators*, pages 241–281. Cambridge University Press, Cambridge, 1999.

[Da Costa and French, 2003] N. C. A. Da Costa and S. French. *Science and Partial Truth*. Oxford University Press, Oxford, 2003.

[Giere, 1988] R.N. Giere. *Explaining Science: A Cognitive Approach*. University of Chicago Press, Chicago, 1988.

[Mayer, 1948] M.G. Mayer. On closed shells in nuclei. *Physical Review*, 74(3):235–239, 1948.

[McMullin, 1985] E. McMullin. Galilean idealisation. *Studies in History and Philosophy of Science*, 16:247–273, 1985.

[Morrison, 1999] M. Morrison. Models as autonomous agents. In Morgan M. and Morrison M., editors, *Models as Mediators*, pages 38–65. Cambridge University Press, Cambridge, 1999.

[Moszkowski, 1957] S.A. Moszkowski. Models of nuclear structure. In S. Flgge, editor, *Encyclopedia of Physics: Structure of Atomic Nuclei. Vol. 39*, pages 411–550. Springer Verlag, Berlin, 1957.

[Portides, forthcoming] D. Portides. Scientific models and the semantic view of scientific theories. *Philosophy of Science*, forthcoming.

[Preston and Bhanduri, 1975] M.A. Preston and R.K. Bhanduri. *Structure of the Nucleus*. Addison-Wesley Publishing, Ontario, 1975.

[Putnam, 1962] H. Putnam. What theories are not. In Suppes Nagel and Tarski, editors, *Logic, Methodology and Philosophy of Science*, pages 240–251. Stanford University Press, Stanford, 1962.

[Segré, 1977] E. Segré. *Nuclei and Particles*. W.A. Benjamin, London, 1977.

[Signell and Marshak, 1958] P.S. Signell and Marshak. Semiphenomenological two-nucleon potential. *Physical Review*, 109(4):1229–1239, 1958.

[Suppe, 1977] F. Suppe, editor. *The Structure of Scientific Theories*. University of Illinois Press, Urbana, 1977.

[Suppe, 1989] F. Suppe. *The Semantic Conception of Theories and Scientific Realism*. University of Illinois Press, Urbana, 1989.

[Van Frassen, 1990] B.C. Van Frassen. *The Scientific Image*. Clarendon Press, Oxford, 1990.

[Woods and Saxon, 1954] R.D. Woods and D.S. Saxon. Diffuse surface optical model for nucleon-nuclei scattering. *Physical Review*, 95:577–578, 1954.

Demetris Portides
University of Cyprus
Email: portides@ucy.ac.cy

On Popper's Logic of Scientific Discovery

ATOCHA ALISEDA

ABSTRACT. The aim of this paper is to elucidate some aspects of Karl Popper's logic of research and place it in the philosophical discussion of scientific discovery. A closer look into his proposal reveals that his work, in the light of recent literature, does point into the direction of some fundamental mechanisms which fall under the study of discovery. The argument for this claim is twofold: one the one hand, when a finer analysis of contexts of research is done, it seems that Popper's logic may be considered as part of the context of discovery, and on the other hand, his account on the growth of knowledge by the method of conjectures and refutations, is in accordance with the "Friends of Discovery" mainstream, at least is so far as the work of Herbert Simon is concerned.

1 Introduction

It is an unfortunate yet an interesting fact that Popper's *Logik der Forschung* first published in German in 1934, was translated into English and published twenty five years later as *The Logic of Scientific Discovery*. An accurate translation would have been: "The Logic of Scientific Research", as found in translations into other languages, such as in Spanish (*La Lógica de la Investigación Científica*).

One reason for its being unfortunate lies in the fact that several renowned scholars (Simon and Laudan amongst others) have accused Popper of denying the very subject matter of what the title of his book suggests, something along the lines of a logical enterprise into the epistemics of scientific theory discovery. These complaints are firmly grounded within the accepted view that for Popper scientific methodology concerns mainly the testing of theories, and this approach clearly leaves outside of its scope issues having to do with discovery. Thus, for those philosophers of science interested in discovery processes as well as in other methods for scientific inquiry outside the realm of justification, it seems natural to leave Popper out of the picture and take the above complaint on the title to be just a confusion originated from its English translation.

However, on a closer look into Popper's philosophy, an additional confusion stirs up when we find (in a publication shortly after) his position beyond justification issues and concerned with the advancement and discovery in science, as witnessed by the following quote: "*Science should be visualized as progressing from problems to problems – to problems of increasing depth. For a scientific theory – an explanatory theory – is, if anything, an attempt to solve a scientific problem, that is to say, a problem concerned with the discovery of an explanation*" [Popper, 1963, p. 222]. This view is in accord with Simon's famous slogan that "*scientific reasoning is problem solving*" made in research in cognitive psychology and artificial intelligence, a claim also put forward by Laudan in the philosophy of science. Moreover, it seems Popper was happy with the English title of his book, for being such an obsessive proof reader of his own work, he made no remarks about it[1].

Additionally, it is also appealing that in the literature of knowledge discovery in science, especially from a computational point of view, some of Popper's fundamental ideas are actually implemented within the simulation of discovery and testing processes in science. For example, the requirement characterizing a theory as 'scientific' when it is subject to refutation, is translated into the '*FITness*' criterion, and plays a role in the process of theory generation, for a proposed theory only survives if it can be refuted within a finite number of instances [Simon, 1977, p. 403].

Therefore, the apparent translation inaccuracy of Popper's *The Logic of Scientific Discovery* is an interesting fact as well, calling to be questioned further. The aim of this paper is to elucidate some aspects of Popper's logic of research and place it in the philosophical discussion of scientific discovery. A closer look into his proposal reveals that his work, in the light of recent literature, does point into the direction of some fundamental mechanisms which fall under the study of discovery. The argument for this claim is twofold: one the one hand, when a finer analysis of contexts of research is done, it seems Popper's logic may be considered as part of the context of discovery, and on the other hand, his account on the growth of knowledge by the method of conjectures and refutations, is in accordance with the "Friends of Discovery" mainstream, at least is so far as the work of Simon is concerned.

---

[1] Popper used every opportunity to clarify his claims and terms put forward in his *Logik der Forschung*, as witnessed in a multitude of footnotes and new appendices to "The Logic of Discovery" and in many remarks found in later publications. In [Popper, 1992] he reports (referring to his "Postscript: After Twenty Years"): "*In this Postscript I reviewed and developed the main problems and solutions discussed in Logik der Forschung. For example, I stressed that I had rejected all attempts at the justification of theories, and that I had replaced justification by criticism.*" (p. 149).

2   Contexts of research: which are they?

The literature on the subject scientific discovery is staggeringly confused by the ambiguity and complexity of the term "discovery". A discovery of an idea leading to a new theory made in science involves a complicated process which goes from the initial conception of an idea throughout its justification and final settlement as a new theory. These two aspects are just the two extremes in a series of intermediate processes – which need not be sequential – including the entertainment of a new idea, its initial evaluation, which may lead to finer ideas in need of evaluation or may be even replaced by other ideas, calling for modification of the original one. Therefore, we must at least acknowledge that scientific discovery is a process subject to division. This fact naturally confronts us with the problem of how to provide a proper division.

An attempt to supply a division is given by the contemporary distinction between the contexts of discovery and the context of justification, originally proposed by Reichenbach in the 30's. This division often presupposes the latter as dealing exclusively with the 'finished research report' of a theory, and thus certainly leaves ample room to the former. In order to bring some order and clarity into the study of the process of scientific discovery at large, several authors identify an intermediate step between the two extremes, the conception and justification of a new idea. While Savary puts forward the phase of "working with ideas" [Savary, 1995], Laudan introduces the "context of pursuit" as an "nether region" between the two contexts [Laudan, 1980, p. 174]. Another dimension in the study of the context of discovery is to distinguish between a narrow and a broad view. While the former view regards issues of discovery as those dealing exclusively with the initial conception of an idea, the latter view is that which deals with the overall process going from the conception of a new idea to its settlement as an idea subject for ultimate justification (a distinction introduced by Laudan in [Laudan, 1980]).

It is however also a matter of choice to extend the boundaries of the context of justification in order to deal with evaluation questions as well, especially when the truth of a theory is not the sole interest to be searched for. A consequence of this move is the proposal to rename the "context of justification" as the "context of evaluation" [Kuipers, 2000, p. 132] or as the "context of appraisal" [Musgrave, 1989, p. 20]. Under the latter view, the context of discovery has in turn been relabeled the "context of invention", in order to avoid the apparent contradiction that arises when we speak of the discovery of hypotheses, as discovery is a "success word" which presupposes that what is discovered, must be true.

Therefore, the original distinction between the contexts of discovery and justification may not only be further subdivided, but also its boundaries may not be so sharply distinguished. A separate issue concerned with all these contexts of research is to inquire whether the "context of discovery" or any other context for that matter, is subject to philosophical reflection and allows for logical analysis.

3 Is there a logic of discovery?

3.1 History

A critical analysis on the question concerning the search of a logic of discovery can be given from a historical point in view as has already been masterly done by Laudan [Laudan, 1980]. In this paper, he is concerned with discovery in its narrow sense, the view that regards issues of discovery as those dealing exclusively with the initial conception of an idea. He identifies three periods in time in the evolution of the overall enterprise, from antiquity to the XIX century. The first and the second period (from antiquity to 1750 and from 1750 to 1820, respectively), are characterized by the search of a logic serving both discovery and theory justification purposes, both holding up to an infallibilistic stance in regard to the problem of the well-foundedness of knowledge. The main difference between these periods is found in the object of science under consideration, something which in turn determines the type of logic to be worked out. In the first period researchers focused mainly on the characterization of empirical laws, such as "all gases expand when heated", in the discovery of universal statements concerning only observable entities. The corresponding logic is inductive. It was until the 1750's that several scientists and philosophers were interested in modeling as well the discovery of explanatory theories, those involving theoretical entities. For this purpose, a much more complex logical mechanism than that of enumerative induction was required. The idea of truth-approximation was behind the conception of these logics, and accordingly were labeled as "self-corrective logics"[2]. Thus, up to this point, the search was for a truth-preserving logic of discovery and justification to warrant infallible knowledge.

The transition from the second to the third period (1820-1830) was given by a major change in view regarding the legitimization of knowledge, namely by a shift from infallibilism to fallibilism. Therefore, around mid XIX century, the enterprise of a logic of discovery was abandoned and replaced by an exclusive search for a logic of justification, a logic of post hoc evaluation.

This is a rough and admittedly not a completely fair reconstruction of

---

[2]Usually, these logics involve a background theory (consisting of laws), initial conditions and a relevant observation. The goal of these logical apparatus was to produce a better and truer theory than the old (background) one.

Laudan's historical analysis, for among other things, he convincingly claims further that recent attempts to revive the search of a logic of discovery have no clear grounds, for the move to a logic of justification made the search of a logic of discovery "epistemologically irrelevant". While I find this argument thought-provoking, and in fact I agree with it, my position is that the relevance of a logic of discovery nowadays is not to be found on epistemic grounds, but rather on heuristic aspects. But here is not the place to develop my argument, but only to complement Laudan's historical critical analysis with my own, based on dividing the question of a logic of discovery into three of them, each of which takes into account one of the following three aspects: its purpose, pursuit and achievement.

Underlying questions

The purpose of a logic of discovery concerns the ultimate goal researches engaged in this enterprise wish to achieve in the end, and the affirmative (negative) response to the question of whether there is such a logic is largely grounded on philosophical ideals. The pursuit of a logic of discovery regards the working activities that researchers in the field are engaged in to attain their goals, and the answer to such a question is given in the form of concrete proposals of logics of discovery. Finally, the achievement of a logic of discovery provides an evaluation between previous aspects, of whether what is actually achieved lives up to its purpose. This three-question division allows to evaluate existing proposals on the logic of discovery as to their coherence, and it also provides a finer grain distinction in order to compare confronting proposals, for they may agree on one question while disagree on the other one.

In the first period identified above (from antiquity to mid XIX century), the search for a logic of discovery was guided by a strong philosophical ideal, that of finding a universal system which would capture the way humans reason in science, including the whole spectrum, from the initial conception of new ideas into their ultimate justification. Following the spirit of Leibniz's "Characteristica Universalis", this ideal was the motor behind the ultimate purpose of finding a logic of discovery. As to the question of pursuit, the central method worked out was the modeling of induction, giving place to proposals such as Bacon's "eliminative induction", which is in fact a method for hypothesis selection. Therefore, regarding the question of achievement of a logic of discovery in this period, the products resulting from its pursuit did captured only to a very small extent what the question of purpose was after, and therefore the project was not coherent in regard to what was searched for and what was found in the end. A similar analysis can be given for the second period. In the third period, the original question of purpose is vanished by the advancement of fallibilism; for it was then clear that a universal calculi in which all ideas could be translated to and by which intellectual arguments were conclusively settled, was an impossible goal to achieve. The question to be searched for instead is concerned with a logic of justification, and accordingly, the question of pursuit focused on developing accounts on these matters. There was complete harmony between the search and the findings, and so the question of achievement rendered this period as coherent.

However, the aim to find a logic dealing with the conception of new ideas was not completely discarded, as witnessed later in the work of Charles S. Peirce and others, but this line of research has remained incoherent as far as the question of achievement is concerned. As absurd as this position seems to be, this is still the ideal to which the "friends of discovery" hold to, but a closer analysis into their approach will help to make sense of their stance. We will limit to the view of Herbert Simon[3].

Simon on discovery

In principle, the pioneering work of Herbert Simon and his team share the ideal on which the whole enterprise of artificial intelligence was initially grounded, namely that of constructing intelligent computers behaving like rational beings, something which resembles the philosophical ideal which guided the search for a logic of discovery in the first period identified above (up to the XIX century). However, it is important to clarify on what terms is this ideal inherited in regard to the question of purpose of a logic of discovery, on the one hand, and put to work with respect to the question of the pursuit of such a logic, on the other.

In his essay *Does scientific discovery have a logic?* Simon sets himself the challenge to refute Popper's general argument, reconstructed for his purposes as follows: "*If 'There is no such thing as a logical method of having new ideas, then there is no such thing as a logical method of having* small *new ideas'* " ([Simon, 1977, p. 327]. My emphasis), and his strategy is precisely to show that an antecedent in the affirmative does not commit to an assessment of the consequent, as Popper seems to suggest. Thus, Simon converts the ambitious aim of searching for a logic of discovery revealing the process of discovery at large, into an unpretentious goal: "*Their modesty [of the examples dealt with] as instances of discovery will be compensated by their transparency in revealing the underlying process*" [Simon, 1977, p. 327].

This humble but brilliant move allows Simon to further draw distinctions

---

[3] Although there are several particularities about Simon's proposal with respect to all others, in general his claims reflect the spirit of the project on the search of a logic of discovery for the Friends of discovery overall enterprise. Our claims are based on the [Simon, 1977] paper.

on the type of problems to be analyzed and on methods to be used. For Simon and his followers, scientific discovery is a problem-solving activity. To this end, a characterization of problems into those which are well structured versus those which are ill structured is provided[4], and the claim for a logic of discovery focuses on the well structured problems. Although there is no precise method by which scientific discovery is achieved, as a form of problem solving, it can be pursued via several methodologies. The key concept in all this is that of *heuristics*, the guide in scientific discovery which is neither totally rational nor absolutely blind. Heuristic methods for discovery are characterized by the use of selective search with fallible results. That is to say, while they provide no complete guarantee to reach a solution, the search in the problem space is not blind, but it is selective according to a predefined strategy[5].

Even though this approach comes from apparently distant disciplines to philosophy of science, namely cognitive psychology and artificial intelligence, they are proposals which suggest the inclusion of computational tools in the philosophy of science research methodology and by so doing claim to reincorporate aspects from the context of discovery within its agenda. However, this approach does not give an account of "the Eureka" moment, even for small ideas.

4 Popper's position

4.1 Discovery

The common view on Popper's position states that issues of discovery cannot be studied within the boundaries of methodology, for he explicitly denies the existence of a logical account for discovery processes, regarding its study a business of psychology. This position is backed up by the following – much

---

[4] A well structured problem is that for which there is a definite criterion for testing, and for which there is at least one problem space in which the initial and the goal state can be represented and all other intermediate states may be reached with appropriate transitions between them. An ill-structured problem lacks at least one of the former conditions.

[5] The authors distinguish between weak and strong methods of discovery. The former is the type of problem solving used in novel domains. It is characterized by its generality, since it does not require in-depth knowledge of its particular domain. In contrast, strong methods are used for cases in which our domain knowledge is rich, and are specially designed for one specific structure. Weak methods include generation, testing, heuristic methods, and means-ends analysis, to build explanations and solutions for given problems. These methods have proved useful in artificial intelligence and cognitive simulation, and are used by several programs. An example is the BACON system which models explanations and descriptive scientific laws, such as Kepler's law, Ohm's law, etcetera. It is however a matter of debate if BACON really makes discoveries, since it produces theories new to the program but not to the world, and its discoveries seem spoonfed rather than created.

too quoted – passage:

> Accordingly I shall distinguish sharply between the processes of conceiving a new idea, and the methods and results of examining it logically. As to the task of the logic of knowledge – in contradistinction to the psychology of knowledge – I shall proceed on the assumption that it consists solely in investigating the methods employed in those systematic tests to which every new idea must be subjected if it is to be seriously entertained. Some might object that it would be more to the purpose to regard it as the business of epistemology to produce what has been called a "rational reconstruction" of the steps that have led the scientist to a discovery – to the finding of some new truth ... But this reconstruction would not describe these processes as they actually happen: it can give only a logical skeleton of the procedure of testing. Still, this is perhaps all that is meant by those who speak of a "rational reconstruction" of the ways in which we gain knowledge.
>
> It so happens that my arguments in this book are quite independent of this problem. However, my view of the matter, for what is worth, is that there is no such thing as a logical method of having new ideas, or a logical reconstruction of this process. [Popper, 1959, p. 31,32]

First of all, it should come at no surprise that Popper's position holds a place in the discussion of discovery issues, as the objects of his analysis are precisely genuinely new ideas. Moreover, we should emphasize that Popper draws a clear division between two processes; amidst the conception of a new idea, and the systematic tests to which a new a new idea should be subjected to, and in the light of this division he advances the claim that not the first but only the second one is amenable to logical examination.

4.2 The growth of scientific knowledge

For Popper the growth of scientific knowledge was the most important of the traditional problems of epistemology. His fallibilist position provided him with the key to reformulate the traditional problem in epistemology, which is focused on the reflection on the sources of our knowledge. Laid down this way, this question is one of origin and begs for an authoritarian answer [Popper, 1963, p. 25], regardless of its answer being placed in our observations or in some fundamental assertions lying at the core of our knowledge. Popper's answer to this question is that we do not know or even hope to know about the sources of our knowledge, since our assertions are only guesses [Popper, 1963, p. 27]. Rather, the question and its answer should be the following: *"How can we hope to detect and eliminate error? Is, I believe, By criticizing the theories or guesses of others"* [Popper, 1963, p. 26]. This is the path to make knowledge grow: *"the advance of knowledge consists, mainly, in the modification of earlier knowledge"* [Popper, 1963, p. 28].

This concern on the growth of knowledge is intimately related to his earlier mentioned view of science as a problem solving activity, and in this respect he writes: *"Thus, science starts from problems, and not from observations; though observations may give rise to a problem, specially if they are unexpected"* [Popper, 1963, p. 222]. Moreover, rather than asking the question 'How do we jump from an observation statement to a theory?', the proper question to ask is the following: *"'How do we jump to from an observation statement to a good theory?'... by jumping first to any theory and then testing it, to find whether it is good or not; i.e. by repeatedly applying the critical method, eliminating many bad theories, and investing many new ones."* [Popper, 1963, p. 55–56].

Thus, the intention in both the title and content of Popper's *The Logic of Scientific Discovery* was to unfold the epistemics of evaluation and selection of newly discovered ideas in science, more in particular, of scientific theory choice. To this end, Popper proposed a rational method for scientific inquiry, the method of conjectures and refutations [Popper, 1963][6]. The motivation behind this method was to provide a criterion of demarcation between science and pseudo-science. Besides its purpose, the logical method of conjectures and refutations is a norm for progress in science, for producing new and better theories in a reliable way. For Popper the growth of scientific knowledge – and even of pre-scientific knowledge – is based on the learning from our mistakes, which according to him is is achieved by the method of trial and error. He did not provided a precise procedure to perform scientific progress, leading to better theories, but rather a set of theory evaluation criteria including a measure for its potential progressiveness (its testability) and the condition of having a greater "empirical content" than the antecedent theory.

5 Popper's position revisited

5.1 On Popper's context of research

According to our previous discussion of contexts of research, where shall we place Popper's position? Taking as basis the original division into contexts

---

[6]This method is proposed as an alternative to induction, in fact it pretends to make induction absolutely irrelevant, and therefore dissolve the problem of induction.

of discovery and justification, if we regard Popper's view on discovery as a narrow one, under which discovery issues concern exclusively the initial conception of an idea, then his position of leaving the "Eureka moment" out of reach for the hand of logical analysis is in complete accord with his motto: *"that there is no such thing as a logical method of having new ideas"*, placing his overall approach as belonging to the context of justification, as it has been interpreted all along. In contrast, if we regard Popper's view as a broad one in regard to discovery (a position argued in [Savary, 1995]), then since he concedes to the second process the possibility of logical examination, that is, to *"the methods employed in those systematic tests to which every new idea must be subjected if it is to be seriously entertained"*, whatever methods these may be, they are certainly previous to the ultimate justification of an idea, and thus the second process falls within the context of discovery as a whole. Moreover, a closer look into Popper's philosophy reveals that his method of conjectures and refutations had no presumptions of assessing a theory as definitively true, as the positivists pretended to be the ultimate goal of justification. Finally, taking into account the finer distinctions proposed for contexts of research, compels us to place Popper's position in regard to the second process into the context of pursuit, and thus falling into the context of discovery when a broad view is acknowledged, but into the context of evaluation or appraisal when the context of justification is enlarged. In any case, what seems to be clear is that while the first process falls into the context of invention, the second one is clearly outside its boundaries.

A first conclusion to draw is that while Popper was genuinely interested in an analysis of new ideas in science, he rendered the very first process of the conception of an idea to be outside the boundaries of the methodology of science, and centered his efforts in giving an account of an ensuing process, that concerned with the methods of analyzing new ideas logically. It is not at all a straightforward matter in which context we shall place Popper's position in regard to his logic of knowledge, and I decide to take no position, as it clearly depends on what we take to be the boundaries and business of the contexts of research, something which holds our discussion under a terminological debate giving reasons both for complaints and for compliments on the English title of his book *The Logic of Scientific Discovery*.

5.2 On Popper's "logic of research"

Popper's logic of scientific discovery embraces a fallibilist stance and has no pretensions to uncover the epistemics of creativity, but focuses on the evaluation and selection of new ideas in science. Thus, his position is clearly akin with the period corresponding to the search for a logic of justification, as far as the question of purpose is concerned. As to the question of pursuit, his account provides criteria for theory justification under his critical rationalism view, by which no theory is finally settled as true, and is also concerned with the advancement of science, characterizing problem situations[7] as well as a method for their solution, the method of conjectures and refutations.

Therefore, as to the question of achievement, it seems that Popper actually found more than what he was looking for, since his account offered not the justification of theories as true, but rather a methodology for having better and stronger theories, something which enters the territory of discovery, in so far as the question of the growth of knowledge is concerned.

To conclude, I think Popper's approach is close to that of Simon, as least in so far as the following basic ideas are concerned: they both hold a fallibilist stance in regard to the well-foundedness of knowledge and view science as a dynamic activity of problem solving in which the growth of knowledge is the main aspect to characterize, as opposed to the view of science as an static enterprise in search of the assessment of theories as true. One reason that allows for the convergence of these two accounts, perhaps obvious by now, is that neither the "Friends of discovery" really account for discovery at large nor Popper neglects its study entirely.

However, a fundamental difference between these approaches is found in the method itself for the advancement of science, in what they regard to be the "logic" for discovery. While for Popper ideas are generated by the method of blind search, Simon and his team develop a full theory to support the view that ideas are generated by the method of "selective search", and clearly this last account allows for a better understanding of how theories and ideas may be generated.

My own view on the issue of discovery is that just as we have to acknowledge discovery as a process subject to division, we have also to acknowledge an inescapable part in the process of discovery, and accordingly recall Popper's motto: *"Admittedly, no creative action can ever be fully explained"* [Popper, 1975, p. 179]. But this sentence naturally implies that creative action may be partially explained (as argued in [Savary, 1995]). The question then, is to identify a proper line drawing a convenient division in order to carry out an analysis of the generation of new ideas in science.

Acknowledgements

This paper is a translation from the original in Spanish, published in [Aliseda, 2004].

---

[7]The notion of a "problem situation" is not explained here. A depth-in analysis may be found in [1995].

BIBLIOGRAPHY

[Aliseda, 2004] A. Aliseda. Sobre la lógica del descubrimiento científico de Karl Popper. *Signos Filosoficos*, VI(11):115–130, 2004.

[Kuipers, 2000] T. Kuipers. *From Instrumentalism to Constructive Realism. On some Relations between Confirmation, Empirical Progress and Truth Approximation*. Kluwer Academic Publishers, Dordrecht, 2000. Synthese Library, vol 287.

[Laudan, 1980] L. Laudan. Why was the logic of discovery abandoned? In T. Nickles, editor, *Scientific Discovery, Logic and Rationality*, pages 173–183, Dordrecht, 1980. Reidel.

[Musgrave, 1989] A. Musgrave. Deductive heuristics. In K. Gavroglu and *et al.*, editors, *Imre Lakatos and Theories of Scientific Change*, pages 15–32, Dordrecht, 1989. Kluwer Academic Publishers.

[Popper, 1959] K.R. Popper. *The Logic of Scientific Discovery*. Hutchinson, London and New York, 1959. Impression 1959. Originally published as *Logik der Forschung*, Springer, 1934.

[Popper, 1963] K.R. Popper. *Conjectures and Refutations. The Growth of Scientific Knowledege*. Routledge and Kegan Paul, London, 1963.

[Popper, 1975] K.R. Popper. *Objective Knowledge. An Evolutionary Approach*. Oxford University Press, Oxford, 1975.

[Popper, 1992] K.R. Popper. *Unended Quest. An Intellectual Autobiography*. Routledge and Kegan Paul, London, 1992.

[Savary, 1995] C. Savary. Discovery and its logic: Popper and the "friends of discovery". *Philosophy of the Social Sciences*, 25(3):318–344, 1995.

[Simon, 1973] H. Simon. Ramsey eliminability and the testability of scientific theories. *British Journal for the Philosophy of Science*, 24:357–408, 1973.

[Simon, 1977] H.A. Simon. Does scientific discovery have a logic? in h. simon, models of discovery [1973]. In H. Simon, editor, *Models of Discovery and Other Topics in the Methods of Science*, pages 326–337, Dordrecht, 1977. Reidel.

Atocha Aliseda
Instituto de Investigaciones Filosoficas
Universidad Nacional Autonoma de Mexico (UNAM)
Mexico City, Mexico
Email: atocha@minerva.filosoficas.unam.mx
URL: http://www.filosoficas.unam.mx/~atocha/home.html

Exemplar-Based Reasoning with the Shortest Derivation
Rens Bod

ABSTRACT. We present an evaluation of the Exemplar-Based Explanation model, termed EBE, which constructs derivations of new phenomena by combining partial derivations of prior phenomena. It turns out that derivational explanation is *massively redundant*: many different derivational explanations exist for the same phenomenon, even if these derivations are subsumed under the same general laws. We argue that EBE can solve this redundancy problem by using a notion of derivational similarity, which favors largest possible derivational chunks. This notion of derivational similarity can be maximized by computing the shortest derivation containing the fewest subtrees. We test EBE on a corpus of phenomena from classical and fluid mechanics that were derived by third-year physics students. Our experiments suggest that humans indeed tend to compute the shortest derivation consisting of fewest subtrees from exemplary derivations. We argue that EBE can also be applied to situations where intermediate models are needed or where laws are difficult to come by.

1 Introduction

This paper deals with the problem of explanatory redundancy, i.e. the problem that there can be different derivational explanations for the same phenomenon, even if these explanations are subsumed under the same general laws. In the worst case, the number of different derivations grows exponentially with the number of terms in the mathematical description of the phenomenon. In other words, derivational explanation is *massively redundant*. This massive redundancy of explanation is problematic not only from an epistemological point of view but also from a cognitive point of view: physics students and experts typically come up with only one and the same derivation for a certain phenomenon.

The main goal of this paper is to create a model that can predict the derivational explanation that humans (i.e. advanced physics students) construct for a certain phenomenon. We argue that the problem of explanatory redundancy can be solved by a notion of "shortest derivation". The shortest derivation of a phenomenon is the derivation consisting of the fewest chunks of previously derived phenomena that function as "exemplars". We show how a computational model of exemplar-based reasoning, known as EBE, can efficiently compute the shortest derivation and that it can accurately select the same derivations as humans do.

The idea that natural phenomena can be explained by modeling them on exemplars is usually attributed to Thomas Kuhn in his account on normal science [Kuhn, 1970]. Kuhn urges that exemplars are "concrete problem solutions that students encounter from the start of their scientific education" [Kuhn, 1970, p. 187] and that "scientists solve puzzles by modeling them on previous puzzle-solutions" [Kuhn, 1970, p. 189]. Instead of explaining a phenomenon from scratch (i.e. all the way down from laws), Kuhn contends that scientists try to match the new phenomenon to one or more previous phenomena-plus-explanations.

In similar vein, Philip Kitcher argues that new phenomena are derived by using the same patterns of derivations ("argument patterns") as used in previously explained phenomena: "Science advances our understanding of nature by showing us how to derive descriptions of many phenomena, using the same patterns of derivation again and again" [Kitcher, 1989, p. 432]. Different from Kuhn, Kitcher proposes a rather concrete account of explanation, known as the "unificationist view", which he still links to Kuhn's view, by interpreting exemplars as derivations [Kitcher, 1989, p. 437–8].

Thomas Nickles relates Kuhn's view to case-based reasoning [Nickles, 2003, p. 171]. Case-based reasoning (CBR) is an artificial intelligence technique that provides an alternative to rule-based problem solving. Instead of solving each new problem from scratch, CBR stores previous problem-solutions in memory as cases. When CBR begins to solve a new problem, it retrieves from memory a case whose problem is similar to the problem being solved. It then adapts the example's solution and thereby solves the problem [Carbonell, 1986; Branting, 1991; Kolodner, 1993].

CBR is congenial to a natural language processing technique known as data-oriented parsing (DOP) [Scha, 1990; Bod, 1998; Collins and Duffy, 2002]. DOP analyzes new sentences on the basis of a corpus of parse trees of previous sentences. In producing and perceiving a new sentence, DOP tries to construct the sentence by combining *subtrees* from parse trees in the corpus. DOP is different from CBR in the way it tackles the problem of derivational redundancy. In case a sentence has more than one possible analysis, DOP selects the parse tree that maximizes probability or minimizes derivation length. DOP can thus be seen as a (probabilistic) extension of CBR. DOP has also been employed to analyzing other perceptual modalities, such as music and vision [Bod, 2002; Bod et al., 2003].

In [Bod, 2005a] and [Bod, 2005b], we created a DOP model for scientific reasoning, which we termed EBE ("exemplar-based explanation"). This EBE model was inspired by Kuhn's and Kitcher's notions of exemplars and by the notion of derivational analogy in CBR [Veloso and Carbonell, 1993]. Similar to the use of parse trees in linguistics, EBE represents derivational explanations in physics by *derivation trees*. Explanations of new phenomena are constructed by combining subtrees from derivation trees of previous phenomena. When a phenomenon has more than one derivation, EBE computes the shortest derivation of the phenomenon consisting of the fewest, and therefore largest derivational chunks in the corpus, just as the DOP model in [Bod, 2000].

In the current paper we provide a first computational and experimental evaluation of EBE. We show that EBE can be implemented by equational reasoning techniques and test EBE on a corpus of phenomena from classical and fluid mechanics that were derived by third-year physics students. Our experiments indicate that EBE assigns to 91% of the phenomena the same derivations as students do, suggesting that humans indeed tend to compute the shortest derivation consisting of fewest subtrees from exemplary derivations.

In the following section, we will first review EBE. Next, in section 3, we will go into the problem of derivational redundancy and demonstrate how it can be solved by EBE. In section 4, we show how EBE can be implemented and discuss a series of experiments with third-year physics students. We discuss related work in section 5 and end with a conclusion in section 6.

2 Review of the Exemplar-Based Explanation (EBE) model

We will illustrate the EBE model in [Bod, 2005a; Bod, 2005b] by a simple idealized example. Consider the following derivation of the Earth's mass from the Moon's orbit in the textbook by [Alonso and Finn, 1996, p. 247]:

> Suppose that a satellite of mass m describes, with a period P, a circular orbit of radius r around a planet of mass M. The force of attraction between the planet and the satellite is $F = GMm/r^2$. This force must be equal to m times the centripetal acceleration $v^2/r = 4\pi^2 r/P^2$ of the satellite. Thus,
>
> $$\frac{4\pi^2 mr}{P^2} = \frac{GMm}{r^2}$$

Canceling the common factor m and solving for M gives

$$M = \frac{4\pi^2 r^3}{GP^2}.$$

By substituting the data for the Moon, $r = 3.84 \times 10^8$ m and $P = 2.36 \times 10^6$ s, Alonso and Finn compute the mass of the Earth: $M = 5.98 \times 10^{24}$ kg. In doing so, Alonso and Finn abstract from many features of the actual Earth-Moon system, such as the gravitational forces of the Sun and other planets, the magnetic fields, the solar wind, etc. Albeit heavily idealized, the derivation provides a concrete problem solution on which various other phenomena can be modeled. In fact, Alonso and Finn reuse parts of this derivation to solve problems such as the velocity of a satellite at a certain distance from the Earth and the escape velocity from the Earth's surface.

To create a formal model of explanation that reuses patterns from previously derived phenomena, we need a formal representation of derivations. As in DOP, EBE proposes to represent derivations by tree structures which indicate how a mathematical description of a phenomenon (or problem) is compositionally derived from laws and antecedent conditions. Figure 1 shows how the derivation for the Earth's mass above may be turned into a tree.

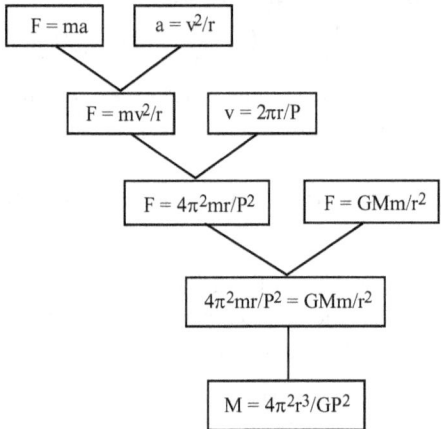

Figure 1. Derivation tree for the Earth's mass.

The tree in Figure 1 represents the various derivation steps (insofar as they are mentioned in the derivation above) from general laws to an equation of the mass M. We will refer to such a tree as a "derivation tree". A derivation tree is a labeled tree in which each node is annotated with a formula; the boxes are only meant as convenient representations of these labels. The formulas at the top of each "vee" (i.e. each pair of binary branches) in the tree can be viewed as premises, and the formula at the bottom of each "vee" can be viewed as a conclusion which is arrived at by simple term substitution. The last derivation step in the tree is not formed by a "vee" but consists in a unary branch which solves the directly preceding formula for a certain variable (in the tree above, for the mass M). Thus, in general, a unary branch refers to a mathematical derivation step that solves an equation for a variable, while a binary branch refers to a physical derivation step which introduces and combines physical laws or conditions (or other knowledge such as approximations and corrections).

Note that a derivation tree captures the notion of covering-law explanation or deductive-nomological (D-N) explanation of [Hempel and Oppenheim, 1948]. In the D-N account, a phenomenon is explained by deducing it from general laws and antecedent conditions. Although there can be various other representations of derivations, we will see that tree structures are exceedingly apt to representing derivations constructed by humans (section 4). Moreover, a major advantage of trees is that they can be straightforwardly decomposed into parts or *subtrees* which can next be recomposed to derive new phenomena.

To give an example, consider the following subtree in Figure 2 which is obtained from the derivation tree in Figure 1 by leaving out the last derivation step (i.e. the solution for the mass M).

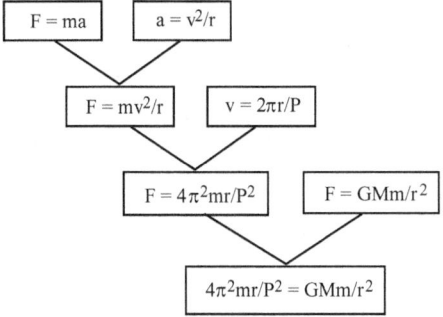

Figure 2. A subtree from the tree in Figure 1.

This subtree can be applied to various other situations. For instance, in deriving the regularity known as Kepler's third law (which states that r^3/P^2 is constant for all planets orbiting around the Sun, or satellites around the

Earth if you wish) the subtree in Figure 2 needs only to be extended with a mathematical derivation step that solves the last equation for r^3/P^2, as represented in Figure 3.

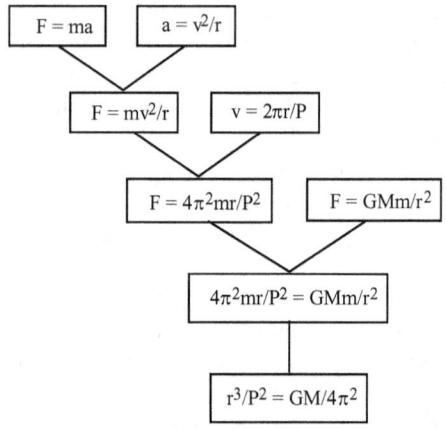

Figure 3. Derivation tree for Kepler's third law.

In a similar way we can also derive the distance of a geostationary satellite, namely by solving the subtree in Figure 2 for r and taking P as the rotation period of the Earth.

It is of course not typically the case that derivations involve only one subtree. In deriving the velocity of a satellite at a certain distance from a planet, we cannot directly use the large subtree in Figure 2, but need to *decompose* the tree in Figure 1 into two smaller subtrees that are *recomposed* by term substitution (represented by the operation "∘") and finally solved for the velocity v in Figure 4.

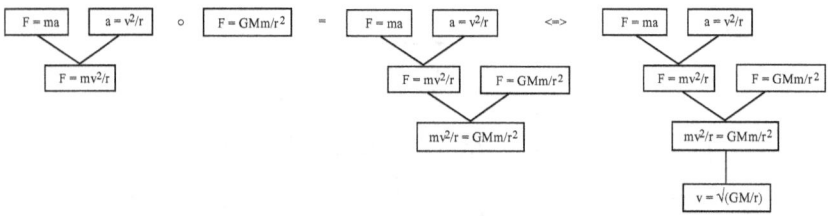

Figure 4. Deriving a phenomenon by combining two subtrees.

Figure 4 shows that we can create new derivation trees by combining subtrees from previous derivation trees. Note that subtrees can be of arbitrary size: from single equations to combinations of laws, up to entire derivations. We shall argue in the next section that the largest subtrees need to be used whenever possible.

The notion of term substitution, though widely used in rewriting systems, may need some further specification. The term-substitution operation "$\circ$" is a partial function on pairs of labeled trees; its range is the set of labeled trees. The combination of tree t and tree u, written as $t \circ u$, is defined iff the equation at the root node of u can be substituted in the equation at the root node of t (i.e. iff the lefthandside of the equation at the root node of u literally appears in the equation at the root node of t). If $t \circ u$ is defined, it yields a tree that expands the root nodes of copies of t and u to a new root node where the righthandside of the equation at the root node of u is substituted in the equation at the root node of t. Note that the substitution operation can be iteratively applied to a sequence of trees, with the convention that $\circ$ is left-associative.

Thus EBE employs a *corpus of derivation trees* representing exemplars together with a *matching procedure* that combines subtrees from the corpus into new derivation trees. This brings us to the following definition for an explanation of a phenomenon with respect to a corpus.

DEFINITION 1 *Given a corpus C of derivation trees $T_1, T_2, \ldots, T_n$ representing exemplars and a term substitution operation $\circ$, an explanation of a phenomenon P with respect to C is a derivation tree T such that (1) there are subtrees $t_1, t_2, \ldots, t_k$ in $T_1, T_2, \ldots, T_n$ for which $t_1 \circ t_2 \circ \ldots \circ t_k = T$, (2) the root node of T is mathematically equivalent to P and (3) the leaf nodes of T are either laws or antecedent conditions or any other equations that cannot be derived from higher-level equations.*

So far, we have only dealt with derivations of *idealized* phenomena that can be deductively constructed from theoretical laws and conditions. It is well known, however, that derivations of *real-world* phenomena may involve *non*-deductive elements such as corrections, normalizations and other adjustments that stand in no deductive relation to laws [Cartwright, 1983; Cartwright, 1999; Giere, 1988].

Yet, it is easy to see that EBE can integrate theoretical laws and phenomenological adjustments within trees as long as such adjustments can be described in terms of mathematical formulas. Definition 1 captures in fact non-deductive elements by referring to them as "equations that cannot be derived from higher-level equations".

It is convenient to give an example of a non-deductive element in a derivation tree, since we will need this in section 4. Consider the problem of deriving the discharge of a jet from a tank in fluid mechanics, which is derived by [Norman et al., 1990, p. 497] from Bernoulli's law as follows:

> Suppose the subscripts 1 and 2 refer to a point in the surface of the liquid in the tank, and a section of the jet just outside the orifice. If the orifice is small we can assume that the velocity of the jet is v at all points in this section.

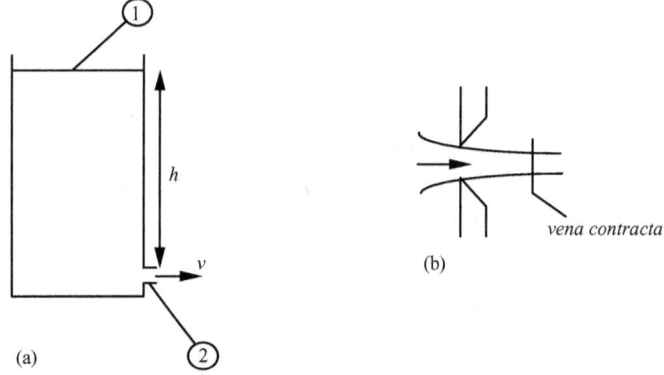

Figure 12.12.

The pressure is atmospheric at points 1 and 2 and therefore $p_1 = p_2$. In addition the velocity v_1 is negligible, provided the liquid in the tank has a large surface area. Let the difference in level between 1 and 2 be h as shown, so that $z_1 - z_2 = h$. With these values, Bernoulli's equation becomes:

$$h = v^2/2g \text{ from which } v = \sqrt{2gh}$$

This result is known as Torricelli's theorem. If the area of the orifice is A the theoretical discharge is:

$$Q(\text{theoretical}) = vA = A\sqrt{2gh}$$

The actual discharge will be less than this. In practice the liquid in the tank converges on the orifice as shown in Figure 12.12b.

The flow does not become parallel until it is a short distance away from the orifice. The section at which this occurs has the Latin name *vena contracta* (*vena* = vein) and the diameter of the jet there is less than that of the orifice. The actual discharge can be written:

$$Q(\text{actual}) = C_d A \sqrt{2gh}$$

where C_d is the coefficient of discharge. Its value depends on the profile of the orifice. For a sharp-edged orifice, as shown in Figure 12.12b, it is about 0.62.

Note that the coefficient of discharge C_d is not derived from higher-level laws in the derivation but is introduced ad hoc. Yet we can still create a derivation tree for this system if we write the coefficient by the rule $Q(\text{actual}) = C_d Q(\text{theoretical})$, which is shown in Figure 5 (in which we also added Bernoulli's law in full):

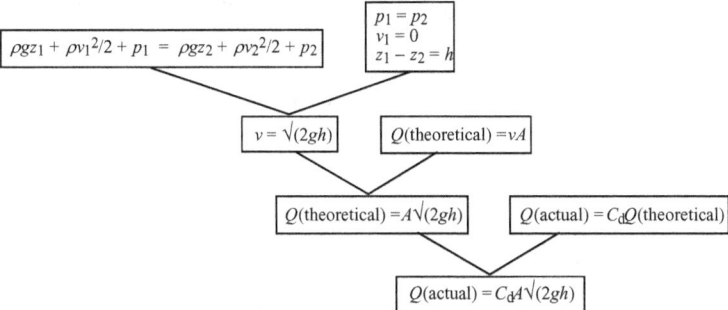

Figure 5. Derivation tree for the discharge of a jet from a tank.

The tree in Figure 5 closely follows the derivation above, where the initial conditions for p_1, p_2, v_1, z_1 and z_2 are represented by a separate label in the tree. The coefficient of discharge is introduced in the tree by the rule $Q(\text{actual}) = C_d Q(\text{theoretical})$. Although this empirical rule does not follow from any higher-level law, it is widely used in fluid mechanics to solve a large number of other systems, ranging from nozzles, notches, nappes, weirs, open channel flows and many pipeline problems – see [Douglas and Matthews, 1996]. The rule is employed in all hydraulic systems with a fluid discharge. Without using it, the predicted discharge of a system can be up to 50% off the mark. By writing the rule in terms of a mathematical equation we incorporate it into a derivation tree and can reuse it wherever needed.

3 Derivational redundancy and the shortest derivation

Given a corpus of derivation trees and a mathematical description of a phenomenon, existing equational reasoning systems, such as *Mathematica* or *TK Solver*, can be employed to derive the phenomenon from the equations in the subtree-roots in the derivation trees (see [Baader and Nipkow, 1998] for an overview on equational reasoning). However, there is a problem that we have not considered so far: there can be many, sometimes extremely many, *different* derivation trees for the same phenomenon. In the worst case, the number of derivation trees grows exponentially with the number of terms in the description of the phenomenon [Baader and Nipkow, 1998]. In other words, derivational explanation is *massively redundant*.

In order to show this, we will first enlarge our corpus used in section 2 with another derivation from Alonso and Finn's textbook. This derivation again provides an exemplary problem solution for the Earth's mass but this time by computing it from the acceleration of an object at the Earth's surface [Alonso and Finn, 1996, p. 246]. This second exemplar can be represented by the derivation tree in Figure 6.

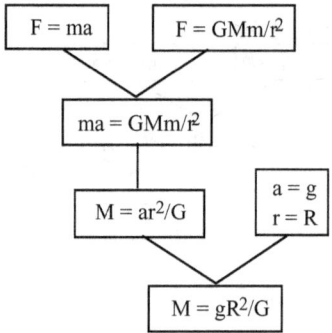

Figure 6. A additional exemplar in the corpus.

By substituting the values for g (the acceleration at the Earth's surface), R (the Earth's radius) and G (the gravitational constant), Alonso and Finn obtain roughly the same value for the Earth's mass as in the previous derivation in Figure 1. They argue that this agreement is 'a proof of the consistency of the theory" [Alonso and Finn, 1996, p. 247]. (Note that the derivation is again idealized: no centrifugal force is taken into account, let alone influences from the Sun or other planets.) Thus the problem of the Earth's mass is derivationally redundant in that it can be solved in at least two different ways. And *both* derivations are used in Alonso and Finn's textbook as exemplars for deriving other phenomena.

When we add the tree in Figure 6 to our corpus, we can model Kepler's regularity also on this exemplar, resulting in an alternative derivation tree given in Figure 7, which uses a large subtree from Figure 6 in combination with two small subtrees from the exemplar in Figure 1.

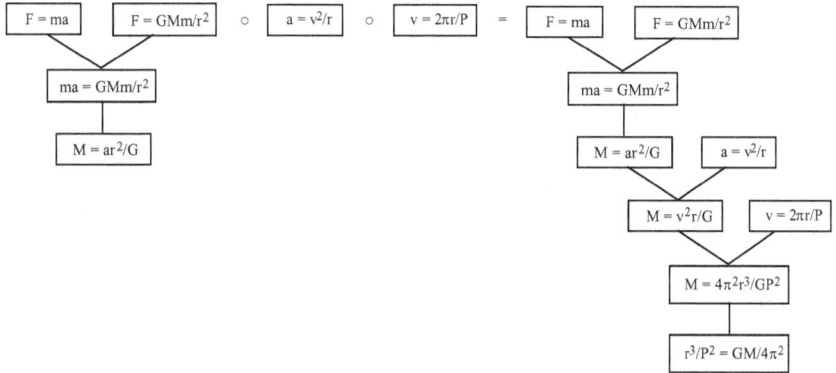

Figure 7. An alternative derivation tree for Kepler's regularity.

Thus there are at least two different derivation trees for Kepler's regularity, represented by Figures 3 and 7, and it is easy to see that many more trees can be constructed by other combinations of laws and subtrees. Note that there is nothing "wrong" with the alternative derivation tree in Figure 7: there are no spurious non-explanatory laws that are irrelevant (as would be e.g. Hook's or Boyle's law – see [Friedman, 1974] for a discussion of non-explanatory laws in derivations). The main difference is that the derivation in Figure 8 is modeled on a *different* exemplar than the derivation in Figure 3. The alternative derivation in Figure 8 is also insightful as it refers to the conceptual equivalence between terrestrial and celestial mechanics in Newtonian dynamics. The fact that Kepler's regularity can be derived not only from Figure 1 but also from Figure 6 suggests that if we bring a satellite down to the Earth's surface it still follows the same regularity.

Yet, it turns out that no physics student comes up with the alternative derivation tree in Figure 7 (see section 4). Why? Apart from the fact that the derivation tree in Figure 3 is slightly smaller, the tree in Figure 3 is more "derivationally similar" to an exemplar in the corpus. That is, the tree in Figure 3 can be constructed by just one large subtree from the corpus, whereas the tree in Figure 7 needs at least 3 subtrees to be constructed from the corpus.

This suggests that if we want to model the way humans derive phenomena, we should mimic as closely as possible the derivations that function as exemplars. A distinctive feature between different derivation trees of a phenomenon is that *some derivation trees are more similar to exemplars than others*. The larger the partial match between a derivation tree and an exemplar, in terms of their largest common subtree, the more "derivationally similar" they are. Since students learn physics not just by memorizing laws, but also by studying exemplary problem solutions, we conjecture that they try to derive a phenomenon by maximizing derivational similarity with previously derived phenomena, or equivalently, by *minimizing derivation length* where the length of a derivation is defined as the number of corpus-subtrees it consists of. We will refer to the derivation of minimal length also as the "shortest derivation". Since subtrees can be of arbitrary size, *the shortest derivation corresponds to the derivation tree which consists of largest partial match(es) with previous derivation trees in the corpus*. This brings us to the following definition of the "best derivation tree" for a phenomenon derived by EBE:

DEFINITION 2 *Let $L(d)$ be the length of derivation d in terms of its number of subtrees, that is, if $d = t_1 \circ \ldots \circ t_k$ then $L(d) = k$. Let d_T be a derivation which results in tree T. Then the best tree, T_{best}, derived by EBE is the tree which is produced by a derivation of minimal length:*

$$T_{\text{best}} = \underset{T}{\operatorname{argmin}}\ L(d_T)$$

It is important to understand the difference between a tree produced by the smallest number of subtrees and an absolute smallest tree. While the tree in Figure 3 is produced by the smallest possible number of corpus-subtrees, it does not correspond to the smallest possible tree, i.e. the tree with the smallest possible number of nodes (or labels). There exists a smaller tree that simply applies all laws at once to arrive at the formula for Kepler's regularity. However, it turns out that no student constructs such minimal derivations (see section 4), and we therefore believe that our notion of shortest derivation being the smallest number of (corpus-)*subtrees* is more appropriate than a notion of shortest derivation defined as the smallest number of *nodes*. Only in case our notion of shortest derivation does not lead to a unique result, i.e. if a phenomenon can be derived by the same number of subtrees, it seems reasonable to choose the tree with the fewest nodes from among the shortest derivations, reflecting a preference for economy if modeling on exemplars does not break ties.

4 Evaluating the EBE model

How can we evaluate EBE? Since we propose EBE to be a cognitive model of human problem solving, it seems appropriate to evaluate our model against manually constructed derivations. To do so, we developed a *test corpus* of manually solved problems (by third-year physics students) and a *training corpus* of exemplary problem solutions from textbooks. Next, we developed an implementation of EBE which computed T_{best} for each test problem by means of subtrees from the training corpus. The performance of EBE on the test problems was compared with the derivations provided by the human participants. Our main goal was to evaluate EBE's conjecture that humans derive new phenomena by constructing the shortest derivation containing the largest possible patterns from previous derivations.

4.1 Method and procedure for the participants

19 third-year physics students from the University of Amsterdam were paid to solve 6 elementary problems from classical mechanics and 5 elementary problems from fluid mechanics. The students had previously followed courses in classical mechanics and fluid mechanics. The 11 problems given to them consisted in deriving a phenomenon from laws, initial conditions and, in the case of fluid mechanics, empirical coefficients. The students were given no other instructions than that they should solve the problems by paper and pencil in class. The first two and the last two of the problems are given below:

Problem nr. 1

Show that the period of the Earth's rotation for which an object at the equator would become weightless is given by $P = 2\pi\sqrt{R/g}$ where R is the Earth's radius and g is the gravitational acceleration at the Earth's surface.

Problem nr. 2

Show that the theoretical velocity which an object attains in free fall from height h is given by $v = \sqrt{2gh}$ where g is the gravitational acceleration at the Earth's surface.

Problem nr. 10

When water flows through a right-angled V-notch, show that the discharge is given by $Q = KH^{5/2}$ in which K is a constant and H is the height of the surface of the water above the bottom of the notch.

Problem nr. 11

Show that the theoretical rate of flow through a rectangular notch is given by $Q = (2/3)B\sqrt{2g}H^{3/2}$ where B is the width of the notch and H is the height of the water level above the bottom of the notch.

After the students had solved the problems on paper, they were given a short, ten-minutes tutorial on the concept of derivation tree, in particular on the difference between binary branches (used for combining laws, conditions etc.), and unary branches (used for mathematical derivation steps). Next, the students were asked to draw derivation trees for their problem solutions. The students had no difficulties with this task and there were no questions during the drawing of the trees. This suggests that derivation trees are comprehensible structures for representing problem solutions.

There was a high agreement among the derivation trees constructed by the participants: on average 91.4% (SD=1.2) of the derivation trees per problem matched. (We abstracted from the difference between right-branching and left-branching trees; all trees given in this paper are right-branching, but they can without loss of generality be represented by left-branching trees, and some of the students actually did this.) Only the most frequent derivation tree for each problem was put in the test corpus.

Next, the students were asked to draw derivation trees for 9 exemplary problem solutions that are used as exemplars in the textbooks by [Alonso and Finn, 1996, chapter 11] and [Douglas and Matthews, 1996, chapter 7]. The three example problems in Figures 1, 5 and 6 were among these exemplars. It should be said, however, that none of the students derived $F = mv^2/r$ from $F = ma$ and $a = v^2/r$, as we did in Figure 1. Instead all students used the equation for centripetal force $F = mv^2/r$ directly as a law. The agreement among the derivation trees for the exemplary solutions was very high: 97.7% (SD=0.3). The most frequent tree for each exemplary solution was put in the training corpus.

All test problems could be solved out of subtrees from the training corpus, but this fact was not told to the students: they first had to solve the test problems after which they were asked to draw trees for the exemplary problem solutions from the textbooks. Each student accomplished the task in less than 2.5 hours (including the tutorial).

4.2 Method and procedure for EBE

The goal for EBE was to construct T_{best} for each of the 11 problems from the test corpus by means of the subtrees from the training corpus of 9 exemplars. To accomplish this, we implemented EBE by using *TK Solver* as a backbone (release 5.0, Universal Technical Systems Inc.). *TK* solves an equation given a list of other equations – provided that there is a solution. To make *TK* suitable for EBE, we programmed a shell around *TK* (written in *C*) which first converted each derivation tree from the training corpus into all its subtrees and next extracted the equations from the subtree-roots. Each equation was indexed to remember the subtree it was extracted from. This resulted in a list L of 148 equations. For each test problem, *TK* derived a set of solutions given the list L of equations. Several problems received more than 100 solutions, which gives an idea of the derivational redundancy if we do not have any mechanism to break ties.

From *TK*'s output, our program selected the shortest solution for each problem that used the fewest equations. The equations of the shortest solution were converted back to their corresponding subtrees, which were combined into the derivation tree T_{best}. Note that in this way T_{best} consisted of the largest partial matches with trees in the training corpus. In case T_{best} was not unique the program chose the tree with fewest nodes among the best trees. A major advantage of *TK* is that it hides the algebra, which is good as this was also asked from the students and which corresponds to our use of unary branches in trees. (We will not go into the equational reasoning techniques used in *TK*, as these are nowadays rather standard – see [Baader and Nipkow, 1998]).

Of course *TK* is by itself not comparable to human problem solving. But what we want to evaluate is not *TK*, but EBE's claim that humans derive new phenomena by constructing (our notion of) the shortest derivation. Thus *TK* is only taken as a tool and should be seen as a "provider" of the set of *possible* solutions from which EBE should predict the *best* solution that humans come up with.

4.3 Results and evaluation

The best trees computed by EBE were compared with the derivation trees constructed by the students in the test corpus. Using an exact match metric, where we only abstracted from the difference between right- and left-branching, the accuracy of EBE was 91%. That is, for 10 out of 11 phenomena, EBE predicted the same derivation tree as assigned by (the vast majority of) the students.

To put our 91% accuracy into perspective, we also computed the accuracy by choosing a *random* derivation tree, T_{random}, from among *all* possible trees that were constructed by EBE (i.e. trees that did not necessarily correspond to the shortest derivation but that could be constructed from *TK*'s output). In this case, the accuracy was 0%. By repeatedly choosing a random tree 100 times for each problem, we still obtained an accuracy of just

0.6%. Although our test set consists of only 11 trees, the difference between 91% accuracy obtained by T_{best} and the 0% (or 0.6%) accuracy obtained by T_{random} – which is however still a "correct" derivation – suggests that T_{best} mimics human problem solving more closely than T_{random}. We also computed the accuracy by choosing the *absolute* smallest tree containing fewest nodes among all proposed trees in *TK*'s output. This resulted in 18% accuracy (2 out of 11). Table 1 gives an overview of the results.

criterion	agreement with trees created by students
random tree	0.6%
smallest tree	18%
shortest derivation	91%

Table 1. Accuracy of EBE compared to students' performance for three different criteria.

These results suggest that if we want to accurately predict the derivations that humans construct for phenomena, we should not just combine laws and initial conditions, neither search for the smallest tree in terms of nodes, but search for the tree which consists of the fewest subtrees from previous derivations, as generated by our notion of shortest derivation. Of course, further experiments on larger corpora are needed to support these results.

Another interesting result was that EBE correctly predicted for each problem whether it could be solved by problem solutions from classical or from fluid mechanics. It could even predict on which exemplar(s) a new problem could best be modeled: namely the exemplar from which the largest subtree could be used. It happened only once that T_{best} was not unique. This occurred with problem nr 2 given above: "Show that the theoretical velocity which an object attains in free fall from height h is given by $v = \sqrt{2gh}$ where g is the gravitational acceleration at the Earth's surface". Although this problem is exceedingly simple, EBE constructed two equally short derivation trees for it with equally many nodes. It is interesting therefore to go with some more detail into this problem.

All students but one derived the formula $v = \sqrt{2gh}$ by equating potential energy mgh with kinetic energy $mv^2/2$ (one student derived it from the Galilean equations of motion, resulting in a considerably longer derivation). EBE, on the other hand, returned two derivations that were equally short. One of these derivations was the same as the (majority of the) students came up with, i.e. by equating potential and kinetic energy – which in fact literally appeared as a subtree in a larger derivation tree in the training set and could thus be constructed by reusing exactly that subtree. The other derivation was taken from the exemplary solution from fluid mechanics for the jet through an orifice, and also consisted of just one subtree, namely a subtree in Figure 5 that has the formula $v = \sqrt{2gh}$ as its root and Bernoulli's equation and the initial conditions as its leaves. Both trees could thus be constructed by exactly one subtree from the corpus and both consisted of three nodes. Thus even after choosing the tree with the fewest nodes from among the shortest derivations, T_{best} was not unique. (Of course, this all hinges on the contents of our training corpus, but remember that we did actually use the exemplary derivations from the relevant chapters of the two textbooks.)

To get out of this impasse, we could add yet another selection criterion to EBE, for example one which chooses the tree with fewest *terms* (variables and constants) in the labels. Because the derivation in classical mechanics contains fewer terms than the derivation in fluid mechanics, this selection criterion would let EBE predict the same derivation as the vast majority of the students did for this problem. But this additional criterion does not solve our impasse in general, since EBE would then predict the wrong derivation if the problem was framed in terms of fluid mechanics, i.e.: "Show that the theoretical velocity of a jet through a small orifice from a tank of height h is $v = \sqrt{2gh}$ where g is the gravitational acceleration at the Earth's surface".

What is actually going on is that EBE constructs derivation trees by *syntactic* reasoning only. It does not distinguish between a point mass and a unit-volume of fluid. But in reading problem nr. 2, we may just as well interpret the word "object" as a part of fluid rather than as a point mass, resulting into two different phenomena. This is not as far-fetched as it seems, since historically Daniel Bernoulli solved the problem of the velocity of water by analogically treating a flow in terms of Newtonian-like particles, which makes the two phenomena "equivalent".

If we want to avoid EBE mixing up derivations from different domains, we could either create different corpora for different domains (which makes EBE much less interesting as we want EBE to find out which exemplar should be used), or introduce different variables for point mass velocity and fluid velocity. The latter can be accomplished by using subcategorizations, e.g. v_p for the velocity of a particle and v_f for a fluid. This way, the two velocities cannot be substituted, and the phenomena $v_p = \sqrt{2gh}$ and $v_f = \sqrt{2gh}$ get each a different derivation. However, as problem nr. 2 is stated above, it is inherently ambiguous and EBE rightly comes up with the somewhat unexpected result of two different best trees. Interestingly, none of the students came up with two best trees, and none came up with the fluid-mechanical interpretation.

This suggests an interesting potential of EBE which may be useful in practice: EBE can provide explanations for phenomena that scientists themselves would perhaps not come up with. By computing for example all candidates for T_best or by slightly loosening the notion of shortest derivation, EBE can investigate an increasingly large space of potential explanations.

5 Related work and discussion

Related work in physics problem-solving has been carried out in Analogical Reasoning and Case-Based Reasoning (CBR) [Carbonell, 1986; VanLehn, 1998; Owensby and Kolodner, 2004]. A major difference with the current work is EBE's focus on solving the problem of derivational redundancy. However, as mentioned in section 1, EBE can be seen as an instantiation of CBR where phenomena are represented by trees and where new phenomena are derived by the *smallest* number of subtrees of previously derived phenomena.

Another related approach, Explanation-Based Learning (EBL) [Mitchell et al., 1986], also memorizes previous cases with the main goal to speed up computation. Moreover, EBL searches for general rules that can cover an entire class of cases. In our approach we already have laws in the exemplars, so explanation-based learning does not seem to be of great use here. Yet, it may be that similar reasoning patterns can be covered by higher-level meta rules, which we will leave for future research.

Our work is different from the field known as Discovery Science [Langley et al., 1992], in that we focus on modeling human problem solving in "normal science" and not on discovering new laws. While there may be possible connections between the discovery of new laws and new explanations by EBE, our aim in this paper was to attempt, if at all, to automate "normal science" rather than "revolutionary science".

There is an important question as to how far EBE stretches. Can we also represent more complex systems by means of derivation trees? EBE hinges on the fact that all knowledge must be represented by equations. But what if such knowledge involves an intermediate model, as is e.g. the case with Prandtl's *boundary layer* model discussed by [Morrison, 1999]? Morrison rightly notes that the boundary layer model is autonomous in that it was not created by some approximation of the general Navier-Stokes equations for fluid dynamics. Yet, Prandtl's model does represent a mathematical structure which approximates the Navier-Stokes equations, which means that it can be represented by a derivation tree. Such a derivation tree does of course not represent the creative act of inventing the boundary-layer model, but once it has been invented it can be reused as an exemplar by EBE to derive a range of new engineering problems.

Note that EBE does not demand that every phenomenon or system be linked to universal laws. As in Kitcher's account, a phenomenon may be derived from a phenomenological model only – without any derivational relation to high-level theory. This is e.g. the case in quantum-chromodynamics, where phenomenological models such as the MIT-bag model are used to decribe certain features of quarks (see [Hartmann, 1999]). Such phenomenological models do serve as exemplars, albeit there is no deductive relation with theory. This kind of situation also occurs in disciplines where universal laws are difficult to come by or where they are not present at all, such as in biology, economics and linguistics (see [Bod, 2005b]). As long as there is a model or regularity representing the phenomenon, we can construct a derivation tree for it. In EBE there is no need to link a phenomenon to general laws, except if the phenomenon can be derived from them. In the "worst" case a derivation tree consists only of the empirical regularity describing the phenomenon.

6 Conclusion

We have given an evaluation of the exemplar-based explanation model, known as EBE. EBE constructs derivations of new phenomena by combining partial derivations of prior phenomena. We argued that the problem of derivational redundancy could be solved by selecting the shortest derivation consisting of fewest partial derivations. We implemented EBE by adapting existing equational reasoning techniques. Experiments on a corpus of phenomena from classical and fluid mechanics indicated that EBE predicts the same derivations as humans come up with a very high degree of accuracy. We argued that EBE can also be applied to situations where intermediate models are needed or where laws are difficult to come by.

BIBLIOGRAPHY

[Alonso and Finn, 1996] M. Alonso and E. Finn. *Physics*. Addison-Wesley, New York, 1996.

[Baader and Nipkow, 1998] F. Baader and T. Nipkow. *Term Rewriting and All That*. Cambridge University Press, Cambridge, 1998.

[Bod *et al.*, 2003] R. Bod, R. Scha, and K. Sima'an. *Data-Oriented Parsing*. University of Chicago Press, Chicago, 2003.

[Bod, 1998] R. Bod. *Beyond Grammar*. CSLI Publications, Stanford, 1998.

[Bod, 2000] R. Bod. Parsing with the shortest derivation. In *Proceedings COLING 2000*, pages 69–76, 2000.

[Bod, 2002] R. Bod. A unified model of structural organization in language and music. *Journal of Artificial Intelligence Research*, 17:289–308, 2002.

[Bod, 2005a] R. Bod. Exemplar-based explanation. In L. Magnani and R. Dossena, editors, *Computing, Philosophy, and Cognition*, volume 4 of *Texts in Philosophy*, pages 329–348. ECAP 2004, Pavia, Italy, College Publications, 2005.

[Bod, 2005b] R. Bod. Towards a general model of applying science. *International Studies in the Philosophy of Science*, 19(3), 2005.

[Branting, 1991] K. Branting. Reasoning with portions of precedents. In *Proceedings Third International Conference on AI and Law*, 1991.

[Carbonell, 1986] J. Carbonell. Derivational analogy: A theory of reconstructive problem solving and expertise acquisition. In Michalski et al., editor, *Machine Learning, Vol. II*, pages 371–392. Morgan Kaufmann, 1986.

[Cartwright, 1983] N. Cartwright. *How the Laws of Physics Lie*. Oxford University Press, Oxford, 1983.

[Cartwright, 1999] N. Cartwright. *The Dappled World*. Cambridge University Press, Cambridge, 1999.

[Collins and Duffy, 2002] M. Collins and N. Duffy. New ranking algorithms for parsing and tagging: Kernels over discrete structures, and the voted perceptron. In *Proceedings ACL 2002*, 2002.

[Douglas and Matthews, 1996] J. Douglas and R. Matthews. *Fluid Mechanics*, volume 1. Longman, 3rd edition, 1996.

[Friedman, 1974] M. Friedman. Explanation and scientific understanding. *Journal of Philosophy*, 71:5–19, 1974.

[Giere, 1988] R. Giere. *Explaining Science: A Cognitive Approach*. University of Chicago Press, Chicago, 1988.

[Giere, 1999] R. Giere. *Science without Laws*. University of Chicago Press, Chicago, 1999.

[Hartmann, 1999] S. Hartmann. Models and stories in hadron physics. In M. Morgan and M. Morrison, editors, *Models as Mediators*. Cambridge University Press, 1999.

[Hempel and Oppenheim, 1948] C. Hempel and P. Oppenheim. Studies in the logic of explanation. *Philosophy of Science*, 15:135–175, 1948.

[Kitcher, 1981] P. Kitcher. Explanatory unification. *Philosophy of Science*, 48:507–531, 1981.

[Kitcher, 1989] P. Kitcher. Explanatory unification and the causal structure of the world. In P. Kitcher and W. Salmon, editors, *Scientific Explanation*. University of Minnesota Press, 1989.

[Kolodner, 1993] J. Kolodner. *Case-Based Reasoning*. Morgan Kaufmann, New York, 1993.

[Kuhn, 1970] T. Kuhn. *The Structure of Scientific Revolutions*. University of Chicago Press, Chicago, 2nd edition, 1970.

[Langley et al., 1992] P. Langley, H. Simon, G. Bradshaw, and J. Zytkow. *Scientific Discovery*. The MIT Press, Cambridge, MA, 2nd edition, 1992.

[Mitchell et al., 1986] T. Mitchell, R. Keller, and S. Kedar-Cabelli. Explanation-based generalization: A unifying view. *Machine Learning*, 1:47–80, 1986.

[Morgan and Morrison, 1999] M. Morgan and M. Morrison. *Models as Mediators*. Cambridge University Press, Cambridge, 1999.

[Morrison, 1999] M. Morrison. Models as autonomous agents. In M. Morgan and M. Morrison, editors, *Models as Mediators*. Cambridge University Press, 1999.

[Nickles, 2003] T. Nickles. Normal science: from logic to case-based and model-based reasoning. In T. Nickles, editor, *Thomas Kuhn*. Cambridge University Press, 2003.

[Norman et al., 1990] E. Norman, J. Riley, and M. Whittaker. *Advanced Design and Technology*. Longman, New York, 1990.

[Owensby and Kolodner, 2004] J. Owensby and J. Kolodner. Case interpretation and application in support of scientific reasoning. In *Proceedings CogSci 2004*, 2004.

[Scha, 1990] R. Scha. Language theory and language technology; competence and performance. In Q.A.M. de Kort and G.L.J. Leerdam, editors, *Computertoepassingen in de Neerlandistiek*. Almer, 1990.

[Suppe, 1977] F. Suppe. *The Structure of Scientific Theories*. University of Illinois Press, Illinois, 2nd edition, 1977.

[VanLehn, 1998] K. VanLehn. Analogy events: How examples are used during problem solving. *Cognitive Science*, 22(3):347–388, 1998.

[Veloso and Carbonell, 1993] M. Veloso and J. Carbonell. Derivational analogy in prodigy: Automating case acquisition, storage, and utilization. *Machine Learning*, 10(3):249–278, 1993.

Rens Bod
Institute for Logic, Language and Computation
University of Amsterdam
1018 TV Amsterdam, The Netherlands
Email: rens@science.uva.nl

The Role of Simulation Models in Visual Cognition

ARTURO CARSETTI

ABSTRACT. Cognitive processes can be considered, in the first instance, as self-organizing and complex processes characterized by a continuous emergence of new categorization forms and by self-referentiality. In order to understand the inner mechanisms of this kind of processes we have to outline a theory of more and more sophisticated forms of organization. We need, for instance, to define new measures of meaningful complexity, new architectures of semantic neural networks, etc. In particular, we have to take into consideration the genetic and "genealogical" aspects that characterize the inner development of cognitive symbolic structures.

However, cognition is not only a self-organizing process. It is also a co-operative and coupled process. If we consider the external environment as a complex, multiple and stratified Source which interacts with the nervous system, we can easily realize that the cognitive activities devoted to the "intelligent" search for the depth information living in the Source, may determine the same change of the complexity conditions according to which the Source progressively expresses its "wild" action. In this sense, simulation models are not neutral or purely speculative. The true cognition appears to be necessarily connected with successful forms of reading, those forms that permit a specific coherent unfolding of the deep information content of the Source. Therefore, the simulation models, if valid, materialize as "creative" channels, i.e., as autonomous functional systems, as the same roots of a new possible development of the entire system represented by brain and its Reality.

As is well known, cognition is not only a self-organizing process, it is also a co-operative and coupled process. If we consider the external environment as a complex, multiple and stratified Source which interacts with the nervous system, we can easily realize that the cognitive activities devoted to the "intelligent" search for the depth information living in the Source, may determine the very change of the complexity conditions according to which the Source progressively expresses its "wild" action. In this sense, simulation models are not neutral or purely speculative. The true cognition appears to be necessarily connected with successful forms of reading, those forms that permit a specific coherent unfolding of the deep information content of the Source. Therefore, the simulation models, if valid, materialize as "creative" channels, i.e., as autonomous functional systems, as the same roots of a new possible development of the entire system represented by brain and its Reality. In other words, cognitive activity is rooted in Reality, but at the same time represents the necessary means whereby Reality can embody itself in an objective way: i.e., in accordance with an in-depth "nesting" process and a surface unfolding of operational meaning. In this sense, the objectivity of Reality is also proportionate to the autonomy reached by cognitive processes.

Within this conceptual framework, reference procedures (as well as simulation processes) thus appear as related to the modalities providing the successful constitution of the channel, of the actual link, first of all, between operations of vision and thought. Such procedures ensure not only a "regimentation" or an adequate replica, but, also, the real constitution of a cognitive autonomy in accordance with the truth. A method thus emerges which is simultaneously project, *telos* and regulating activity: a code which becomes process, positing itself as the foundation of a constantly renewed synthesis between function and meaning. In this sense, reference procedures act as guide, mirror and canalization with respect to primary information flows and involved selective forces. They also constitute a precise support for the operations which "imprison" meaning and "inscribe" the "file" considered as an autonomous generating system. In this way, they offer themselves as the actual instruments for the constant renewal of the code, for the invention and the actual articulation of an ever new incompressibility. Hence the possible definition of new axiomatic systems, new measure spaces, the real displaying of processes of continuous reorganization at the semantic level. Indeed, it is only through a complete, first-order "reduction" and a consequent non-standard second-order analysis that new incompressibility will actually manifest itself. Therefore, the reference procedures appear to be related to a process of multiplication of minds, as well as to a process of unification of meanings which finally emerges as vision *via* models. Here also the possibility emerges of a connection between things that are seen and those that are unseen, between visual recognition of objects and thought concerning their secret interconnections. In other words, this is the connection between the eyes of the mind and those of operational meaning, a meaning which is progressively enclosed within generative thinking and manages to express itself completely through the body's intelligence.

The functional analysis reveals even more clearly, if possible, the precise awareness that, at the level of a cognitive system, in addition to processes of rational perception, we also face specific ongoing processes of semantic categorization. It is exactly when such processes unfold in a coherent and harmonious way that the "I" not only manages to emerge as an observation system, but is also moulded by the simultaneous display of the structures of intentionality. Through the intentional vision, the "I" comes to sense the Other's thought-process emerging at the level of its interiority. The drawing thus outlined, however, is meant for the Other, for the Other's autonomy, for its emerging as objectivity and action. This enables me to think of the autonomy of the Nature that "lives" (within) me.

At the level of intuition-based categorization processes, the file is selected from the ongoing morphogenesis. When the original meaning manages to express new lymph through a renewed production of forms, the self-inscribing file might express its unification potentialities through the successive individuation of concepts which, however, are selected and moulded at an intuitive level. Hence the possibility of an actual "inscription" to the same extent as the morphogenesis, but also the realization of a reduction process, the very laying down of an original creativity within a mono-dimensional and dynamic framework. It is exactly when the reduction is carried out, though, that the procedures of reflection, the identification of limits and completion can be completely performed on the basis of the constant support of the *telos'* activity, of the primary regulation activities proper to the organism, taken as ongoing projectuality.

From an informational point of view, life can be characterized in terms of a concrete answer to three difficult questions: "how is information generated?", "how is information transmitted?" and "how is information assimilated?". With respect to this last interrogative, we have immediately to realize that the assimilation-process of external information implies the existence of specific forms of determination at the neural level as well as the continuous development of a specific cognitive synthesis. Actually, information relative to the system stimulus is not a simple amount of neutral sense-data to be ordered, it is linked to the "unfolding" of the selective action proper to the optical sieve, it articulates through the imposition of a whole web of constraints, possibly determining alternative channels at, for example, the level of internal trajectories. Depth information grafts itself on (and is triggered by) recurrent cycles of a self-organizing activity characterized by the formation and a continuous *compositio* of multi-level attractors. The possibility of the development of new systems of pattern recognition, of new modules of reading will depend on the extent to which new successful "garlands" of the functional patterns presented by the optical sieve are established at the neural level in an adequate way. The afore-mentioned self-organizing activity thus constitutes the real support for the effective emergence of an autonomous cognitive system and its consciousness. Insofar as an "I" manages to close the "garland" successfully, and imprison the thread of meaning, thereby harmonizing with the ongoing "multiplication" of mental processes at the visual level, it can posits itself as an adequate grid-instrument for the "reading-reflection" on behalf of the Source of itself (but in accordance with the metamorphosis in action), for its self-generating and "reflecting" as *Natura naturata*, a Nature which the very units (monads) of multiplication (the final result of this very metamorphosis) will actually be able to read and see through the eyes of mind.

If we take into consideration, for instance, visual cognition we can easily realize that vision is the end result of a construction realized in the conditions of experience. It is "direct" and organic in nature because the product of neither simple mental associations nor reversible reasoning, but, primarily, the "harmonic" and targeted articulation of specific attractors at different embedded levels. The resulting texture is experienced at the conscious level by means of self-reflection; we actually sense that it cannot be reduced to anything else, but is primary and self-constituting. We see visual objects; they have no independent existence in themselves but cannot be broken down into elementary data. Grasping the information at the visual level means managing to hear, as it were, inner speech. It means first of all capturing and "playing" each time, in an inner generative language, through progressive assimilation, selection and real metamorphosis (albeit partially and roughly) and according to "genealogical" modules, the articulation of the complex semantic apparatus which works at the deep level and moulds and subtends, in a mediate way, the presentation of the functional patterns at the level of the optical sieve.

What must be ensured, then, is that meaning can be extended like a thread within the file, constructing a "garland"; only on the strength of this construction can an "I" posit itself together with a sieve: a sieve in particular related to the world which is becoming visible. In this sense, the world which then comes to "dance" before my eyes is impregnated with meaning. The "I" which perceives it realizes itself as the fixed point of the garland with respect to the "capturing" of the thread inside the file and the genealogically-modulated articulation of the file which manages to express its invariance and become "vision" (visual thinking which is also able to inspect itself), anchoring its generativity at a deep semantic dimension. The model can shape itself as such and succeed in opening the eyes of the mind in proportion to its ability to permit the categorial to anchor itself to (and be filled by) intuition (which is not, however, static, but emerges as linked to a continuous process of metamorphosis). And it is exactly in relation to the adequate constitution of the channel that a sieve can effectively articulate itself and cogently realize its selective work at the informational level. This can only happen if the two selection processes meet, and a *telos* shapes itself autonomously so as to offer itself as guide and support for the task of both capturing and "ring-threading". It is the (anchoring) rhythm-scanning of the labyrinth by the thread of meaning which allows for the opening of the eyes, and it is the truth, then, which determines and possesses them. Hence the construction of an "I" as a fixed point: the "I" of those eyes (an "I" which perceives and which exists in proportion to its ability to perceive). What they see is a generativity in action, its surfacing rhythm being dictated intuitively. What this also produces, however, is a file that is incarnated in a body that posits itself as "my" body, or more precisely, as the body of "my" mind: hence the progressive outlining of a meaning, "my" meaning which is gradually pervaded by life.

Vision as emergence aims first of all to grasp (and "play") the paths and the modalities that determine the selective action, the modalities specifically relative to the revelation of the afore-mentioned semantic apparatus at the surface level according to different and successive phases of generality. These paths and modalities thus manage to "speak" through my own fibres. It is exactly through a similar self-organizing process, characterized by the presence of a double-selection mechanism, that the mind can partially manage to perceive (and assimilate) depth information in an objective way [Carsetti, 2004]. The extent to which the simulation model succeeds, albeit partially, in encapsulating the secret cipher of this articulation through a specific chain of programs determines the model's ability to see with the eyes of the mind as well as the successive irruption of new patterns of creativity. To assimilate and see, the system must first "think" internally (at the imagination level) the secret structures of the possible, and then posit itself as a channel (through the precise indication of forms of potential coagulum) for the process of opening and anchoring of depth information. This process then works itself gradually into the system's fibres, *via* possible selection, in accordance with the coagulum possibilities and the meaningful connections offered successively by the system itself.

The revelation and channeling procedures thus emerge as an essential and integrant part of a larger and coupled process of self-organization. In connection with this process we can ascertain the successive edification of an I-subject conceived as a progressively wrought work of abstraction, unification, and emergence. The fixed points which manage to articulate themselves within this channel, at the level of the trajectories of neural dynamics, represent the real bases on which the "I" can reflect and progressively constitute itself. The I-subject can thus perceive to the extent in which the single visual perceptions are the end result of a coupled process which, through selection, finally leads the original Source to articulate and present itself as true invariance and as "harmony" within (and through) the architectures of reflection, imagination, computation and vision, at the level of the effective constitution of a body and "its" intelligence: the body of "my" mind. These perceptions are (partially) veridical, direct, and irreducible. They exist not in themselves, but, on the contrary, for the "I", but simultaneously constitute the primary departure-point for every successive form of reasoning perpetrated by the observer. As an observer I shall thus witness *Natura naturata* since I have connected functional forms at the semantic level in accordance with a successful and coherent "score".

It is precisely through a coupled process of self-organization of the kind that it will finally be possible to manage to define specific procedures of reconstruction and representation within the system, whereby the system will be able to identify a given object within its context, together with its *Sinn*. The system will thus be able to perceive the visual object as immersed within its surroundings, as a self-sustaining reality, and, at the same time, feel it living and acting objectively within its own fibres. In this way it will be possible for the brain to perceive (and assimilate) depth information according to the truth (albeit partially). As Kanizsa [1979] maintained, the world perceived at the visual level is constituted not by objects or static forms, but by processes appearing "imbued" with meaning. The line *per se* does not exist: only the line which enters, goes behind, divides, etc.: a line evolving according to a precise holistic context, in comparison with which function and meaning are indissolubly interlinked. The static line is in actual fact the result of a dynamic compensation of forces. Just as the meaning of words is connected with a universe of highly-dynamic functions and functional processes which operate syntheses, cancelations, integrations, etc. (a universe which can only be described in terms of symbolic dynamics), in the same way, at the level of vision we must continuously unravel and construct schemata, simulate and assimilate, make ourselves available for selection by the coordinated information penetrating from external reality, we have, in particular, to continuously adjust our action in accordance with the internal selection mechanisms through a precise "journey" into the regions of intensionality.

In accordance with these considerations, the brain actually appears as a self-organizing measuring device (as the effective and self-organizing articulation of a biological evaluation "space") [Grossberg, 2000]. This device articulates progressively through a manifold of processing stages characterized by patterns of continuous interaction and integration. At the level of the brain, the computation unit is not furnished by a single processing stage but by an ensemble of processing streams. In this sense, the brain aims first of all to constitute itself as a grid capable of partially recovering in its interior, at the simulation level, the meaningful unity (the irradiating and unifying warp) living at the level of the semantic dimension by means of an adequate texture of self-organizing programs.

This can be understood if we start from a number of simple considerations. The visual process, as stated above, occurs within a coupled system equipped with self-reflection, in which a precise distinction obtains between vision and thought, although they maintain a constant and indissoluble functional exchange. A system of the kind subsumes the articulation of a series of specific processes: a process of mirroring, a process of simulation, a process of assimilation, a process of "irruption", a process of intentional observation, etc. It also subsumes the successive outlining of functions which self-regulate, as well as the progressive construction of increasingly forms of real autonomy. That function which self-organizes with its meaning, and which posits itself as emergent, is "experienced" as vision insofar as it manages to establish itself, at the network level, as a specific modulation and integration of biological circuits capable of realizing a partial engraving of the original *Sinn* (of the deep process of "scanning" and unification articulating at the level of the system-body of meaning). The resulting picture is of a world characterized by continuous emergence and by a constant composition and restructuring of schemes. This composition works at the horizontal and vertical level with a functional and constant internal "thickening" of the processing streams involved in accordance with a precise *bricolage*.

In this sense, vision extends within a coupled system characterized by the presence of a double selection: external and internal , the latter regarding the universe of meaning (this is, actually, a point of fact we are now ready to examine in the light of current achievements in contemporary theoretical Biology). Within the process, meaning reveals itself (albeit partially) in (and through) the effected emergence. Only in this way can a real assimilation process articulate, on the basis first of all of a coherent construction of possible schemes, falsification acts, and so on. This process can then gradually recognize itself in the realized emergence as an act of vision concerning the emergence itself.

In self-organizing emergence, then, we find, simultaneously, a process of assimilation, one of growth, one of "inscription" and one of stabilization and reduction through fixed points. It is therefore not surprising that, as soon as the assimilation (and the unfolding by unification) of meaning occurs correctly, vision appears veridical. What this particularly presupposes as an essential component of the process is also the articulated presence of definite capacities of self-reflection and precise replication-mechanisms at the level of vision by models. Actually, if it is obvious that no thought can exist which has not first filtered through the senses, it is equally clear that there can be no effective vision, at the level of simulation model, unless specific elaboration has taken place able to "coagulate" the activity of internal selection. The outline offered by the model serves first of all to propose possible integration schemes able to support and prime the nesting proper to the internal selection. At the moment of the realization of the embodiment, new vision by models emerges, and the outline as independent instrument is abandoned because superseded. In this sense it is true that at the level of the eyes of mind we have visual cognition, and not intellectual reading. Function and meaning articulate together, but in accordance with the development of a process of *adequatio*, and not of autonomous and direct creation. I will be unable to think of vision during emergence, but will be able to use it, once realized, to construct further simulation models. Growth, modulation, and successive integration thus exist "within and among" the channels together with specific differentiation processes.

It is far from easy to determine mathematics for processes of the kind, since it is clearly impossible to restrict the processes of self-reflection and assimilation totally within the limits of a mechanistic reductionism. Actually, internal and external selection are based on principles and on choices which are articulated on a deep, creative level. Insofar as these principles and choices enter the scene, for example, at the second order level, they cannot be previously determined at the first order level; they are produced by the ongoing dialectics, by the symbolic dynamics in action and are revealed in emergence, i.e. when they really constitute me as the subject which sees and thinks. As for self-reflection, the space occupied by these choices, too, cannot be reductively determined: yet the thread must be untangled and the space explored. The mind has to function as a bridge between internal and external selection. This is the *Via*-Method, relying on the continuous invention of new mathematics, new geometry, new formal axioms, etc. Hence the importance of the eye of the phenomenologist, and in particular of the perceptologist, s/he who listens to the channels, and hence, at the same time, the importance of the eye of the mathematician, s/he who explores the thread of simulation in the regions of pure abstraction. Amodal completion in this context emerges as a privileged window opened on a microcosm which is largely articulated according to the fibres proper to the architecture of mind. Objects are identified through the qualities elaborated and calculated along and through the channels. I neither colonize nor occupy, to use Freeman's words: I offer myself as a gridiron and I am selected. What remains on my flesh, the operative selection, is the inscription by means of cancelations and negative engravings of the deep functional patterns according to which the real processes "pulsate".

The simulation model thus constructed permits a more coherent integration and articulation of the channels, laying the foundation for the self-organized synthesis of ever-new neural circuits. Objects, in their quality of being immersed in the real world, then emerge as related to other objects possessing different features, and so on. Through and beyond these interrelations, holistic properties and dimensions then gradually reveal themselves, which I must grasp in order to see the objects with their meaning, if I am to understand the meaning of things. Apples exist not in isolation, but as objects on a table, on a tree: they are, for instance, in Quine's words, "immersed in red", a reality I can only grasp by means of a complicated second-order process of analysis, elaboration, and comparison which can thereafter be reduced, through concatenations of horizontal and vertical constraints, specific rules and the successive determination of precise fixed points, to the first-order level. I thus need constant integration of channels and formal instruments to grasp information of the kind, i.e. to assimilate structural and holistic relations and relative ties in an adequate way.

In other words, I will understand the meaning of things only if I am able to give the correct coagulum recipes with a view to being selected so as to grasp and capture not only the superficial aspects of objects in the world, but their mutual relations as they interact in depth, in obedience, for instance, to a specific intensional dimension. Only if I provide the correct coagulum, and select the right languages, will these relations emerge through the trigger operated by the "creative" procedures proper to depth information. Information about the outside world and the "genealogical" apparatus "feeding" it is thus extended: hence the need for an "internal" guide to the first articulation and growth of the mechanisms of vision, the need for a particular "thread": the thread of operational meaning, Ariadne's thread, primarily. The rhythm-scanning of the original labyrinth operated by Ariadne together with the operations concerning the "replica" in action allow the eyes of mind to open: herein we can recognize the progressive opening of the eyes of the Minotaur led by the hand through the process of metamorphosis. Ariadne is a lesson in how to "think by forms": how to order and unify (according to a semantic unification and representation process) generative thoughts and functional patterns in order to see, while the Minotaur represents the far-flung multiplicity of channels, the pure creativity in action that progressively constitutes itself as a model, thus allowing for the emergence of a specific (and multiple) mental activity. Selected and guided by Ariadne, and beginning mentally to perceive her (at the level of the inner "replica" effected on the basis of the blueprint of the original scanning), he becomes aware of a new process of self-organization articulating within his channels. Besides thinking of Ariadne, he will also be able to see an external world to the extent to which he himself has been selected by it and filled by intuitions (and he will also be able to see himself as a part of the world-Nature: i.e., as an observer). The self-organization of the channels coincides with the successive stages in his metamorphosis, with his own gradual cognitive development and with his very achievement of a form of effective, intellectual autonomy. It is exactly within the secret paths of this process of metamorphosis that we can ascertain the objective articulation, at a deep (and mental) level, of the specific procedures proper to knowledge construction.

BIBLIOGRAPHY

[Atlan, 1992] H. Atlan. Self-organizing networks: weak, strong and intentional, the role of their underdetermination. *La Nuova Critica*, 19/20:51–155, 1992.

[Carnap and Hillel, 1950] R. Carnap and Y. Bar Hillel. An outline of a theory of semantic information. Report, Massachussets Institute of Technology, 1950.

[Carsetti, 1993] A. Carsetti. Meaning and complexity: the role of non-standard models. *La Nuova Critica*, 22:57–86, 1993.

[Carsetti, 2000] A. Carsetti, editor. *Functional Models of Cognition. Self-organizing Dynamics and Semantic Structures in Cognitive Systems*. Kluwer Academic Publishers, Dordrecht, 2000.

[Carsetti, 2004] A. Carsetti, editor. *Seeing, Thinking and Knowing. Meaning and Self-Organisation in Visual Cognition and Thought*. Kluwer Academic Publishers, Dordrecht, 2004.

[Chaitin, 1987] G. Chaitin. *Algorithmic Information Theory*. Cambridge University Press, Cambridge, 1987.

[Grossberg, 2000] S. Grossberg. Linking mind to brain: the mathematics of biological intelligence. *Notices of AMS*, pages 1358–1374, 2000.

[Herken, 1988] R. Herken, editor. *The Universal Turing Machine. A Half Century Survey*, Oxford, 1988. Oxford University Press.

[Hoffmann, 1998] G. Hoffmann. *Visual Intelligence: How We Create What We See*. WW. Norton, New York, 1998.

[Kanizsa, 1979] G. Kanizsa. *Organisation in Vision: Essay on Gestalt Perception*. Praeger, New York, 1979.

[Kohonen, 1984] R. Kohonen. *Self-organization and Associative Memories*. Springer, Berlin, 1984.

[Talmy, 2000] L. Talmy. *Toward a Cognitive Semantics*. MIT Press, Cambridge, 2000.

Arturo Carsetti
Department of Philosophy
University of Rome "Tor Vergata"
Rome, Italy
Email: `art-car@iol.it`

Thought Experiments and Imagery in Expert Protocols

JOHN J. CLEMENT

ABSTRACT. This paper focuses on case studies from think-aloud protocols with expert scientists solving explanation problems in which they appear to make predictions for novel systems they have never seen by "imagining what will happen" or "running" the system in their heads. Nersessian [2002] has proposed, based on her reading of historical records of investigations in scientific work such as Maxwell's work on electromagnetic field theory, that thought experiments can play a role in scientific theory formulation. Such thought experiments are intriguing because (1) they appear to play a powerful role in science; and (2) the subject appears to gain something like empirical information without making any new observations. This raises what I call the *fundamental paradox of thought experiments*, expressed as: "How can findings that carry conviction result from a new experiment conducted entirely within the head?" Here I will analyze examples of thought experiments from think aloud protocols in order to examine some approaches to resolving the paradox and to begin to explore the breadth of circumstances in which thought experiments (TE's) of different types are used. Scholars have long been intrigued with the nature of thought experiments in science but the definition and scope of the term thought experiment has remained controversial, as have theories of the mechanisms by which they work. This motivates developing a taxonomy and theory of thought experiment processes based on observations from expert protocols. Findings from such studies may also give us a more systematic way to analyze the types of TE's used in the history of science and in instruction.

Expert case study

The data base for this study comes from ten professors and advanced graduate students in scientific fields who were recorded while thinking aloud about the following "Spring Problem":

A weight is hung on a spring. The original spring is replaced with a spring made of the same kind of wire, with the same number of coils, but with coils that are twice as wide in diameter. Will the spring stretch from its natural length more, less, or the same amount under the same weight? (Assume the mass of the spring is negligible.) Why do you think so?

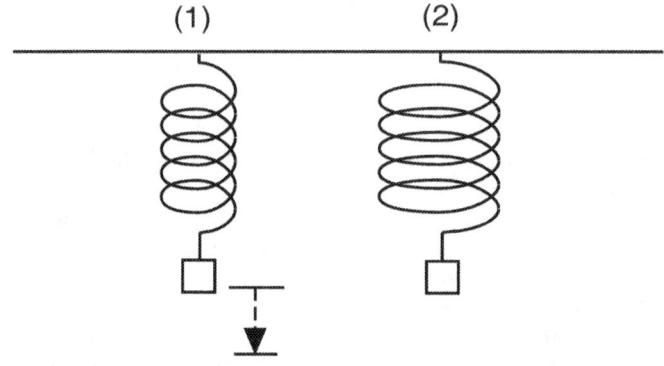

Figure 1.

This data base has yielded previous findings on analogies and creative model construction cycles [Clement, 1988; Clement, 1989; Clement, 2004]. For example, the 1988 paper coded frequencies for different types of analogies across all subjects in the data base, and the 1998 and 2004 papers extended this analysis to other tasks. The 1989 paper documented creative analogies, Aha! insights and cyclical model evaluation and revision processes. Griffith, Nersessian, and Goel [2000], developed a computational model for the generation and revision of analogies from these protocols. The present paper uses newly analyzed hand motions and other indicators in these video tapes to examine thought experiments.

Unfortunately, there is no consensus on a definition of 'thought experiment'. I will present some examples of TE's from one of these subjects and then give a definition. In the following episode subject S2 simply imagined what would happen to wide and narrow springs:

> **Protocol Section 1:** S: "I'm *going to try to visualize it* to imagine what would happen–my guess would be that it [wider spring] would stretch more–this is *a kind of kinesthetics sense* that somehow a bigger spring is looser – Umm, that's high uncertainty."

I will use italicized type to identify observations that have potential as evidence for imagery (both kinesthetic and visual) and simulation use. This appears to be a TE in the sense that he makes a prediction for a concrete situation, and it is one he has not previously observed. In this case his visualization, as he puts it, gives him the correct answer, but he is not very confident in the result, so it is a high–uncertainty TE.

S2 also generated an analogy in which he predicted that a long horizontal rod fixed at one end would bend more than a short one (with the same sized weight attached to the other end of each rod), inferring that segments of the wider spring would bend more and therefore stretch more. However, he was concerned about the appropriateness of this analogy at a deeper level because of the apparent lack of a match between (1) bending producing an increasing slope in the rod and (2) a lack of increasing slope in the wire in a stretched spring. One can visualize this discrepancy here by thinking of the increasing slope a bug would experience walking down a bending rod and the constant slope the bug would experience walking down the helix of a stretched spring. (The latter is my own descriptive analogy for purposes of clarity – not the subject's.) This discrepancy led him to question whether the bending rod was an adequate model for the spring.

> **Section 2**: "But then it occurs to me that there's something clearly wrong with that [bending rod] metaphor, because ...it would (raises hands together in front of face) droop (*moves r. hand to the right in a downward curve*) like that, its slope (*retraces curved path in air* with l. hand) would steadily increase, whereas in a [real] spring, the slope of the spiral is constant..."

Imagining the spring with an increasing slope appears to be a thought experiment in which he "runs" the idea of bending taking place in the spring as it stretches. (This implies that the coils would be farther apart at the bottom of the spring than the top.) This can be seen as a novel thought experiment in which he examines the consequences of running the "bending model" of the spring. This anomaly of a spring with an increasing slope produces a mismatch with his prior knowledge that spring coils should not become wider apart at one end (when the mass of the wire is negligible). This anomaly appears to bother him considerably and drives further work on the problem, eventually discrediting the bending model. This TE also appears to be *generative* in that it generates a situation with new and novel properties, namely a spring with increasing slope. The passages in parentheses indicate *depictive hand motions* which suggest the use of imagery.

Definition

These examples motivate the following definition. Performing an *(untested) thought experiment* (in the broad sense) is the act of considering an untested, concrete system (the "experiment" or case) and attempting to predict aspects of its behavior. Those aspects of behavior must be new and untested in the sense that the subject has not observed them before nor been informed about them. I use the phrase "untested" thought experiment to emphasize that this does not include cases where the subject simply replays a previously observed event. Still, the above definition is intentionally very broad. Later I will give a second narrower definition for referring to a more specialized use of TE's. However, an advantage of the broad definition above is that it appears to designate cases that raise the fundamental paradox.

Polygonal springs

After spending nearly 30 minutes considering this and other analogies, he generates the polygonal coil cases in Figure 2a and 2b. While analyzing the hexagon in terms of bending effects below, it occurs to him in an Aha episode that there will also be *twisting* effects in the segments.

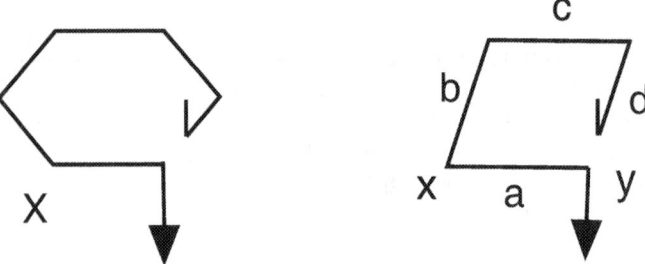

Figure 2. a and b.

Section 3 - *Running an analogy and activating a schema:*

"Just looking at this it occurs to me that when force is applied here (at arrow in Figure 2a), you not only get a bend on this segment, but because there's a pivot here (point x), you get a torsion [strain produced by twisting movement of the wire] effect. Aha!! Maybe the behavior of the spring has something to do with twist (*makes twisting motion with right hand*) forces as well as bend forces. That's a real interesting idea."

Twisting of the wire and the resulting torsion is in fact a key element in the analysis of spring behavior as understood by engineers. Its discovery here represents a major insight in finding a new causal mechanism. The torsion discovery and Aha! phenomenon above is an interesting process in itself and is discussed in Clement [1989]. However in this paper I want to focus on examining the possibility that imagistic simulation plays a role in performing and making inferences for thought experiments rather than on how they are generated. The subject continues:

> **Section 4**: "Let me accentuate the torsion force by making a square where there's a right angle. I like that, a right angle. That unmixes the bend from the torsion. Now I have two forces introducing a stretch. I have the force that bends this segment a (Figure 2b) and in addition I have a torsion (*makes twisting motion with right hand*) force which twists [rod b] at vertex, um, x" [as if side a were a wrench acting to twist side b].
>
> **Section 5**: "Now let's assume that torsion and bend (*makes bending motion with hands together*) don't interact...does this (points to square) gain in slope–toward the bottom? Indeed, we have a structure here which does not have this increasing slope as you get to the bottom. It's only if one looks at the fine structure; the rod between the Y and the X, that one sees the flop (*moves left hand horizontally in a downward curve*) effect". "Now I feel I have a good model of sp- of a spring."

Because bending and twisting still allow the slope to "start over from zero" at each corner, the square coil is a new model in which the accumulating slope difficulty does not occur, suggesting a way to resolve his previous anomaly. He goes on to ask about the effect of coil width for the square coil model.

> **Section 6**: "Now making the sides longer certainly would make the [square] spring stretch more... The longer the segment, (*holds hands up in front as if holding something between them*) the more (*makes bending motion with right hand*) the bendability."
>
> **Section 7**: "Now the same thing would happen to the torsion I think, because". "If I have a longer rod (*moves hands apart*), and *I put a twist* on it (*moves hands as if twisting a rod*), it seems to me–again, physical intuition–that it will twist more... I'm (*raises hands in same position as before and holds them there*

Figure 3. a, b, and c.

continuously until the next motion below)... *imagining holding something that has a certain twistyness to it, a-and twisting it...*"

Section 8: Now I'm confirming (*moves right hand slowly toward left hand*) that by using this method of limits. As (*moves right hand slowly toward left hand until they almost touch at the word "closer"*) I bring my hand up closer and closer to the original place where I hold it, *I realize very clearly that it will get harder and harder to twist.* So that confirms my intuition so I'm quite confident of that... (see Figure 3).

(The reader may wish to try this thought experiment with images of coat hanger wire, bent to have "handles" at each end of the wire.) Later the subject distinguishes between confidence in the answer to the spring problem and confidence in his understanding of it, and indicates that the torsion analysis has increased his subjective feeling of understanding from "way, way down" to "like, 80%". At this point S2 appears to have a mental model of the spring as working like a square coil that contains elements that both bend and twist. S2 uses thought experiments to predict correctly that the wider spring will stretch more. These also suggest that the slope of the stretched spring will be constant throughout (also correct), resolving S2's previous anomaly about increasing slope.

Traditional descriptions of expertise focus on practiced skills and domain specific knowledge about systems that are very familiar, whereas in this setting, the subject exhibits something more like adaptive expertise wherein he tries to use more general heuristics to invent and evaluate new models of an unfamiliar system. Unlike practiced homework problems, the spring problem challenges S2 in this way, and appears to put him on the "frontier" of his own personal knowledge.

Under the definition proposed, each of the eight numbered sections above can be considered to contain an untested thought experiment. The epistemological status of thought experiments is controversial in terms of whether they can "officially" support or discount scientific theory in the realm of publicly tested theory. Real experiments are ordinarily preferred when available. Nevertheless, scientists such as Galileo, Newton, and Maxwell have included powerful thought experiments in their published works. Perhaps a more interesting question is whether they can play an important role in the generation and initial evaluation of theories by a scientist during the challenging process of theory construction. Nersessian [2002] discusses the way in which Maxwell realized through a TE that adjacent electromagnetic vortices in the ether would in effect "jam" like meshing gears trying to rotate in the same direction and the way that this TE led him to reject and improve upon this early model. Related processes that can be interpreted as utilizing thought experiments have been documented in the case of Michael Faraday by Ippolito and Tweney [1995] and in the case of astronomers by Trickett and Trafton [2002]. In the present case we appear to have evidence that (1) TE's have been part of the subject's generating and considering important new hypotheses (e.g. protocol sections 1 and 3); (2) a TE (section 2) has raised serious doubts about one hypothesis; (3) TE's have boosted his confidence in other hypotheses about certain aspects of system behavior (section 8) and have increased his feeling of understanding how the system works. Thus TE's appeared to play a central, not just peripheral, role within the thinking of this subject in helping to generate, cast serious doubt on, or support hypotheses.

Use of imagistic simulation

In this section I will put forward an initial explanation of how TE's work by introducing the concept of imagistic simulation. Italicized type above in sections 7 and 8 identifies examples of several imagery-related observation categories, in the following order: *personal action* projections (spontaneously redescribing a system action in terms of a human action, consistent with the use of kinesthetic imagery), *depictive hand motions*, and *imagery reports*. The latter occurs when a subject spontaneously uses terms like "imagining", "picturing", a situation, or "feeling what it's like to manipulate" a situation. In this case it is a *dynamic imagery report* (involving movement or forces). None of these observations are infallible indicators on their own, but I take them as evidence for imagery, and this is reinforced when more than one appear together. Such indicators also appear alongside new predictions at many earlier points indicated by italicized type in the protocol segments. (There is not space for a review here, but an increasing variety of studies of depictive gestures suggest that they are expressions of core meanings or reasoning strategies and not simply translations of speech.

Others indicate that the same brain areas are active during real actions and corresponding imagined actions.)

One can draw on the precedent of motor schema theory [Schmidt, 1982] in hypothesizing that perceptual motor knowledge structures that can control real actions over time (e.g. a schema for "twisting" objects) are involved here. The observations in sections 7 and 8, for example, can be explained via what I have called schema–driven *imagistic simulations* wherein: (1) the subject has activated a somewhat general and permanent perceptual motor schema that can control the action of twisting real objects; (2) the schema assimilates an image of two rods of different lengths that is more specific and temporary; (3) the action schema "runs through its program" vicariously without touching real objects, generating a simulation of twisting the two rods, and the subject compares the anticipated effort required for each. Such a simulation may draw out implicit knowledge in the schema that the subject has not attended to before – e.g. in this case the simulation may draw out knowledge embedded in analog tuning parameters of a motor schema to anticipate differences in the effort required to twist a long and short rod. In other words, a hypothesis can be made, with initial grounding in data such as that in protocol sections 7 and 8, that the subject is going through a process wherein a general action schema assimilates the image of a particular object and produces expectations about its behavior in a subsequent dynamic image, or simulation [Clement, 1994; Clement, 2003]. The knowledge being used there is "embodied" in this sense.

Such domain specific schemas that generate simulations can work in concert with more general spatial reasoning skills – reasoning operations that embody spatio-temporal constraints on *any* system of objects, such as the constraint that solid objects may not occupy the same space, or that the face of an object turning on a vertical axis will disappear and reappear (cf. [Shepard and Cooper, 1982]). Given these elements the ability to run thought experiments is most plausibly explained using a framework that includes (a) flexible perceptual motor schemas that can run imagistic simulations via the extended application of the schema outside of its normal domain of application ("outside" means that the schema is being applied to either: an unfamiliar situation; or a familiar situation along with question that has never been asked before about the outcome); (b) converting implicit into explicit knowledge and/or (c) spatial reasoning such as that involved in section 5 in imagining whether the contributions to stretching in each side of the square coil add or cancel [Clement, 1994]. This last case also involves a (d) compound simulation wherein multiple simulations (of bending or twisting in each side) are performed on the same image and the results added together imagistically. One can point to the above sources

(a) through (d) as potential origins of conviction in thought experiments, to help us begin to explain the fundamental paradox. They can also explain the effectiveness of the extreme case at the end of the transcript above as an example of what I call "imagery or simulation enhancement", a phenomenon difficult to explain in other ways [Clement, 1994]. The extreme case leads to the same prediction with a much higher degree of confidence. I infer that this comes from increasing the differences between the two images being compared and making that difference more detectable under inspection of the images – here the kinesthetic difference in the torque or twisting force applied to a "normal" rod and a very short rod in order to put a certain amount of twist in it. Thus, I hypothesize that this is a case of "simulation enhancement" and that the role of this extreme case is *to enhance the subject's ability to run and compare imagistic simulations with high confidence.* In this case the main source of conviction in the simulation is the tapping of implicit knowledge and its conversion into explicit knowledge. The ability of the theory to explain why the extreme case helps in this case lends some initial support to the theory.

Another subject, S6, attempted to imagine the direction in which the wire would twist in a spring coil when it was stretched, but found it difficult. By adding "little paint dots" to his image of thick spring wire, he was eventually able to do this. This appears to be another kind of imagery enhancement strategy.

In summary, there is evidence that the TE's documented here involved the use of imagery. The successful use of TE's above can be explained by hypothesizing that they involve an imagistic simulation process wherein a perceptual motor schema generates dynamic imagery that can be interrogated. Four possible sources of conviction in TE's were suggested that can begin to explain the fundamental paradox.

Evaluative Gedanken experiments

Band spring. A use of the term "thought experiment" in a *narrower sense* is what I call an *evaluative Gedanken experiment.* Performing an *evaluative Gedanken experiment* is the act of considering an untested, concrete system designed to help *evaluate* a *scientific concept, model,* or *theory* – and attempting to predict aspects of its behavior. An evaluative Gedanken experiment is usually more complex than a simple thought experiment. In the cases discussed here the subject considers a new system for which the present model is predictive, but for which another source of knowledge is also predictive, giving the potential for conflict or consonance between the two prediction methods. An example from subject S2 is the case of a spring made of a vertically oriented band of material shown at the bottom of Figure 4a. (the reader might imagine the metal strip unwound from a coffee can, with coils that are reshaped to make a spring, say, 3" wide.)

> **Section 9**: "How about a spring made of something that can't bend. And if you showed that it still behaved like a spring you would be showing that the bend isn't the most important part. Or isn't particularly relevant at all maybe somehow... How *could I imagine such a structure*?... I'm thinking of something that's made of a band... *we're trying to imagine configurations* that wouldn't bend. Since its cross section is like that (see Figure 14.4) ... it can't bend in the up/down (*indicates up/down directions with hands*) direction like that because it's too tall. But it can easily twist (*motions as if twisting an object*)."

Given the initial imagery reports here, I interpret this to mean that the subject imagined that such a spring would still be quite stretchable even though the band "cannot bend in the up–down direction", challenging the necessity of bending as not "particularly relevant at all" to stretching. In this type of evaluatory Gedanken experiment he designs a special case where the bending model yields a prediction, (predicts no stretch) but where he also has some other independent source of information that can evaluate that prediction (physical intuition predicts that it will stretch), as shown in Figure 4a. This is a Gedanken experiment because it is designed to help test a model. It may actually involve two thought experiments, so it is a more complex reasoning pattern than a simple thought experiment. Either of the competing predictions just mentioned and represented by vertical arrows in Figure 4a may occur via a simple thought experiment.

Galileo. A similar pattern can be seen in the Dropping the Cannonball Through a Hole in the Earth experiment shown in Figure 4b, used by Galileo [Galilei, 2001] in his *Dialogues on the Two Chief World Systems*. (Actually this thought experiment can be traced back at least as far as his predecessor Tartaglia.) The larger issue at hand is whether Aristotle's model of "Natural" and "Violent" motion, as two different types of motion, is valid. Here "Natural" motion is motion toward the center of the Earth, where as throwing a ball upward is "Violent" (unnatural) motion. Drilling a hole through the Earth and dropping a cannonball into it appears to produce a smooth transition from natural to violent motion at the center of the Earth. And yet the cause of upward motion seems eminently "natural".

> Salviati: But [the cannonball] having arrived at the center is it your belief that it would pass on beyond..?
>
> Simplicio: I think it would keep on going a long way.

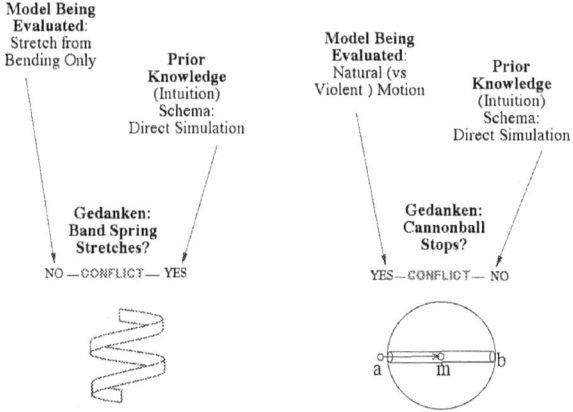

Figure 4. a and b.

Salviati: Now wouldn't this motion beyond the center be upward, and, according to what you have said, preternatural and constrained? Let me see you find an external thrower who shall overtake it.. to throw it upward.

The inference that the cannonball appears to pass through the center absolutely smoothly without any discontinuous change attacks Aristotle's central distinction between Natural and Violent motion and helps discredit the model in the dialogues. Again, as shown in Figure 4b, the Gedanken case evokes one prediction from the model and a conflicting prediction from another source, here one's intuitions about the strong momentum of a speeding cannonball. Thus the pattern of reasoning in the band spring experiment from the expert protocol appears to have the same form as an evaluative Gedanken experiment from Galileo's dialogues.

Shear forces in spring. A similar pattern was seen when another subject, S15 (interviewed after the original set of subjects discussed at the beginning of this paper) generated a microscopic model of shear forces and displacements in the spring wire that shows one element being displaced downward relative to the previous element. He drew these wire element displacements as a kind of slowly descending, spiral staircase. However, by running a Gedanken experiment he discovers a contradiction: this model predicts that stretch in a loosely coiled wire and a tightly coiled wire of the same length should be the same. But this is counter to his intuition since in running this comparison macroscopically without analyzing it, he imagines that the tightly coiled wire would stretch less. This case is not the same as the original problem because the wire length in the newly proposed experiment is the same for both springs. It is a Gedanken experiment that pits the shear model against strong intuitions about the behavior of loosely and tightly coiled material, and it yields a contradiction to his model. Multiple instances of depictive hand motions in this episode provide evidence for imagistic thought.

> **S15**: "(Draws spring like coil with elements descending in steps so that each element is slightly displaced vertically with respect to the next) The mechanism of force communication is this sort of shearing-like property (h*olds both hands side by side oriented vertically, palms open and facing each other*) But now the next think I want to try to understand is, ok,... given that mechanism, explain why then that the very loosely wound coil (*makes a wide tracing of a large coil with his right hand*) and the tightly wound coil have different elasticities. The net displacement is proportional to how much relative motion (*holds palms flat, face down, so index fingers are touching and slides one hand upward and one down*) the communication of force... calls for...
>
> But now I could take this thing, [the same piece of wire] and I could wind it tightly, or I could wind it loosely, and I have the sense that if my model's correct, the rate that I wind it is gonna be (pause) ok. Now I'm extremely unhappy – The stiffness of the spring is just a function of the length of the material – (*holds palms flat, face down, so index fingers are touching and slides hands in opposite directions*) – the contradiction [is]... that I could wind this thing as tight as a spring... or as a very loose spring and it would still have the same displacement".

Here again in this case, his current model predicts one result (equal stretching of the two coils) while a powerful intuition schema predicts the opposite result, discrediting the model strongly.

Conclusion

In this paper I have attempted to show that it is possible to document the use of thought experiments in think aloud protocols. We saw that for the three subjects discussed here, thought experiments can be either supportive of or conflicting with an explanatory model. By this I do not mean confirming or rejecting a hypothesized model with certainty. However, the present paper argues that at the very least, TE's can add or detract support for a model that may be competing with other models in the mind of the scientist as a plausible theory.

The Fundamental Paradox of TE's: "How can findings that carry conviction result from a new experiment conducted entirely within the head?" is a long standing problem for philosophy of science as well as cognitive science. It appears to apply to all Untested TE's as defined here in the broad sense as the act of predicting the behavior of an untested, concrete, system (the "experiment"). Since this definition for untested thought experiments is quite broad, this suggests that the paradox applies surprisingly broadly. And in fact, the examples of TE's presented in this paper occurred within a broad variety of types of reasoning: Elemental TE's, Analogy, Extreme Case Use, Running a Model, and Gedanken Experiments.

Toward explaining the paradox. Instead of one source of conviction that might explain the fundamental paradox, a number of sources were hypothesized to account for the transcript data. However, a central theme is that these all depend on imagistic representations. There was evidence that the TE's in the protocols used imagistic simulations. Multiple indicators were identified, including: predicting changes in a system, imagery reports, personal action projections, depictive hand motions, and dynamic imagery reports. The hypothesized sources of conviction in thought experiments utilizing such imagistic simulations included: (a) flexible perceptual motor schemas that can run imagistic simulations via the extended application of the schema outside of its normal domain of application; (the Bending Spring, and Vertical Band Spring cases) (b) converting implicit into explicit knowledge (twisting rod case); (c) spatial reasoning (increased stretch in wider square coil case; and non–cumulative slope in square coil); and (d) compound imagistic simulation (also used in the these two square coil cases).

Are thought experiments primarily based on prior experience or creative reasoning? In an early paper on thought experiments, Gooding [1992] took a position that was intermediate between the empiricist and rationalist ends of the spectrum, concluding that: "What is needed is a combination of empirical knowledge and the ability to reason with it." The present analysis fleshes out details in Gooding's general position and provides evidence that supports it. The first two sources above are developmentally experiential to the extent that they have historical roots in experience. In (a) using an old schema in a new situation outside its normal domain of application can mean that there is some uncertainty in the application, but there may also be a fair amount of conviction. In (b) one uses stored knowledge, possibly from an old empirical source, that is encoded in analog form and needs to be made explicit. Both of these processes use prior knowledge schemas but must extend them via reasoning since, by definition, a TE deals with an unfamiliar aspect of a case.

Sources c and d above primarily utilize additional reasoning capabilities to make further extensions of thought. In (c) spatial reasoning, about perceptual transformations or the way movements add together for example, is a type of plausible reasoning that is presumably ubiquitous but largely unrecognized in philosophy of science. And in (d) the subject runs two schemas on the same case in a way that involves sequencing or coordinating and predicting the effects of a new combination of actions, a related form of generative reasoning via imagistic simulation. Thus it can be hypothesized that TE's can utilize specific experiential knowledge; but this imagistic knowledge is extended via reasoning. They can use both prior knowledge schemas and reasoning processes which extend the range and types of inferences made from using the schemas in new or novel situations. Thus I believe that analyzing the contribution of the above four aspects of imagistic simulation can begin to give us a theoretical picture of how thought experiments work. This may not be the only mechanism used, but it appears to be a very important one.

Gedanken experiments. I also introduced the concept of Evaluatory Gedanken Experiments: (TE's in the narrow sense): these are designed to help evaluate a theory, model, or concept [Clement, 2003]. These are more similar to real experiments, and can include something like the control of variables. Their argument structures can be more complex than that of a simple untested thought experiment. The structure of evaluative Gedanken experiments observed in the expert protocols here appears to be the same as that of a thought experiment discussed by Galileo. This leads us to conjecture that the concepts derived here may be applicable to certain thought experiments in the history of science. It is hoped that the descriptors generated here will help us to develop a theory of TE's based on evidence from naturalistic observations of behaviors, and that this can interact fruitfully with the analysis of historical TE's.

Educational experiments. A relatively small number of specific Gedanken experiments (in the narrow sense) in the history of science have been recognized for their pedagogical value wherein the scientist uses the experiment to convince others of the validity of his or her theory (cf. [Reiner and Gilbert, 2000]). Because TE's in the broad sense appear in many types of reasoning, they are more ubiquitous than previously thought. This suggests that they may play a more widespread role in instruction. Therefore I believe understanding the mechanisms for TE's in scientists will lead to important payoffs for education as well.

In the framework developed here, expert TE's can make use of embodied analog perceptual–motor systems and schemas used in mental simulations. Both the broad and narrow concepts of thought experiment as defined here appear to be useful, and both can be documented in think aloud protocols. The broad concept is appropriate for expressing the fundamental paradox. The narrower concept of an evaluative Gedanken experiment encompasses Galileo's use of the "hole through the earth" thought experiment, impressive in that it was designed to contribute to eliminating an established theory.

BIBLIOGRAPHY

[Clement, 1988] J. Clement. Observed methods for generating analogies in scientific problem solving. *Cognitive Science*, 12:563–586, 1988.

[Clement, 1989] J. Clement. Learning via model construction and criticism: Protocol evidence on sources of creativity in science. In J. Glover, R. Ronning, and C. Reynolds, editors, *Handbook of Creativity: Assessment, Theory and Research*, New York, 1989. Plenum.

[Clement, 1994] J. Clement. Use of physical intuition and imagistic simulation in expert problem solving. In D. Tirosh, editor, *Implicit and Explicit Knowledge*, Norwood, NJ, 1994. Ablex.

[Clement, 1998] J. Clement. Expert novice similarities and instruction using analogies. *International Journal of Science Education*, 30(10):1271–1286, 1998.

[Clement, 2003] J. Clement. Imagistic simulation in scientific model construction. In *Proceedings of the Twenty-Fifth Annual Conference of the Cognitive Science Society*, 25, Mahwah, NJ, 2003. Erlbaum.

[Clement, 2004] J. Clement. Imagistic processes in analogical reasoning: Conserving transformations and dual simulations. In K. Forbus, D. Gentner, and T. Rieger, editors, *Proceedings of the Twenty-Sixth Annual Conference of the Cognitive Science Society*, pages 233–238, Mahwah, 2004. Erlbaum.

[Galilei, 2001] G. Galilei. *Dialogues Concerning the Two Chief World Systems*. Random House, New York, 2001.

[Gooding, 1992] D. Gooding. The procedural turn: or, why do thought experiments work? In R. Giere, editor, *Cognitive Models of Science*, pages 45–76, Minneapolis, 1992. University of Minnesota Press.

[Gooding, 1998] D. Gooding. Thought experiment. In E. Craig, editor, *The Encyclopaedia of Philosophy*, London, 1998. Routledge.

[Griffith et al., 2000] T. W. Griffith, N. J. Nersessian, and A. Goel. Function-follows-form transformations in scientific problem solving. In *Proceedings of the Cognitive Science Society 22*, pages 196–201, Mahwah, N.J., 2000. Erlbaum.

[Ippolito and Tweney, 1995] M. F. Ippolito and R. D. Tweney. The inception of insight. In R. J. Sternberg and J. E. Davidson, editors, *The Nature of Insight*, Cambridge, MA, 1995. MIT Press.

[Kuhn, 1977] T. W. Kuhn. A function for thought experiments. In T. Kuhn, editor, *The Essential Tension*, Chicago, 1977. University of Chicago Press.

[Magnani, 1999] L. Magnani. Model-based creative abduction. In L. Magnani, N. J. Nersessian, and P. Thagard, editors, *Model-Based Reasoning in Scientific Discovery*, pages 219–238, New York, 1999. Academic/Plenum Publishers.

[Nersessian, 2002] N. J. Nersessian. Maxwell and "the method of physical analogy": model-based reasoning, generic abstraction, and conceptual change. In D. Malament, editor, *Essays in the History and Philosophy of Science and Mathematics*, pages 129–166, Lasalle, IL, 2002. Open Court.

[Reiner and Gilbert, 2000] M. Reiner and J. Gilbert. Epistemological resources for thought experimentation in science learning. *International Journal of Science Education*, 22(5):589–506, 2000.

[Schmidt, 1982] R. A. Schmidt. *Motor Control and Learning*. Human Kinetics Publishers, Chmpaign, IL, 1982.

[Shepard and Cooper, 1982] R. Shepard and L. Cooper. *Mental Images and their Transformations*. MIT Press, Cambridge, MA, 1982.

[Trickett and Trafton, 2002] S. Trickett and J. G. Trafton. The instantiation and use of conceptual simulations in evaluating hypotheses: Movies-in-the-mind in scientific reasoning. In W. Gray and C. Shunn, editors, *Proceedings of the Twenty-Fourth Annual Conference of the Cognitive Science Society*, pages 878–883, Mahwah, NJ, 2002. Erlbaum.

John J. Clement
University of Massachusetts
Amherst, USA
Email: clement@srri.umass.edu

Technological Thinking and Moral Imagination

MICHAEL E. GORMAN

ABSTRACT. This paper begins with an overview of the methods used to study scientific thinking, then applies two of these approaches to distinct case-studies of invention. The first involves Alexander Graham Bell's work on the telephone; the second involves a modern design team developing a textile that they intended to embody environmental intelligence. In both cases, analogies to nature are used to form mental models of a new technological system; in the second case, this mental model has an explicitly moral component. A computational model of Bell's network of enterprises has been created by Marin Simina at Georgia Tech, but as yet, no computational model exists for the textile case, though it might be possible to adapt Gooding and Addis' Belief 2.0.

1 Introduction

Kevin Dunbar and Jonathan Fugelsang provide a taxonomy of approaches that one can take when studying scientific thinking [Dunbar and Fugelsang, 2005]. They begin with a distinction (first made by [Bruner et al., 1956]) between *in vitro* studies of scientific thinking, which involve abstract tasks like the 2-4-6 [Gorman, 1992] and *in vivo* studies, which involve observing and analysing scientific practice. [Dunbar, 1999] has previously employed an *in vitro* task to study how participants reasoned about a genetic control mechanism, and also conducted *in vivo* studies of molecular biology laboratories. Dunbar and Fugelsang label four more approaches in the same style:

Ex Vivo research, in which a scientist is taken out of her or his laboratory and investigated using in vitro problems similar to those they would use in their research. John Clement's work is a good example (see his paper in this volume).

In Magnetico research, using techniques like MRI to study brain patterns during problem-solving, including potentially both *in vitro* and *in vivo* research.

In Silico research, involving computational simulation and modelling of the cognitive processes underlying scientific thinking. Shrager and Langley provide a classic overview of this work [Shrager and Langley, 1990].

Sub Specie Historiae research, focusing on detailed historical accounts of scientific and technological problem-solving. Nancy Nersessian's work on Maxwell [Magnani et al., 1999] and Gooding and Tweney's work on Faraday serve as good examples [Gooding, 2005; Tweney et al., 2005]. These *sub specie historiae* studies can serve as data for *in silico* simulations.

This chapter will extend this literature into invention, using two of the approaches outlined above: a sub specie historiae case-study and an in vivo case-study of a modern design team. The latter will introduce a new element: inventions motivated by ethical concerns.

2 Alexander Graham Bell's path to the telephone

2.1 Bell's network of enterprises

Howard Gruber coined the term 'network of enterprises' to refer to the way in which Darwin pursued multiple research projects study of barnacles, pigeon-breeding and geology, for example that came together in his theory of evolution [Gruber, 1989]. Bell's network was not as broad as Darwin's, but it did include two main enterprises: teaching the deaf and multiple telegraphy. Bell's family was very involved with teaching the deaf. As part of that, Bell wanted to find a way for deaf people to be able to see the sound patterns they were making, so they could improve them. He experimented with a device called the manometric flame capsule; it had a speaking tube and membrane on the other side of which was a chamber through which gas was fed to a small flame. As one spoke, the gas was alternately compressed and decompressed by the vibration of the membrane, changing the height of the flame, which was showed in mirrors. Bell noted that, "when we speak to the apparatus, an undulatory band of light makes its appearance in the mirror. The upper edge of the luminous band appears to be carved into beautiful waves of various shapes and sizes, and when we sing different vowel sounds into the mouth-piece of the instrument, retaining the voice on a uniform level, the form or shape of the undulations visible in the mirror changes with every vowel. I thought that if I could discover the shape or form of vibration that was characteristic of the elements of English speech, I could depict these upon paper by photographic means for the information of my deaf pupil" [Bell, 1908, pp. 24–25].

Since Bell could not physically record the manometric flame patterns, he concentrated on another device, the phonautograph, that consisted of a cone and membrane with a lever attached to the membrane. When one spoke into the cone, the lever vibrated. At the end of the lever was a bristle brush which traced the shape of the sound wave on a piece of smoked glass that was moved in a direction perpendicular to the motion of the lever. "I proposed to use these glass plates as negatives, and by photographic means, print off copies of the tracings for the use of my pupils" [Bell, 1908, p. 26].

Considering the phonautograph's geometry–with its thin, light membrane and the relatively heavy wooden lever and style–Bell was struck by the resemblance between the device and the structure of the human ear. The ear analogy suggested the sorts of modifications he might undertake to successfully replicate the way in which the ear translated sound into mechanical motion. Bell sought to duplicate "the shape of the membrane of the human ear, the shapes of the bones attached to it, the mode of connection between the two, etc." [Bell, 1908, p.29].

Bell built an ear phonautograph in the summer of 1874, roughly the same time as he was conceiving his harp apparatus. This grisly device included the actual bones from a human ear that were set into vibration when one spoke into a cone and diaphragm to which they were attached.

From the phonautograph, Bell gained a kinesthetic understanding of the process by which speech was translated into an undulating mechanical motion in the ear. This understanding would lead to a unique mental model for a device that would transmit speech.

Multiple telegraphy

Thomas Hughes emphasizes the importance of reverse salients in technology [Hughes, 1987]. A reverse salient is the part of a technological frontier that lags behind, holding up progress. In the early 1870s, the reverse-salient in communication was the same as it has been many times during the history of communications: the problem of increasing bandwidth. At the time Bell began his multiple telegraphy work, two messages could be sent down a single wire at the same time. This meant that sending multiple stock quotes simultaneously from New York to Chicago would require multiple uninsulated wires to run alongside the railroad an unacceptable solution. Everyone in the communications industry knew that whoever found a way to get more messages down a single wire at the same time would be rich and famous.

Transmission of speech would do nothing to solve this bandwidth problem: you could at most get two people, one on each end, exchanging information over a single wire. So the transmission of speech was thought to be a minor application, useful for certain individuals who liked to have private telegraph lines connecting home to office.

Teaching the deaf was a wonderful profession, but it was not a route to fame and fortune, and like many young people, Alec Bell was ambitious. So in addition to his goal to make speech visible to the deaf, he added another enterprise to his network. His first idea for a multiple telegraph came from musical tones. Why not take two forks that produced exactly the same tone and turn one into a telegraph transmitter and the other into a receiver? If one could do this with one pair of forks, why not do it with four, eight or even sixteen distinct tones, all carrying information down the same wire? From J. Baille's *The Wonders of Electricity*[1] Bell got the idea of substituting a steel reed for a tuning fork; the reed's pitch could be precisely tuned simply by adjusting its length, whereas the end of a tuning fork had to be filed to change its pitch.

Bell was developing what Bernie Carlson and I call a 'mechanical representation' an artifact that represents, for the inventor, the solution to part of a problem she encounters frequently. Mechanical representations are a kind of mechanistic mental model [De Kleer, 1983] that is realized in an artifact and is used preferentially by a particular inventor. Edison, for example, used a drum cylinder in both the phonograph and the kinetoscope, and that mechanical representation structured both devices [Carlson and Gorman, 1989]. "Any creative technologists possesses a mental set of stock solutions from which he draws in addressing problems" [Jenkins, 1984, p. 153].

In Bell's case, the initial problem was sending tones down a wire, and the steel reeds continued from this point to be his preferred solution. When he was imagining or sketching a telegraph circuit, he could put this reed and coil combination into it and know exactly what he intended.

But Bell had great difficulty putting this reed-and-coil combination into an effective multiple telegraph circuit. In November of 1873 he concluded that, "with my own crude workmanship, and with the limited time and means at my disposal, I could not hope to construct any better models. I therefore from this time devoted less time to practical experiment than to the theoretical development of the details of the invention" [Bell, 1876, p. 8].

2.2 The error that led to the first speaking telegraph

In the summer of 1874, Bell put reeds on either pole of a horseshoe magnet, and experimented with sending the sound of either reed, separately or in combination. Bell's goal was to magnetize the reed itself and therefore avoid distortions that occurred when he used an non-magnetic reed in combination with an electromagnet exactly the sorts of distortions that other multiple telegraph inventors like Bell's rival Elisha Gray were able to avoid by using dampers and electromagnets of different resistances [Gorman et al., 1993].

---

[1] J. Baille, *The Wonders of Electricity*, New York, Charles Scribner, 1872, pp. 140–143.

Bell's horseshoe magnet experiment was partly successful, enough to lead him to imagine a device he called the 'harp apparatus'. In multiple telegraphy, the goal was to use a series of reeds far enough apart in tone to be able to be sent simultaneously down the wire and clearly distinguished at the other end. But Bell now imagined a device that did the opposite, where reeds almost indistinguishably different in tone were aligned along a single electromagnet. Such a device might function like the strings on a piano or a harp and vibrate in response to any tone made near them; these vibrations would then induce a current which could be carried to a receiving harp. Bell did not know how many reeds would be needed, and he was sure the device would not induce a current strong enough to carry a signal over the wire. Therefore, this imaginary device functioned as a mental model for how speech might be transmitted, adding a new enterprise to his network.

Bell had obtained support from Gardiner Hubbard, father of one of his pupils, Mabel. Hubbard wanted to break Western Union's virtual monopoly on what we would now call information services [Carlson, 1994]. One way was through the development of new technologies like the multiple harmonic telegraph system proposed by his daughter's teacher.

To complicate matters, Bell was courting Mabel. Therefore, out of deference to his primary backer and potential father-in-law, Gardiner Hubbard, Bell suspended the goal of creating a speaking telegraph and got back to multiple telegraphy, working with Tom Watson, the assistant his backers had enabled him to hire.

On June 2, 1875, Bell and Watson were testing set up three multiple telegraph stations, A, B and C, each with three of the tuned reeds. Bell wanted to be able to pluck reed A1 and have the reeds B1 and C1 vibrate and only those reeds. When Bell depressed the telegraph key corresponding to A1, B1 vibrated, but Watson, who was in another room with C1, noticed it was stuck. To release it, Watson plucked it. Bell rushed in to Watson's room to see what he was doing. Plucking C1 had caused all three of the reeds at B to vibrate, carrying both the pitch and the overtones of C1.

Seen from the standpoint of multiple telegraphy, this result was an error– one stuck reed caused three reeds at the other station to vibrate, and one could hear overtones, whereas what one really wanted was a pure tone. But this error suggested a route to the transmission of speech. His familiar reed-and-coil mechanical representation could perform the function of the entire harp. "These experiments at once removed the doubt that had been in my mind since the summer of 1874, that magneto-electric currents generated by the vibration of an armature in front of an electro-magnet would be too feeble to produce audible effects that could be practically utilized for the purposes of multiple telegraphy and of speech-transmission" [Bell, 1908, p.59].

Bell suspended the goal of pursuing multiple telegraphy and immediately asked Watson to build a telephone in which a reed relay was attached to a diaphragm or membrane with a speaking cavity over it. As one spoke into the cavity, the membrane would vibrate; these vibrations would be translated into an electrical current by his reed-and-coil mechanical representation, which would send them to the same device on the other end. Bell and Watson heard a kind of mumbling that suggested they were on the right track. Bell then wrote an application for a patent that focused especially on the undulating current Bell felt was they key to the transmission of speech. The patent was submitted by Gardiner Hubbard on February 14th, 1876.

2.3 ALEC: A computational simulation of Bell's network of enterprises

Marin Simina built a computer model of the way in which Bell suspended and activated goals, and applied it to the June 2nd experiment [Simina, 1999]. This simulation, called ALEC, engaged in opportunistic reasoning [Simina and Kolodner, 1995], posing and adopting new enterprises. The June 2nd experiment triggered and evaluate and critque function in ALEC, which led it to suspend the telegraph goal and activate the speech goal.

ALEC differs from KEKADA, another program which could take advantage of surprising results. KEKADA simulated Hans Krebs' discovery of the Ornithine cycle. But while KEKADA could propose experiments to follow up on surprising results, it lacked ALEC's ability to activate a suspended goal and use that to direct further research. ALEC is, in a sense, more stubborn than KEKADA because it does not eliminate its suspended goals, it simply suspends them.

What ALEC demonstrated was that a network of enterprises model could be translated into a set of functions and representations that would simulate Bell's decision to suspend the multiple telegraphy goal on June 2nd and pursue the telephone. This kind of in silico simulation is a valuable check on the plausibility and rigor of the sub specie historiae account. The problem with such simulations is that they can also be constructed from different assumptions: they may address the consistency and rigor of a qualitative model of an inventor's processes, but they cannot address its validity.

2.4 Bell's ear mental model

On June 30, 1875, Bell wrote Gardiner Hubbard: "I shall have ready tomorrow afternoon an instrument modeled after the human ear–by means of which I hope... to transmit a vocal sound... I am like a man in a fog who is sure of his latitude and longitude. I know I am close to the land for which I am bound and when the fog lifts I shall see it right before me". The instrument was a second version of the Gallows telephone, constructed by Watson; it worked little better than the first. Bell's reference to 'an instrument modeled after the human ear' shows a link back to his experiments with the ear phonautograph: he knew exactly how the ear translated sound into an undulating mechanical motion. This ear mental model made Bell "sure of his latitude and longitude".

When studying Bell's papers in the Library of Congress, I discovered a page in his experimental notebook[2] where he actually sketched his ear mental model. On February 18, 1876, Bell drew an ear with two different mechanisms for inducing a current next to the bone (see Figure 1). Above the sketch, Bell wrote, "Make transmitting instrument after the model of the human ear. Make armature after the shape of the ossicles. Follow out the analogy of nature" [Bell, 1876, p.13].

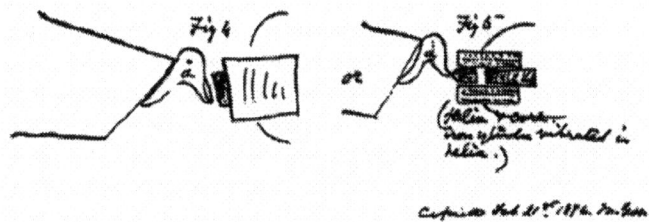

Figure 1. Sketch in Bell's notebook showing the bones of the middle ear inserted between a speaking tube and two arrangements of coils and magnets that would induce a current.

The left (Figure 4) and right (Figure 5) sketches by Bell shown in Figure 1 show the bones of the middle ear attached to a speaking tube and two ways of inducing a current, if an armature can be found that will serve a function similar to the steel reed in his mechanical representation.

Above this entry in the notebook, Bell sketched how the iron core and helix, or coil, of wire shown in the rightmost part of Figure 1 could be used to transmit an undulatory current to the ear and also to vibrate a flame so that the deaf could see the undulatory pattern. He will adopt an invention

---

[2] Bell began this experimental notebook at the urging of Gardiner Hubbard, who told him, "Whenever you recall any fact connected with your invention, jot it down on paper, as time will be essential to us, and the more things you actually performed by you at an earlier date, the better for our case". Gardiner G. Hubbard to Alexander Graham Bell, November 19, 1874, Bell Family Papers, Library of Congress, Box 80. The notebook begins with a recap experiments done in 1875 that Bell cannot date precisely; the first dated entry is the one that culminates in the sketch of his mental model.

heuristic often used by inventors "follow the analogy of nature" [Gorman, 1998] and build a telephone after the model of the human ear.

But it is not enough to propose such an analogy. The inventor needs a good understanding of both the domain of origin for the analogy, and the target domain [Gentner, 1980]. Bell had built the ear phonautograph, a device that showed how the ossicles translated sound into an undulating wave. His June 2nd experiment showed that his reed-and-coil mechanical representation could generate the undulating current because of the mumbling he and Watson had heard. His multiple enterprises connect in his ear mental model.

The ear analogy also provided Bell with a model that suggested where to concentrate his subsequent efforts. I refer to these as slots in a mental model, following frame theory [Weber and Perkins, 1989], which proposed that frames for a problem include slots whose values can vary. In this case, Bell had a slot for an armature that would serve the function of the ossicles, and knew he had to concentrate most of his energies there. He also had a slot for varying arrangements of magnets and coils to translate vibrations into the undulating current. This mental model structured his approach to the problem of actually transmitting speech.

I will bracket, at this point, the fascinating story of Bell's patent application and his competition with his rival Elisha Gray, who thought harmonic telegraphy was a much more important application than speaking telegraphy [Gorman et al., 1993]. In the space here, I also cannot go through the whole story of how Bell actually created a device that transmitted speech. But I want to show how his processes can be subjected to cognitive analysis in a way that would be amenable to modeling.

Bell received his patent on March 7th, and began immediately to work with Watson on a device to transmit speech. It is possible to develop a problem-behavior graph of the variations they tried, relying on Bell's notebook as a source. Problem-behavior graphs are a way of translating protocols obtained from problem-solvers, talking aloud, into graphs that show the path they took towards a goal, including wrong-turns and side branches [Ericsson and Simon, 1984]. A problem behavior graph can also be constructed based on a detailed historical record like an inventor or scientists' notebook [Tweney, 1989].

In Bell's case, I developed a version of this method that started from slots and showed each substitution Bell made. He followed a VOTAT, or vary-one-thing-at-a-time, strategy [Farris and Revlin, 1991] for the most part, occasionally making two substitutions at the same time. The distinguished historian Robert Bruce described Bell's series of experiments on March 8th as "random" (Bruce, 1973), but the problem-behavior graph shows he was trying to establish what Dunbar has called a baseline control [Dunbar and Fugelsang, 2005]. As shown in the slot diagram, he begins with the same arrangement he used in the Gallow's telephone, two of his reed-and-coil mechanical representations connected in a circuit except that in the transmitting one, he goes back to the tuning fork (see Figure 2).

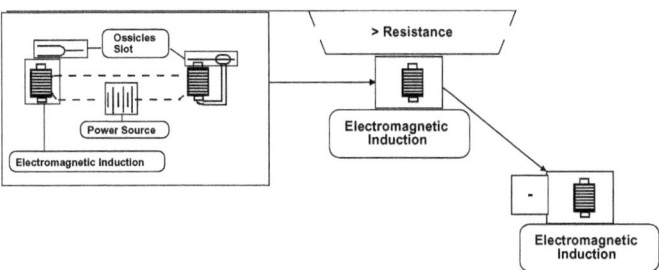

Figure 2. Slot diagram and first two experiments conducted by Bell to accomplish the transmission of speech, March 8, 1876.

The arrangement in the slot diagram transmits to a clearly audible tone, so the arrow goes to the right, indicating positive progress towards the goal of transmitting sounds. Bell then substitutes a coil of higher resistance on the transmitting end in what I have called the electromagnetic induction slot and the resulting sound is fainter so the arrow diagonals downward, indicating this route does not lead as directly toward the goal. Bell next takes out the transmitting coil, which looks nonsensical of course this will not transmit unless one considers the baseline control function. Bell has both transmitter and receiver in the same room; he needs to be sure he can tell the difference in the receiver when the tuning fork alone is vibrated, without any means of inducing the undulating current.

He next adds the coil back in, and removes the reed on the receiving end (see Figure 3). The result is no signal, indicated by the downward arrow. He is now sure he can tell the difference between a transmitted signal and the background noise of the fork in the air.

Bell switches next to the armature slot, substituting first an armature of soft iron on the receiving end, which has a positive result, shown by the arrow leading to the right. Then his final step is to try removing the power source as expected, this leads to a negative result, indicated by the arrow pointing downward.

Bell next switches to experiments in which water is used as a variable resistance medium in the ossicles slot. The account of these experiments

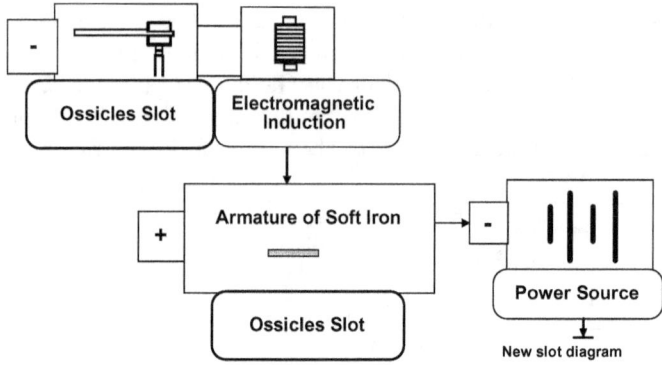

Figure 3. Further experiments by Bell on March 8th, 1876, following slot diagram in Figure 2.

goes beyond the scope of this paper [Gorman, 1995]. Suffice to say that this baseline control has prepared Bell to embark on a new line of research, confident that he has been able to replicate effects using familiar devices.

The process shown by the problem-behavior graphs theoretically could be modeled using the same simulation environment created by Gooding and Addis and applied to Faraday [Gooding and Addis, 1999]. This kind of simulation would also allow the exploration of alternate invention paths Bell or his competitors might have taken.

When I presented this work to modern engineers and engineering students, they challenged the generality of a *sub species historiae* result, and asked how it was relevant to modern teams pursing technological breakthroughs. Bell appears to be a solitary 'heroic' inventor, conducting his experiments in isolation with Watson. In contrast, Edison ran what amounts to the first R&D lab at Menlo Park, where he and his assistants produced an improved telephone transmitter [Carlson and Gorman, 1989].

In fact, Bell was networked into the scientific community of his day [Hounshell, 1979], and in Gardiner Hubbard, had a promoter who made sure Bell's patent was submitted in the nick of time and that he got on the program at key exhibitions. Invention always occurs in a social context, and that context is often easier to see in hindsight.

But I agreed that future studies of invention should incorporate teams. Since one of my tasks was to introduce engineering students to ethics, I was particularly interested in cases where a team invention was spurred by societal concerns.

3 Embodying environmental intelligence

Artifacts can embody mental models, as we saw with Bell's telephone design, which was a kind of electromechanical ear. Similarly, a modern design team set out to create a high-end furniture textile that would serve as an exemplar of a new, environmentally-sustainable industrial revolution. I gained entry to this team through William McDonough, our Dean of Architecture and the source of the team's mental model (see below). My then PhD student Matthew M. Mehalik, who had worked with me on the telephone, took on the daunting task of following a global design team through interviews and occasional site visits [Gorman and Mehalik, 2002] [Mehalik, 2000b] [Mehalik, 2000a]. Matt and I could revisit the Bell notebooks at any time, digging deeper without annoying the inventor. Modern invention teams need to be followed in real-time; they will not sit still for protocols, and their design notebooks often represent valuable intellectual property they are not willing to share.

This environmental design began with a New York fashion designer named Susan Lyons, Vice President at DesignTex, a firm that specializes in creating and manufacturing textiles for the high-end commercial market. Customers in the late 1990s had been asking about the environmental impact of DesignTex's product. Ms. Lyons remembered that her mother "put her money where her mouth was before it was hip to do so,"[3] teaching her children to rinse and re-use plastic bags and to compost. So for business and personal reasons, Ms. Lyons sought a way to incorporate an environmental theme into a new line of furniture fabric.

One of her potential allies was Albin Kaelin, Managing Director of Rohner Textil AG, a mill located in Switzerland, a few hundred yards from the Rhine. The mill had to treat and dispose of its wastewater, which posed a potential threat to Lake Constance, the largest drinking water reservoir in Europe. Also, according to Swiss regulations, the end-trimmings from Kaelin's fabrics were too toxic to put in a landfill. Trimmings could be burned in the regional incinerator to generate electricity, but the air pollutants had to be scrubbed before being released into the air.

Like Lyons, Kaelin had personal reasons for pursuing an environmental agenda. Like Lyons, he also saw a business opportunity, in his case, involving reduction of the costs mandated by regulation. Kaelin and his colleagues developed a fabric called Climatex that could be certified by the German-based association, Eco-Tex. The institute, concerned with human ecology issues, tested Climatex for pH value, content of free and partially releasable formaldehyde, residues of heavy metals, residues of pesticides,

---

[3]Unless indiacted otherwise, quoatations are from interviews with the participants in this network.

Braungart, a PhD in Chemistry who had been the head of Greenpeace's chemistry department. At one point, Braungart lowered himself down a smokestack at Ciba Geigy, to halt production until the company's emissions permits were renewed. The CEO Alex Krauer asked if they could work together, instead of fighting. Braungart founded the Environmental Protection Agency (EPEA) in 1987 so he could consult with companies like Ciba Geigy on improving their processes and products.

Braungart traveled to Kaelin's mill in December of 1993. His evaluation required him to examine all stages of the fabric-construction process. Because the mill was involved with the fabric weaving, he also inspected the mill's suppliers: farmers, yarn spinners, dyers and twisters. Yarn spinners created a cord of yarn/thread from the pieces of individual material fibers, such as wool. Yarn twisters take two or more cords of thread/yarn and twist them together, producing a much thicker and stronger piece of yarn. Dyers added the colors to the yarn. Finishers added chemicals to the finished weave to make it more durable, flame resistant, static resistant, and stain resistant, if such qualities were required.

By the end of January 1994, Kaelin had provided Braungart access to all aspects of the manufacturing processes involved in the creation of the new wool-ramie blend, which Kaelin called Climatex Lifecycle. But the fabric failed the EPEA's tests because of the composition of the dyes. Braungart took advantage of his relationship with Ciba-Geigy to gain access to their manufacturing processes. .Braungart found that only 16 out of the 1800 available dyes passed the protocol. Any color could be created from a combination of these 16 dyes, but when they were combined to create black, the result did not pass the protocol.

At this point, the design team included DesignTex, Rohner Textil, the EPEA, William McDonough and Ciba-Geigy. Take away any of these participants, and the fabric could not be created.

What was missing were customers. Lyons, McDonough, Kaelin and Braungart thought it was important that customers share the waste equals food mental model. The end result was a product, Climatex Lifecycle, that was accompanied by a green booklet authored by McDonough that explained how the fabric embodied waste equals food. The fabric won numerous awards and sold well at the high end of the furniture textile market. Kaelin kept the supply chain aligned with the principles and procedures adopted by the design team. For example, when Rohner Textil's dyemaster started using a less expensive alternative to the dyes approved by the EPEA, but one that fulfilled all the environmental constraints, Kaelin stopped him. The EPEA had to be consulted first.

Climatex Lifecycle is an artifact that embodies a set of ethical design principles. It was produced by an interdisciplinary global team that had to navigate frequent challenges, only a few of which were mentioned in this brief account. It is not possible to construct a problem-behavior graph of any parts of the cognitive processes involved in the fabric invention, because the details are still considered confidential. A future sub species historiae study may be able to get access to these records, just as the Bell notebooks are now available to the public. Whoever does this study will be glad to have the interviews and other information we obtained from the participants when they were working together.

4 Modeling, moral imagination and environmental management

The information we now have might be sufficient to create a computational model of the design process. For example, one could perhaps take the Belief 2.0 system developed by Addis and Gooding and use it to represent each of the five key players and at least simplified versions of their roles on the team: an agent like McDonough advocating uncompromising adherence to the waste equals food principle, one like Braungart with a system to check for adherence, one like Kaelin managing a complex supply chain and one like Lyons looking to sell the final product. One could simulate dilemmas actually experienced by the team, e.g., the time they had to decide whether to modify the fabric design to accommodate robot assemblers at Steelcase, a potential customer. The modifications meant adding a chemical that barely passed the EPEA's protocol. The team decided to make the change, but committed to finding a better alternative in the future. A simulation might help us explore alternate invention paths this team could take in response to current and future dilemmas.

The Climatex Lifecycle team dealt with only part of the global environmental system. Participants focused on certain kinds of environmental waste, and ignored others like the effect of the energy used in manufacturing the fabric on the carbon cycle. Brad Allenby has proposed a comprehensive Earth Systems Engineering Management (ESEM) approach that looks for improvement across the whole global ecosystem, not just a part [Allenby, 2005]. A core principle of ESEM is entering into a dialogue with the environmental system, making reversible changes and monitoring the effects. Mental models in this approach are dynamic, changing as a team's perception of the system is altered by new data and perspectives.

This kind of adaptive management with multiple human and natural stakeholders will require moral imagination [Werhane, 1999]. The first step in moral imagination involves being aware not just of one's own model of the system, but also the implicit ethical assumptions one is making. Environmental debates often have ideological and religious overtones; those interested in restoring an area like the Everglades to a pristine state may conflict with those who see the Everglades primarily as a reservoir that can be used to fuel further development. A first step in a dialogue is to make these implicit assumptions explicit, and recognize that they are views which can be changed. McDonough's waste equals food is a good invention heuristic, based on an analogy to nature, but if it hardens into an ideological view of what constitutes nature, it will fail to adapt to our growing understanding of these systems.

The second step involves sharing these views at a deep level involves being willing to "inhabit the worlds" of others, seeing how the system looks given their mental models [Johnson, 1993]. The third step then involves considering alternative goals for managing the system and mental models for its behavior, based on the consideration of alternate perspectives. Moral imagination is an ethical equivalent to modeling: be aware of one's assumptions, construct alternate models based on other assumptions, evaluate the alternatives and come up with a more comprehensive and imaginative model. This process is not easy, because it is not merely a superficial consensus it involves deep dialogue coupled with the best possible scientific monitoring of the system.

For moral imagination to work, stakeholders managing a system like the Everglades or Phoenix will have to use models to visualize future system states. Obviously, such an approach will depend on a variety of models of the complex system being managed as long as it is remembered that the models are simplifications, to be used for heuristic purposes. These models have to include human actors, including the members of teams proposing new technological solutions. Then the model can be used as a heuristic by the team to improve its own decision-making, especially if combined with Jeff Shrager's reflective diary method, in which each team member keeps a record of her or his problem-solving processes [Shrager, 2005]. These reflections could be used to build a model of the group's own decision processes, which would be altered by the model, leading to more reflections and an iteration of the model. The point is to improve the decision-making process, especially in situations where the team has to be interdisciplinary, engage in moral imagination and be constantly on the alert for information suggesting they have misunderstood or failed to incorporate parts of the system.

5 Conclusions

Sub species historiae studies of inventor's processes are amenable to modeling at the level of goals and also potentially at the level of problem behavior graphs. An interesting challenge is to link the two to show how shifts in mental models and goals can be modeled at the problem behavior level.

The problem with sub species historiae studies is that they are dependent on good records. Studies of modern inventors are therefore a necessary complement, and also mean that we will be studying cases whose outcomes are unknown, eliminating any bias towards inventors known to be successful. Intellectual property issues raise problems for modern studies. One way to overcome this is to conduct real-time studies, but seal the records until appropriate protection has been obtained for the IP involved.[4]

Design teams focused on ethical principles are an especially interesting problem. Modeling their potential courses of action in real time may be helpful to the teams themselves, especially if combined with Shrager's diary method.

BIBLIOGRAPHY

[Allenby, 2005] B. Allenby. Technology at the global scale: Integrative cognitivism and earth systems engineering management. In M.E. Gorman, R.D. Tweney, D.C. Gooding, and A. Kincannon, editors, *Scientific and technological thinking*, pages 303–344. Lawrence Erlbaum Associates, 2005.

[Bell, 1876] A.G. Bell. *Experiments made by A. Graham Bell*, volume I. Bell Family Papers, Library of Congress, 1876.

[Bell, 1908] A.G. Bell. *The Bell Telephone: Deposition of Alexander Graham Bell*. American Bell Telephone Co, Boston, 1908.

[Bruce, 1973] R.V. Bruce. *Bell: Alexander Graham Bell and the Conquest of Solitude*. Little, Brown, Boston, 1973.

[Bruner et al., 1956] J. Bruner, J. Goodnow, and Austin G. *A Study of Thinking*. John Wiley, New York, 1956.

[Carlson and Gorman, 1989] W.B. Carlson and M.E. Gorman. Thinking and doing at menlo park: Edison's development of the telephone, 1876–1878. In W. Pretzer, editor, *Thomas Edison's Menlo Park Laboratory*. Wayne Stat University Press, 1989.

[Carlson and Gorman, 1990] W.B. Carlson and M.E. Gorman. Understanding invention as a cognitive process: The case of thomas edison and early motion pictures, 1888–1891. *Social Studies of Science*, 20:387–430, 1990.

[Carlson, 1994] W.B. Carlson. Entrepreneurship in the early development of the telephone: How did william orton and gardiner hubbard conceptualize this new technology? *Business and Economic History*, 23(2):161–192, 1994.

[De Kleer, 1983] J.B. De Kleer. Assumptions and ambiguities in mechanistic mental models. In D.S. Gentner, editor, *Mental Models*, pages 155–190. Lawrence Erlbaum Associtates, 1983.

[Dunbar and Fugelsang, 2005] K.N. Dunbar and J.A. Fugelsang. Causal thinking in science: How science and students interpret the unexpected. In M.E. Gorman, R.D. Tweney, D.C. Gooding, and A. Kincannon, editors, *Scientific and technological thinking*, pages 57–80. Lawrence Erlbaum Associates, 2005.

[Dunbar, 1999] K. Dunbar. How scientists build models: In vivo science as a window on the scientific mind. In L. Magnani, N. Nersessian, and P. Thagard, editors, *Model-based reasoning in scientific discovery*, pages 85–100. Kluwer Academic Press, New York, 1999.

---

[4]This kind of study is problematic in a publish or perish environment, as the time-delay may be significant.

[Ericsson and Simon, 1984] K.A. Ericsson and H.A. Simon. *Protocol Analysis: Verbal Reports as Data.* MIT Press, Cambridge, MA, 1984.
[Farris and Revlin, 1991] H. Farris and R. Revlin. Rule discovery strategies: Falsification without disconfirmation (replye to gorman). *Social Studies of Science,* 21(3):565–567, 1991.
[Gentner, 1980] D. Gentner. The structure of analogical models in science. Technical Report 4451, Bolt, Beranek and Newman Report, 1980.
[Gooding and Addis, 1999] D.C. Gooding and T. Addis. A simulation of model-based reasoning about disparate phenomena. In L. Magnani, N. Nersessian, and P. Thagard, editors, *Model-based reasoning in scientific discovery,* pages 103–123. Kluwer Academic Press, New York, 1999.
[Gooding, 2005] D.C. Gooding. Seeing the forest for the trees: Visualization, cognition and scientific inference. In M.E. Gorman, R.D. Tweney, D.C. Gooding, and A. Kincannon, editors, *Scientific and technological thinking,* pages 173–218. Lawrence Erlbaum Associates, 2005.
[Gorman and Mehalik, 2002] M.E. Gorman and M.M. Mehalik. Turning good into gold: A comparative study of two environmental invention networks. *Science, Technology & Human Values,* 27(4):499–529, 2002.
[Gorman et al., 1993] M.E. Gorman, Mehalik, W.B. M.M., Carlson, and M. Oblon. Alexander graham bell, elisha gray and the speaking telegraph: A cognitive comparison. *History of Technology,* 15:1–56, 1993.
[Gorman, 1992] M.E. Gorman. *Simulating Science: Heuristics, Mental Models and Technoscientific Thinking.* Indiana University Press., Bloomington, 1992.
[Gorman, 1995] M.E. Gorman. Confirmation, disconfirmation and invention: The case of alexander graham bell and the telephone. *Thinking and Reasoning,* I(3):31–53, 1995.
[Gorman, 1998] M.E. Gorman. *Transforming nature: Ethics, invention and design.* Cambridge University Press, Cambridge, 1998.
[Gruber, 1989] H.E. Gruber. Networks of enterprise in creative scientific work. In B. Gholson, W.R. Shadish, R.A. Neimeyer, and A.C. Houts, editors, *Psychology of Science: Contributions to Metascience,* pages 246–266. Cambridge University Press, 1989.
[Hounshell, 1979] D.A. Hounshell. Bell and gray: Contrasts in style, politics and etiquette. In *Proceedings of the IEEE, 64,* pages 1305–1314, 1979.
[Hughes, 1987] T.P. Hughes. The evolution of large technological systems. In W.E. Bjiker, T.P. Hughes, and T.J. Pinch, editors, *The Social Construction of Technological Systems,* pages 51–82. MIT Press, 1987.
[Jenkins, 1984] R.V. Jenkins. Elements of style: Continuities in edison's thinking. *Annuals of the New York Academy of Sciences,* 424:149–162, 1984.
[Johnson, 1993] M. Johnson. *Moral imagination.* University of Chicago Press, Chicago, 1993.
[Magnani et al., 1999] L. Magnani, N. Nersessian, and P. Thagard, editors. *Model-based reasoning in conceptual change.* Kluwer Academic Pres., New York, 1999.
[McDonough, 1992] W. McDonough. *The Hannover principles: Design for sustainability.* William McDonough Architects, 1992.
[McDonough, 1995] W. McDonough. *The William McDonough Collection.* DesignTex, Inc., New York, 1995.
[McDonough, 1997] W. McDonough. Waste equals food: desigining for material and ethical prosperity, 1997. Paper presented at the Environmental Challenges to Business. Ruffin Lectures in Business Ethics.
[Mehalik, 2000a] M.M. Mehalik. Sustainable network design: A commercial fabric case study. *Interfaces: Special edition on ecologically sustainable practices,* 33(3), 2000.
[Mehalik, 2000b] M.M. Mehalik. Technical and design tools: The integration of iso 14001, life cycle development, environmental design and cost accounting. In Hillary, editor, *ISO 14001 Case Studies and Practical Experience.* Greenleaf, 2000.

[Shrager and Langley, 1990] J. Shrager and P. Langley. *Computational Models of Scientific Discovery and Theory Formation.* Morgan Kaufmann Publishers, Inc, San Mateo, CA, 1990.

[Shrager, 2005] J. Shrager. Diary of an insane cell mechanic. In M.E. Gorman, R.D. Tweney, D.C. Gooding, and A. Kincannon, editors, *Scientific and technological thinking.* Lawrence Erlbaum Associates, 2005.

[Simina and Kolodner, 1995] M.D. Simina and J.L. Kolodner. Opportunistic reasoning: A design perspective, 1995. Paper presented at the 17th Annual Cognitive Science Conference.

[Simina, 1999] M. Simina. *Enterprise-directed reasoning: Opportunism and deliberation in creative reasoning.* PhD thesis, Georgia Institute of Technology, Atlanta, GA, 1999.

[Tweney et al., 2005] R. D. Tweney, R. P. Mears, and C. Spitzmuller. Replicating the process of discovery: Michael faraday and the interaction of gold and light. In M.E. Gorman, R.D. Tweney, D.C. Gooding, and A. Kincannon, editors, *Scientific and technological thinking*, pages 137–158. Lawrence Erlbaum Associates, 2005.

[Tweney, 1989] R.D. Tweney. A framework for the cognitive psychology of science. In B. Gholson, W.R. Shadish, R.A. Neimeyer, and A.C. Houts, editors, *Psychology of Science: Contributions to Metascience.* Cambridge University Press, 1989.

[Weber and Perkins, 1989] R.J. Weber and D.N. Perkins. How to invent artifacts and ideas. *New Ideas in Psychology*, 7:49–72, 1989.

[Werhane, 1999] P.H. Werhane. *Moral imagination and management decision making.* Oxford University Press, Oxford, 1999.

Michael E. Gorman
School of Engineering and Applied Science
University of Virginia
Email: meg3c@virginia.edu

Disembodying Minds, Externalizing Minds
How Brains Make Up Creative Scientific Reasoning

LORENZO MAGNANI

ABSTRACT. The first part of the paper illustrates that at the roots of scientific creativity there is a process of disembodiment of mind that presents a new cognitive perspective on the distributed scientific reasoning on concepts and models. Taking advantage of Turing's comparison between "unorganized" brains and "logical" and "practical machines", and of some paleoanthropological results on the birth of material culture, I will illustrate the centrality to cognition and scientific reasoning of the disembodiment of mind from the point of view of the cognitive interplay between internal and external representations, both mimetic and creative. I consider this interplay critical in making up distributed scientific reasoning through the dynamical interactions with the environment. I also think the disembodiment of mind can nicely account for low-level cognitive processes of cognition, bringing up the question of how could higher-level processes like the ones occurring in science be comprised and how would they interact with lower-level ones. To the aim of explaining these higher-level mechanisms I provide the analysis of model-based and manipulative abduction, and of those external representations and epistemic mediators that furnish "prostheses" able to support the creation of a real efficacious distributed scientific reasoning.

1 From the prehistoric brains to environmental scenarios

In an interesting article of 1950, Computing machinery and intelligence" [Turing, 1950], which adopts an evolutionary perspective, Turing maintains that a big cortex can provide an evolutionary advantage only in presence of a massive storage information and knowledge on external supports that only an already developed small community of human beings can possess. Evidence from paleoanthropology seems to support this perspective. Some research in cognitive paleoanthropology teaches us that high level and reflective consciousness in terms of thoughts about our own thoughts and about

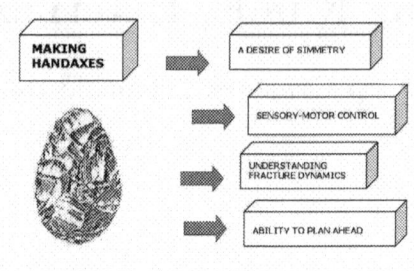

Figure 1.

our feelings (that is consciousness not merely considered as raw sensation) is intertwined with the development of modern language (speech) and material culture. After 250.000 years ago several hominid species had brains as large as ours today, but their behavior lacked any sign of art or symbolic behavior. If we consider high-level consciousness as related to a high-level organization – in Turing's sense – of human cortex, its origins can be related to the active role of environmental, social, linguistic, and cultural aspects.

Handaxes were made by Early Humans and firstly appeared 1,4 million years ago, still made by some of the Neanderthals in Europe just 50.000 years ago. The making of handaxes is strictly intertwined with the development of consciousness. Many needed capabilities constitute a part of an evolved psychology that appeared long before the first handaxed were manufactured. It seems humans were pre-adapted for some components required to make handaxes [Mithen, 1999] (cf. Figure 1):

1. imposition of symmetry (already evolved through predators escape and social interaction). It has been an unintentional by-product of the bifacial knapping technique but also deliberately imposed in other cases. It is also well-known that the attention to symmetry may have developed through social interaction and predator escape, as it may allow one to recognize that one is being directly stared at [Dennett, 1991].

2. understanding fracture dynamics (for example evident from Oldowan tools and from nut cracking by chimpanzees today);

3. ability to plan ahead (modifying plans and reacting to contingencies,

such unexpected flaws in the material and miss-hits), still evident in the minds of Oldowan tool makers and in chimpanzees;

4. high degree of sensory-motor control: "Nodules, preforms, and near finished artifacts must be struck at precisely the right angle with precisely the right degree of force if the desired flake is to be detached" [Mithen, 1999, p. 285]. The origin of this capability is usually tracked back to encephalization – the increased number of nerve tracts and of the integration between them allows for the firing of smaller muscle groups – and bipedalism – that requires a more complex integrated highly fractionated nervous system, which in turn presupposes a larger brain.

The combination of these four resources produced the birth of what Mithen calls technical intelligence of early human mind, that is consequently related to the construction of handaxes. Indeed they indicate high intelligence and good health. They cannot be compared to the artifacts made by animals, like honeycomb or spider web, deriving from the iteration of fixed actions which do not require consciousness and intelligence.

1.1 Private speech and fleeting consciousness

Two central factors play a fundamental role in the combination of the four resources above:

- the exploitation of private speech (speaking to oneself) to trail between planning, fracture dynamic, motor control and symmetry (also in children there is a kind of private muttering which makes explicit what is implicit in the various abilities);

- a good degree of fleeting consciousness (thoughts about thoughts).

In the meantime these two aspects played a fundamental role in the development of consciousness and thought:

> So my argument is that when our ancestors made handaxes there were private mutterings accompanying the crack of stone against stone. Those private mutterings were instrumental in pulling the knowledge required for handaxes manufacture into an emergent consciousness. But what type of consciousness? I think probably one that was fleeting one: one that existed during the act of manufacture and that did not the endure. One quite unlike the consciousness about one's emotions, feelings, and desires that were associated with the social world and that probably were

part of a completely separated cognitive domain, that of social intelligence, in the early human mind. [Mithen, 1999, p. 288]

This use of private speech can be certainly considered a "tool" for organizing brains and so for manipulating, expanding, and exploring minds, a tool that probably evolved with another: talking to each other. Both private and public language act as tools for thought and play a fundamental role in the evolution "opening up our minds to ourselves" and so in the emergence of cognitive chances for new meanings.

1.2 Material culture as distributed cognition

Another tool appeared in the latter stages of human evolution, that played a great role in the evolutions of primitive minds, that is in the organization of human brains. Handaxes also are at the birth of material culture, so as new cognitive chances can co-evolve:

- the mind of some early humans, like the Neanderthals, were constituted by relatively isolated cognitive domains, Mithen calls different intelligences, probably endowed with different degrees of consciousness about the thoughts and knowledge within each domain (natural history intelligence, technical intelligence, social intelligence). These isolated cognitive domains became integrated also taking advantage of the role of public language;

- degrees of high level consciousness appear, human beings need thoughts about thoughts; – social intelligence and public language arise.

It is extremely important to stress that material culture is not just the product of this massive cognitive chance but also cause of it. "The clever trick that humans learnt was to disembody their minds into the material world around them: a linguistic utterance might be considered as a disembodied thought. But such utterances last just for a few seconds. Material culture endures" [Mithen, 1999, p. 291].

In this perspective we acknowledge that material artifacts are tools for thoughts as is language: tools for exploring, expanding, and manipulating our own minds. Moreover, in this perspective the evolution of culture is inextricably linked with the evolution of consciousness and thought. Early human brain becomes a kind of universal "intelligent" machine, extremely flexible so that we did no longer need different "separated" intelligent machines doing different jobs. A single one will suffice. As the engineering problem of producing various machines for various jobs is replaced by the office work of "programming" the universal machine to do these jobs, so the different intelligences become integrated in a new universal device endowed with a high-level type of consciousness.

From this perspective the expansion of the minds is in the meantime a continuous process of disembodiment of the minds themselves into the material world around them. In this regard the evolution of the mind is inextricably linked with the evolution of large, integrated, material cognitive systems. In the following section I will illustrate this extraordinary interplay between human brains and the new cognitive systems they make in the light of the the development of scientific reasoning.

2 Disembodiment of mind

A wonderful example of cognition productive for the discovery of new concepts - through disembodiment of mind - is the carving of what most likely is the mythical being from the last ice age, 30.000 years ago, a half human/half lion figure carved from mammoth ivory found at Hohlenstein Stadel, Germany. We can consider this case an ancestor of the distributed reasoning that is peculiar to scientific mentality.

> An evolved mind is unlikely to have a natural home for this being, as such entities do not exist in the natural world, the mind needs new chances: so whereas evolved minds could think about humans by exploiting modules shaped by natural selection, and about lions by deploying content rich mental modules moulded by natural selection and about other lions by using other content rich modules from the natural history cognitive domain, how could one think about entities that were part human and part animal? Such entities had no home in the mind. [Mithen, 1999, p. 291]

A mind consisting of different separated intelligences cannot come up with such entity (Figure 2). The only way is to extend the mind into the material word, building in the environment primitive *scenarios*[1] made by rocks, blackboards, paper, ivory, and writing, painting, and carving: "artifacts such as this figure play the role of anchors for ideas and have no natural home within the mind; for ideas that take us beyond those that natural selection could enable us to possess" [Mithen, 1999, p. 291].

---

[1] Ohsawa proposed this term to explain some aspects of change discovery. Cf. for example [Nara and Ohsawa, 2004] that stresses the attention on the activity of extracting chances through communication of scenarios: given that a scenario is a time series of events under a given context, during a communication scenario drawings are generated, externalized (that is disembodied) and subsequently analyzed in the process of chance discovery.

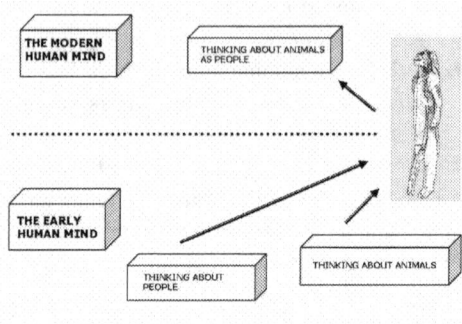

Figure 2.

In the case of our figure we face with an anthropomorphic thinking created by the material representation serving to anchor the cognitive representation of supernatural being. In this case the material culture disembodies thoughts (and provides chances potentially significant to adopt new inferences and new actions), that otherwise will soon disappear, without being transmitted to other human beings. The early human mind possessed two separated intelligences for thinking about animals and people. Through the mediation of the material culture the modern human mind can arrive to internally think about the new concept of animal and people at the same time. But the new mean-ing occurred over there, in the external material world where the mind picked up it.

Artifacts as external objects allowed humans to loosen and cut those *chains* on our unorganized brains imposed by our evolutionary past. Chains that always limited the brains of other human beings, such as the Neanderthals. Loosing chains and securing ideas to external objects was also a way to creatively re-organize brains as universal machines for thinking.

I think the disembodiment of mind can nicely account for low-level cognitive processes of conceptual change, bringing up the question of how could higher-level processes – like the ones occurring in scientific reasoning - be comprised and how would they interact with lower-level ones. To the aim of explaining these higher-level mechanisms I have provided a cognitive framework where model-based and manipulative abduction together with external representations and epistemic mediators play a central role (cf. the following sections).

3 Mimetic and creative representations

3.1 External and internal representations

We have said that through the mediation of the material culture the modern human mind can creatively arrive to internally think about animals and people at the same time. We can also account for this process of disembodiment from an interesting cognitive point of view.

I maintain that representations can be external and internal. We can say that - external representations are formed by external materials that express (through reification) concepts and problems that do not have a natural home in the brain;

- internalized representations are internal re-projections, a kind of recapitulations, (learning) of external representations in terms of neural patterns of activation in the brain. They can sometimes be "internally" manipulated like external objects and can originate new internal reconstructed representations through the neural activity of transformation and integration.

  This process explains why human beings seem to perform both computations of a connectionist type such as the ones involving representations as

- (I Level) patterns of neural activation that arise as the result of the interaction between body and environment - that interaction that is extremely fruitful for creative results - (and suitably shaped by the evolution and the individual history): pattern completion or image recognition, and computations that use representations as

- (II Level) derived combinatorial syntax and semantics dynamically shaped by the various external representations and reasoning devices found or constructed in the environment (for example geometrical diagrams in mathematical creativity); they are neurologically represented contingently as pattern of neural activations that "sometimes" tend to become stabilized structures and to fix and so to permanently belong to the I Level above.

The I Level originates those sensations [they constitute a kind of "face" we think the world has], that provide room for the II Level to reflect the structure of the environment, and, most important, that can follow the computations suggested by these external structures. It is clear we can now conclude that the growth of the brain and especially the synaptic and dendritic growth are profoundly determined by the environment.

When the fixation is reached the patterns of neural activation no longer need a direct stimulus from the environment for their construction. In a certain sense they can be viewed as fixed internal records of external structures that can exist also in the absence of such external structures. These patterns of neural activation that constitute the I Level Representations always keep record of the experience that generated them and, thus, always carry the II Level Representation associated to them, even if in a different form, the form of memory and not the form of a vivid sensorial experience. Now, the human agent, via neural mechanisms, can retrieve these II Level Representations and use them as internal representations or use parts of them to construct new internal representations very different from the ones stored in memory (cf. also [Gatti and Magnani, 2005]).

Human beings delegate cognitive features to external representations because in many problem solving situations the internal computation would be impossible or it would involve a very great effort because of human mind's limited capacity. First a kind of alienation is performed, second a recapitulation is accomplished at the neuronal level by re-representing internally that which was "discovered" outside. Consequently only later on we perform cognitive operations on the structure of data that synaptic patterns have "picked up" in an analogical way from the environment. Internal representations used in cognitive processes have a deep origin in the experience lived in the environment. I think there are two kinds of artifacts that play the role of external objects (representations) active in this process of disembodiment of the mind: creative and mimetic. Mimetic external representations mirror concepts and problems that are already represented in the brain and need to be enhanced, solved, further complicated, etc., so they sometimes can creatively give rise to new concepts, models, and perspectives. Following my perspective it is at this point evident that the "mind" transcends the boundary of the individual and includes parts of that individual's environment[2].

---

[2] A different perspective also considers the dynamics of the distinction between internal and external aspects of cognition stressing the attention to the role of thoughts about thoughts in human consciousness [Brenner, 2005]. In this view a critique of the classical idea of an independent existence of a human agent is given: it coincides with the energetic actualization of representations, intentionalities and processes, in a sense that also echoes the Peircian deep semiotic conclusion about "thinking" I have emphasized in a recent paper on what I call "semiotic brains": "It is sufficient to say that there is no element whatever of man's consciousness which has not something corresponding to it in the word; and the reason is obvious. It is that the word or sign which man uses is the man himself. For, as the fact that every thought is a sign, taken in conjunction with the fact that life is a train of thoughts, proves that man is a sign; so, that every thought is an external sign, proves that man is an external sign. That is to say, the man and the external sign are identical, in the same sense in which the words homo and man are identical. Thus my language is the sum total of myself; fore the man is the thought" [Peirce, 1931-1958,

4 Theoretical and manipulative reasoning

Science is one of the most explicitly constructed, abstract, and creative forms of human knowledge. Today researchers concentrated on the concept of *abduction* pointed out by C.S. Peirce as a fundamental mechanism by which it is possible to account for the introduction of new explanatory hypotheses in science.

Abduction is the process of *inferring* certain facts and/or laws and hypotheses that render some sentences plausible, that *explain* or *discover* some (eventually new) phenomenon or observation; it is the process of reasoning in which explanatory hypotheses are formed and evaluated. There are two main epistemological meanings of the word abduction [Magnani, 2001]: 1) abduction that only generates "plausible" hypotheses ("selective" or "creative") and 2) abduction considered as inference "to the best explanation", which also evaluates hypotheses. An illustration from the field of medical knowledge is represented by the discovery of a new disease and the manifestations it causes which can be considered as the result of a creative abductive inference. Therefore, "creative" abduction deals with the whole field of the growth of scientific knowledge. This is irrelevant in medical diagnosis where instead the task is to "select" from an encyclopedia of pre-stored diagnostic entities.

Theoretical abduction[3] certainly illustrates much of what is important in creative abductive reasoning, in humans and in computational programs, but fails to account for many cases of explanations occurring in science when the exploitation of environment is crucial. It fails to account for those cases in which there is a kind of "discovering through doing", cases in which new and still unexpressed information is codified by means of manipulations of some external objects (*epistemic mediators*). The concept of *manipulative abduction*[4] captures a large part of scientific thinking where the role of action is central, and where the features of this action are implicit and hard to be elicited: action can provide otherwise unavailable information that enables the agent to solve problems by starting and by performing a suitable abductive process of generation or selection of hypotheses.

Many attempts have been made to model abduction by developing some formal tools in order to illustrate its computational properties and the relationships with the different forms of deductive reasoning [Bylander *et al.*,

---

5.314].

[3]Magnani [2001; 2002] introduces the concept of theoretical abduction. He maintains that there are two kinds of theoretical abduction, "sentential", related to logic and to verbal/symbolic inferences, and "model-based", related to the exploitation of internalized models of diagrams, pictures, etc., cf. below in this paper.

[4]Manipulative abduction and epistemic mediators are introduced and illustrated in [Magnani, 2001].

1991]. Some of the formal models of abductive reasoning are based on the theory of the *epistemic state* of an agent [Boutilier and Becher, 1995], where the epistemic state of an individual is modeled as a consistent set of beliefs that can change by expansion and contraction (*belief revision framework*). These kinds of logical models are called sentential [Magnani, 2001].

They exclusively deal with selective abduction (diagnostic reasoning)[5] and relate to the idea of preserving *consistency*. Exclusively considering the sentential view of abduction does not enable us to say much about creative processes in science, and, therefore, about the nomological and most interesting creative aspects of abduction. It mainly refers to the *selective* (diagnostic) and merely *explanatory* aspects of reasoning and to the idea that abduction is mainly an inference *to the best explanation* [Magnani, 2001].

5 Internal, external, hybrid scenarios

If we want to provide a suitable framework for analyzing the most interesting cases of conceptual changes in science we do not have to limit ourselves to the *sentential* view of theoretical abduction but we have to consider a broader *inferential* one: the *model-based* sides of creative abduction (cf. below).

From Peirce's philosophical point of view, all thinking is in signs, and signs can be icons, indices or symbols. Moreover, all inference is a form of sign activity, where the word sign includes "feeling, image, conception, and other representation" [Peirce, 1931-1958, 5.283], and, in Kantian words, all synthetic forms of cognition. That is, a considerable part of the thinking activity is model-based. Of course model-based reasoning acquires its peculiar creative relevance when embedded in abductive processes, so that we can individuate a *model-based abduction*.

Hence we must think in terms of *model-based abduction* (and not in terms of sentential abduction) to explain complex processes like scientific conceptual change. are Different varieties of *model-based abductions* [Magnani, 1999] are related to high-level types of scientific conceptual change [Thagard, 1992]. Following Nersessian [1999], the term "model-based reasoning" is used to indicate the construction and manipulation of various kinds of representations, not mainly sentential and/or formal, but mental and/or related to external mediators.

Manipulative abduction [Magnani, 2001] happens when we are thinking through doing and not only, in a pragmatic sense, about doing. So the idea of manipulative abduction goes beyond the well-known role of experi-

---

[5] As previously indicated, it is important to distinguish between *selective* (abduction that merely selects from an encyclopedia of pre-stored hypotheses), and *creative* abduction (abduction that generates new hypotheses).

ments as capable of forming new scientific laws by means of the results (the nature's answers to the investigator's question) they present, or of merely playing a predictive role (in confirmation and in falsification). Manipulative abduction refers to an extra-theoretical behavior that aims at creating communicable accounts of new experiences to integrate them into previously existing systems of experimental and linguistic (theoretical) practices. The existence of this kind of extra-theoretical cognitive behavior is also testified by the many everyday situations in which humans are perfectly able to perform very efficacious (and habitual) tasks without the immediate possibility of realizing their conceptual explanation.

In the following section manipulative abduction will be considered from the perspective of distributed scientific reasoning. The power of model-based reasoning and abduction (both theoretical and manipulative) mainly depends on their ability to extract and render explicit a certain amount of important information, unexpressed at the level of available data. They have a fundamental role in the process of transformation of knowledge from its *tacit* to its *explicit* forms, and in the subsequent elicitation and use of knowledge. It is in this process that conceptual discovery, promotion, and production is central. Let us describe how this happens in the case of "external" model-based processes.

5.1 Building distributed scientific reasoning through manipulative abduction

For example, in the simple case of the construction and examination of diagrams in elementary geometrical reasoning, specific experiments serve as states and the implied operators are the manipulations and observations that transform one state into another. The geometrical outcome depends upon practices and specific sensory-motor activities performed on a non-symbolic object, which acts as a dedicated external representational medium supporting the various operators at work. There is a kind of an epistemic negotiation between the sensory framework of the problem solver and the external reality of the diagram [Magnani, 2002]. It is well-known that in the history of geometry many researchers used internal mental imagery and mental representations of diagrams, but also self-generated diagrams (external) to help their thinking.

This process involves an external representation consisting of written symbols and figures that for example are manipulated "by hand". The cognitive system is not merely the mind-brain of the person performing the geometrical task, but the system consisting of the whole body (cognition is *embodied*) of the person plus the external physical representation. In geometrical discovery the whole activity of cognition is located in the system consisting of a human together with diagrams.

An external representation can modify the kind of pre-existent internal computation that a human agent uses to reason about a problem: the Roman numeration system eliminates, by means of the external signs, some of the hardest parts of the addition, whereas the Arabic system does the same in the case of the difficult computations in multiplication. The capacity for inner reasoning and thought results from the internalization of the originally external forms of representation [Zhang, 1997].

In the case of the external representations we can have various objectified knowledge and structures (like physical symbols – e.g. written symbols, and objects – e.g. three-dimensional models, shapes and dimensions), but also external rules, relations, and constraints incorporated in physical situations (spatial relations of written digits, physical constraints in geometrical diagrams and abacuses) (see [Zhang, 1997]). The external representations are contrasted to the internal representations that consist in the knowledge and the structure in memory, as propositions, productions, schemas, neural networks, models, prototypes, images.

The external representations are not merely memory aids: they can give people access to knowledge and skills that are unavailable to internal representations, help researchers to easily identify aspects and to make further inferences, they constrain the range of possible cognitive outcomes in a way that some actions are allowed and others forbidden. They increase the chance discoverability of new concepts. The "mind" is limited because of the restricted range of information processing, the limited power of working memory and attention, the limited speed of some learning and reasoning operations; on the other hand the environment is intricate, because of the huge amount of data, real time requirement, uncertainty factors. The interplay between external environment and internal representations lead to the creation of suitable reasoning scenarios we call call *hybrid*.

5.2 The extra-theoretical dimension of scientific reasoning: templates of epistemic acting and epistemic mediators

Peirce gives an interesting example of model-based abduction related to sense activity: "A man can distinguish different textures of cloth by feeling: but not immediately, for he requires to move fingers over the cloth, which shows that he is obliged to compare sensations of one instant with those of another" [Peirce, 1931-1958, 5.221]. This surely suggests that abductive movements have also interesting extra-theoretical characters and that there is a role in abductive reasoning for various kinds of manipulations of external objects. *All* knowing is *inferring* and inferring is not instantaneous, it happens in a process that needs an activity of comparisons involving many kinds of models in a more or less considerable lapse of time. All these considerations suggest, then, that there exist a creative form of thinking through doing[6], fundamental as much as the theoretical one: *manipulative abduction* (see [Magnani, 2001] and [Magnani, 2002]). As already said *manipulative* abduction happens when we are thinking *through* doing and not only, in a pragmatic sense, about doing.

Various templates of manipulative behavior exhibit some regularities. The activity of manipulating external things and representations is highly conjectural and not immediately explanatory: these templates are hypotheses of behavior (creative or already cognitively present in the scientist's mind-body system, and sometimes already applied) that abductively enable a kind of epistemic "doing". Hence, some templates of action and manipulation can be selected in the set of the ones available and pre-stored, others have to be created for the first time to perform the most interesting creative cognitive accomplishments of manipulative abduction.

Some common features of the tacit templates of manipulative abduction, that enable us to manipulate things and experiments in science to extract chances are related to: 1. sensibility towards the aspects of the phenomenon which can be regarded as *curious* or *anomalous*; manipulations have to be able to introduce potential inconsistencies in the received knowledge and so to open new possible reasoning opportunities (Oersted's report of his well-known experiment about electromagnetism is devoted to describing some anomalous aspects that did not depend on any particular theory of the nature of electricity and magnetism); 2. preliminary sensibility towards the *dynamical* character of the phenomenon, and not to entities and their properties, common aim of manipulations is to practically reorder the dynamic sequence of events into a static spatial one that should promote a subsequent bird's-eye view (narrative or visual-diagrammatic), fruitful for further outcomes; 3. referral to experimental manipulations that exploit *artificial apparatus* to free new possible stable and repeatable sources of information about hidden knowledge and constraints (Davy set-up in term of an artifactual tower of needles showed that magnetization was related to orientation and does not require physical contact); 4. various contingent ways of epistemic acting: *looking* from different perspectives, *checking* the different information available, *comparing* subsequent events, *choosing*, *discarding*, *imaging* further manipulations, *re-ordering* and *changing relationships* in the world by implicitly *evaluating* the usefulness of a new order (for instance, to help memory). As I have illustrated more in detail in [Magnani

---

[6] In this way the cognitive task is achieved on *external* representations used in lieu of internal ones. Here action performs an *epistemic* and not a merely performatory role, relevant to abductive reasoning.

et al., 2002] The whole activity of manipulation is devoted to building various external *epistemic mediators* that function as an enormous new source of information and knowledge.

Gooding [1990] refers to this kind of concrete manipulative reasoning when he illustrates the role in science of the so-called "construals" that embody tacit inferences in procedures that are often apparatus and machine based. The embodiment is of course an expert manipulation of objects in a highly constrained experimental environment, and is directed by abductive movements that imply the strategic application of old and new *templates* of behavior mainly connected with extra-theoretical components, for instance emotional, esthetical, ethical, and economic.

The hypothetical character of construals is clear: they can be developed to examine further chances, or discarded, they are provisional creative organization of experience and some of them become in their turn hypothetical *interpretations* of experience, that is more theory-oriented, their reference is gradually stabilized in terms of established observational practices. Step by step the new interpretation – that at the beginning is completely "practice-laden" – relates to more "theoretical" modes of understanding (narrative, visual, diagrammatic, symbolic, conceptual, simulative), closer to the constructive effects of theoretical abduction. When the reference is stabilized the effects of incommensurability with other stabilized observations can become evident. But it is just the construal of certain phenomena that can be shared by the sustainers of rival theories. Gooding [1990] shows how Davy and Faraday could see the same attractive and repulsive actions at work in the phenomena they respectively produced; their discourse and practice as to the role of their construals of phenomena clearly demonstrate they did not inhabit different, incommensurable worlds in some cases. Moreover, the experience is constructed, reconstructed, and distributed across a social network of negotiations among the different scientists by means of construals.

The whole activity of manipulation is devoted to building various external *epistemic mediators*[7]. that function as an enormous new source of information and knowledge. Therefore, manipulative abduction represents

---

[7]This expression is derived from the cognitive anthropologist Hutchins [1995], who coined the expression "mediating structure" to refer to various external tools that can be built to cognitively help the activity of navigating in modern but also in "primitive" settings. Any written procedure is a simple example of a cognitive "mediating structure" with possible cognitive aims, so mathematical symbols and diagrams: "Language, cultural knowledge, mental models, arithmetic procedures, and rules of logic are all mediating structures too. So are traffic lights, supermarkets layouts, and the contexts we arrange for one another's behavior. Mediating structures can be embodied in artifacts, in ideas, in systems of social interactions [...]" [Hutchins, 1995, pp. 290–291].

a kind of redistribution of the epistemic and cognitive effort to manage objects and information that cannot be immediately represented or found internally (for example exploiting the resources of visual imagery).[8]

From the point of view of everyday situations manipulative abductive reasoning and epistemic mediators exhibit other very interesting features critical to scientific reasoning: 1. action elaborates a *simplification* of the reasoning task and a redistribution of effort across time, when we need to manipulate concrete things in order to understand structures which are otherwise too abstract, or when we are in presence of *redundant* and unmanageable information; 2. action can be useful in presence of *incomplete* or *inconsistent* information - not only from the "perceptual" point of view - or of a diminished capacity to act upon the world: it is used to get more data to restore coherence and to improve deficient knowledge; 3. action enables us to build *external artifactual models* of task mechanisms instead of the corresponding internal ones, that are adequate to adapt the environment to the agent's needs. 4. action as a *control of sense data* illustrates how we can change the position of our body (and/or of the external objects) and how to exploit various kinds of prostheses (Galileo's telescope, technological instruments and interfaces) to get various new kinds of stimulation: action provides some tactile and visual information (e.g., in surgery), otherwise unavailable. The external artifactual models are endowed with functional properties as components of a memory system crossing the boundary between person and environment (for example they are able to transform the tasks involved in allowing simple manipulations that promote further visual inferences at the level of model-based abduction). The cognitive process is *distributed* between a person (or a group of people) and external representation(s), and so obviously *embedded* and *situated* in a society and in a historical culture[9].

As we have already stressed, the manipulations of external representations and mediators, together with the internal processes, aims at building

---

[8] It is difficult to preserve precise spatial and geometrical relationships using mental imagery, in many situations, especially when one set of them has to be moved relative to another.

[9] Magnani [Magnani, 2001, chapter 6] stresses the importance of the so-called preinventive forms in abductive reasoning. Intuitively an anomaly is something surprising, as Peirce already knew "The breaking of a belief can only be due to some novel experience" [Peirce, 1931-1958, 5.524] or "[...] until we find ourselves confronted with some experience contrary to those expectations" [Peirce, 1931-1958, 7.36]. Therefore it is not strange that something anomalous can be found in those kinds of structures the cognitive psychologists call *preinventive*. Cognitive psychologists have described many kinds of preinventive structures (typically unstable and incomplete and their desirable properties, that constitute particularly interesting ways of "irritating" the mind and stimulating creativity [Finke *et al.*, 1992]: they are certainly of interest for scientific discovery.

what we have called *hybrid scenarios*.

6 Conclusion

It is clear that the manipulation of external objects helps human beings in scientific reasoning and so in its creative aspects. I have illustrated the strategic role played by low-level cases of disembodiment of mind also relating it to the high-level cognitive processes of manipulative abduction, considered as a particular kind of abduction that exploits external models and epistemic mediators endowed with delegated explanatory cognitive roles and attributes which depict interesting features of distributed scientific reasoning.

BIBLIOGRAPHY

[Boutilier and Becher, 1995] C. Boutilier and V. Becher. Abduction as belief revision. *Artificial Intelligence*, 77:43–94, 1995.

[Brenner, 2005] J.E. Brenner. Knowledge as system: a logic of epistemology. In *6th European Congress on Systems Science*, Paris, 2005. Forthcoming.

[Bylander et al., 1991] T. Bylander, D. Allemang, M. C. Tanner, and J.R. Josephson. The computational complexity of abduction. *Artificial Intelligence*, 49:25–60, 1991.

[Dennett, 1991] D. Dennett. *Consciousness Explained*. Little, Brown, and Company, New York, 1991.

[Finke et al., 1992] R.A. Finke, T.B. Ward, and S.M. Smith. *Creative Cognition. Theory, Research, and Applications*. MIT Press, Cambridge, MA, 1992.

[Gatti and Magnani, 2005] A. Gatti and L. Magnani. On the representational role of the environment and on the cognitive nature of manipulations. In L. Magnani and R. Dossena, editors, *Computing, Philosophy, and Cognition*, pages 227–242, London, 2005. King's College Publications.

[Gooding, 1990] D. Gooding. *Experiment and the Making of Meaning*. Kluwer, Dordrecht, 1990.

[Hutchins, 1995] E. Hutchins. *Cognition in the Wild*. MIT Press, Cambridge, MA, 1995.

[Magnani et al., 2002] L. Magnani, M. Piazza, and R. Dossena. Epistemic mediators and chance morphodynamics. In A. Abe, editor, *Proceedings of PRICAI-02 Conference, Working Notes of the 2nd International Workshop on Chance Discovery*, pages 38–46, Tokyo, 2002.

[Magnani, 1999] L. Magnani. Inconsistencies and creative abduction in science. In *AI and Scientific Creativity. Proceedings of the AISB99 Symposium on Scientific Creativity*, pages 1–8, Edinburgh, 1999. Society for the Study of Artificial Intelligence and Simulation of Behaviour, University of Edinburgh.

[Magnani, 2001] L. Magnani. *Abduction, Reason, and Science. Processes of Discovery and Explanation*. Kluwer Academic/Plenum Publishers, New York, 2001.

[Magnani, 2002] L. Magnani. Epistemic mediators and model-based discovery in science. In L. Magnani and N.J. Nersessian, editors, *Model-Based Reasoning: Science, Technology, Values*, pages 305–329, New York, 2002. Kluwer Academic/Plenum Publishers.

[Mithen, 1999] S. Mithen. Handaxes and ice age carvings: hard evidence for the evolution of consciousness. In A.R. Hameroff, A.W. Kaszniak, and D.J. Chalmers, editors, *Toward a Science of Consciousness III. The Third Tucson Discussions and Debates*, pages 281–296, Cambridge, 1999. MIT Press.

[Nara and Ohsawa, 2004] Y. Nara and Y. Ohsawa. Knowing melting pot. emerging topics based on scenario communication on technical foresights. In A. Abe and R. Oehlmann, editors, *The First European Workshop on Chance Discovery*, pages 114–121, Valencia, 2004.

[Nersessian, 1999] N. J. Nersessian. Model-based reasoning in conceptual change. In N.J. Nersessian, L. Magnani, and P. Thagard, editors, *Model-based Reasoning in Scientific Discovery*, pages 5–22, New York, 1999. Kluwer Academic/Plenum Publishers.

[Peirce, 1931-1958] C.S. Peirce. *Collected Papers of Charles Sanders Peirce*. Harvard University Press, Cambridge, MA, 1931–1958. vols. 1–6, Hartshorne, C. and Weiss, P., eds.; vols. 7–8, Burks, A. W., ed.

[Thagard, 1992] P. Thagard. *Conceptual Revolutions*. Princeton University Press, Princeton, 1992.

[Turing, 1950] A.M. Turing. Computing machinery and intelligence. *Mind*, 49:433–460, 1950.

[Zhang, 1997] J. Zhang. The nature of external representations in problem-solving. *Cognitive Science*, 21(2):179–217, 1997.

Lorenzo Magnani
Department of Philosophy and
Computational Philosophy Laboratory
University of Pavia, Pavia, Italy; and
Department of Philosophy, Sun Yat-sen University
Guangzhou (Canton), P.R. China.
Email: lmagnani@unipv.it

Analogical Reasoning with Animal Models in Biomedical Research

CAMERON SHELLEY

ABSTRACT. Critics of the use of animal models in biomedical research have questioned the validity of this ubiquitous practice. One general argument against the validity of animal models centers on doubts about the analogical relationship between model and human. The argument states that since disanalogies threaten the coherence of the model-human analogy, animal models are always nonvalid. In this article, I argue that, on a proper understanding of analogy and disanalogy, disanalogies do not necessarily undermine the coherence of the model-human analogy. Thus, no general conclusions about validity follow; the validity of any given model depends upon the facts of the case in question.

1 Introduction

What makes animal models valid? In other words, what assures us that experiments performed on animals can tell us *anything* about conditions in human beings? In addition to being an interesting question in the philosophy of science about a crucial scientific research practice, the question of the validity of animal models looms large in the continuing debates concerning animal rights. One sort of argument for the abandonment of animal modeling attacks its validity as a scientific research practice. The general idea behind this sort of argument is that if animal modeling can be shown to be non-valid, then it is obviously morally unacceptable. After all, if we cannot show how claims about human conditions can be tested by experiments on animals, then those experiments ought not to be undertaken at all. So, the validity or non-validity of animal models becomes a pressing question for a variety of reasons.

Biomedical researchers generally agree that the validity of animal models depends significantly on, among other things, the existence of *analogies* between animal models and their human targets. In short, the validity of an animal model depends on whether it is analogous to the human condition it is supposed to represent. In their recent book *Brute Science*, philosophers

[LaFollette and Shanks, 1996] have attacked the practice of animal modeling by attacking this source of its validity. Specifically, LaFollette and Shanks argue that *even when* models are analogous to their targets, there remain *disanalogies* between model and target that undercut *any* such claims for the validity of an animal model. So, they conclude, animal modeling is also an immoral practice.

My first goal in this article is to defend the validity of animal models from LaFollette and Shanks's critique. I will argue first that LaFollette and Shanks's attack is ineffective because they construe the concepts of analogy and of disanalogy in a way that is largely irrelevant to the validity of animal models. However, I will also argue that if we *re-construe* these concepts appropriately, then we can construct an argument that is potentially devastating to the validity of animal models. This reconstructed argument involves employing the concepts of analogy and disanalogy developed in recent cognitive science research [Shelley, 2002a; Shelley, 2002b]. Then, I will explain why I think that even the reconstructed argument does not establish the general non-validity of animal modeling.

So, in the first instance, I aim to rescue the validity of animal models from LaFollette and Shanks's attack. However, what I ultimately aim to accomplish is to show how recent research on analogical reasoning in cognitive science can tell us, with far greater clarity than before, what makes animal models valid.

2 Analogy and face validity

As I mentioned above, LaFollette and Shanks argue for the abandonment of animal models on the grounds that the practice is irretrievably non-valid. In this part of the paper, I will present LaFollette and Shanks's argument as they give it, and point out why the argument falls short of establishing the non-validity of animal models. My criticisms of this argument will help to motivate my reconstruction of it afterwards.

LaFollette and Shanks's argument can be very briefly summarized. It begins with the concession that animal models derive a kind of *prima facie* validity from relevant analogies between animal model and human target. Rats and humans, for example, are indeed alike in many ways; so we might reasonably conclude that it makes sense to compare them. However, the argument continues, there are always disanalogies between animal models and human targets, and these disanalogies have the effect of irretrievably undermining the validity of animal-human comparisons. Rats and humans are, after all, unlike in many ways. So, animal models are, in general, scientifically non-valid, and therefore immoral.

There are a number of flaws in LaFollette and Shanks's argument. A crucial flaw concerns the concepts of analogy and disanalogy that they apply. Specifically, LaFollette and Shanks understand analogy to be the similarity or *resemblance* between model and target. This understanding becomes evident when they begin their argument by defining an analogical inference as a sort of syllogism in the tradition of J.S. Mill [1872]:

X has properties $a, \ldots, e, f$. (X = model)
Y has properties $a, \ldots, e$. (Y = target)
Probably, Y has property f.

In other words, two things X and Y are analogous insofar as they have the same properties, such as color, shape, mass, or constitution. Since some resemblance between X and Y has been established, it is probable that X and Y share some further property f in common as well. LaFollette and Shanks then go on to discuss animal models in terms of what they call a *Causal Analog Model* (CAM). They define a CAM by adding three extra constraints on the definition just given. These constraints are:

- the common properties $a, \ldots, e$ must be causal properties that
- are causally connected with the property f we wish to project, and
- there must be no causally relevant disanalogies between X and Y.

The term *causal* simply means that the properties $a, \ldots, e$ give rise to each other through some means. More importantly, note how the third constraint does the damage to animal modeling. It asserts that the presence of *any* causally relevant disanalogy completely undermines the comparison between model and target. Now, LaFollette and Shanks do not define what they mean by disanalogy, but the obvious and conventional meaning in the same tradition is that a disanalogy is a *dissimilarity*. In other words, a model X and a target Y are disanalogous if there is some property, call it p, possessed by either X or Y but not by both. Of course, there are almost countless dissimilarities between human beings and any given variety of animal, some of which are sure to be causally relevant. Given this third constraint on animal modeling, this fact means that any animal model is sure to be non-valid.

Let us accept for the sake of argument that there will always be causally relevant dissimilarities between animal models and human conditions. Does this claim truly imply that animal models are always non-valid? The answer, I think, is *no*. Very little depends upon the resemblance as such between model and target. In the biomedical literature, the validity that an animal model derives from its similarity to its target is known as *face validity*

[Willner, 1991]. For example, a rat that moves less than usual for its kind, or that consumes less food than usual for its kind might be considered as a model of depression in humans [Sarter and Bruno, 2002]. These properties, less movement and less food consumption than typical, are characteristic symptoms of depression in people. Because of these shared properties, this rat model of human depression enjoys some face validity. However, rats, as [Sarter and Bruno, 2002] point out, are not known to write poetry about their experiences in life. Depressed humans are known to do this, so a rat model of human depression would seem to be lacking somewhat in face validity.

However, face validity counts for very little in the evaluation of animal models. As biomedical researchers are well aware, dissimilarities between species are often just the result of differences in the behavioral repertoires of the animals concerned. Two sorts of behavior, for example, may be analogous even though they are dissimilar in causally relevant ways. For instance, giving stimulants to rats induces a behavior of stereotyped rearing on the hind legs. In contrast, giving stimulants to primates induces a behavior of stereotyped scratching [Willner, 1991, p. 14]. Now, rearing and scratching are dissimilar in the sense that they do not resemble one another to the eye. In other words, they are easily told apart. Nevertheless, they are analogous in the sense that both behaviors are the results of treatment with stimulants.

So, the central issue when evaluating comparisons between animals and animals, or between animals and humans, is whether or not the conditions being compared are functionally equivalent. Stereotyped rearing and stereotyped scratching are functionally equivalent as responses to stimulants.

It is because of considerations like these ones that biomedical researchers set little store in the face validity of animal models [Davidson et al., 1987]. This being the case, LaFollette and Shanks's critique of animal models concerning their face validity is of little consequence. However, we now have a concern, namely functional equivalence, that indeed a crucial consideration.

3 Analogy and construct validity

We now have good reasons to reject LaFollette and Shanks's argument against the validity of animal models in general. However, we now have an issue, namely functional equivalence, that is far more important to the validity of animal models. It is now time to show how recent research on analogical reasoning in cognitive science can help us to clarify what makes animal models valid. I will then be in a position to reconstruct LaFollette and Shanks's argument in a way that makes it far more penetrating.

When considering functional equivalence, biomedical researchers refer to the *construct validity* of an animal model. In terms of animal models, construct validity concerns how well the condition of a model *represents* the condition of interest in the target [Willner, 1991]. In other words, a model must correspond to its target in some appropriate way. In general, biomedical researchers have had no explicit way of conceptualizing and assessing this form of correspondence. They simply rely on their intuitions. From the preceding discussion, it is clear that similarity is not the right way to think about this correspondence. Instead of capturing correspondences in terms of shared properties, we need to capture correspondences of equivalent functions. I claim that the concept of analogy as it has developed recently in the cognitive science literature captures exactly this idea. That is to say, cognitive scientists have come to view analogies as a particular sort of functional correspondence and have supplied tests for the assessment of analogies. So, I think that we can make substantial progress in clarifying and assessing the construct validity of animal models by applying this work on analogies.

Consider, for example, a common mouse model for the action of antidepressants in human beings. In the Porsolt Forced-Swim Test [Porsolt *et al.*, 1977], a mouse is placed in a cylinder of water and watched to see how long it swims until it gives up trying to climb out. (The mouse is not drowned at that point but assumes a static position with its hind feet on the cylinder bottom and its nose out of the water.) It turns out that mice that are treated with antidepressants tend to swim longer than normal mice do. This model of the action of antidepressants enjoys construct validity in the sense that the increased time that mice treated with antidepressants spend trying to extricate themselves from the cylinder corresponds to the increased hope for success in life that depressed people treated with antidepressants feel in the pursuit of their goals. The analogy between mouse and human that is present in this case may be captured as in table 1.

This table represents the analogy in accord with the principles of the *structure-mapping theory* [Gentner, 1983]. The information in the mouse-human analogy is laid out according to a few simple principles. All the information concerning the model domain, that is, the mouse, is located in the left-hand column. All the information concerning the target domain, that is, the depressed human patient, is located in the right-hand column. Each row in the table represents a correspondence between the model and target. For example, the mouse and the human patient are both placed in the same row (at the top) because the mouse and patient correspond to each other in the analogy. Now, there are three different kinds of correspondences that may be found in an analogy, and these are captured as three groups of rows in the table. The topmost area in the table, consist-

Mouse	Human
mouse	patient
antidepressant	antidepressant
safety	goals
longer-time	further-extent
look-for(mouse,safety)	hope-for(patient,goals)
persist(mouse,longer-time)	persist(patient,further-extent)
receive(mouse,antidepressant)	receive(patient,antidepressant)
because(persist,look-for&receive)	because(persist,hope-for&receive)

Table 1. The analogy representing the construct validity of the Porsolt Forced-Swim test.

ing of the top four rows, represents correspondences between the simple properties or *attributes* of each domain, such as *safety* and *goals* in row three. The middle area, consisting of the next three rows, represents correspondences between *relations* among the attributes or facts regarding the attributes, such as *look-for(mouse,safety)* ↔ *hope-for(patient,goals)* in row five. In plain English, what this row states is that the search for safety by the mouse corresponds to the hope for achievement of goals by the patient. The bottom area, consisting of the last row, represents correspondences between the *system relations* of each domain, such as *because(persist,look-for&receive)* ↔ *because(persist,hope-for&receive)*. This row states that just as the persistence of the test mouse for a longer time is due to its search for safety and its treatment with antidepressants, the persistence of the psychiatric patient to a further extent is due to his or her hope for achievement of life goals and treatment with antidepressants.

We can see from looking at the representation in the table that the content of the analogy is unified by the system relations. The *because* relations, in this case, summarize the whole analogy when rendered in plain English. A look at the table shows that all the basic attributes and relations in each domain are related to each other by falling under the system relations. It is these high-level relations that provide the overall structure of the analogy and thereby determine the mutual relevance of each piece of information contained within it.

This way of representing the corresponding relations in an analogy, anchored by the system relations, captures the functional equivalence between model and target. In doing so, it captures the construct validity of the model. So, we can evaluate the construct validity of this animal model by evaluating this analogy. The multiconstraint theory of analogy provides three criteria for performing this evaluation [Holyoak and Thagard, 1995]. The better the analogy satisfies these constraints, the stronger or more coherent the analogy is considered.

Structural consistency: This constraint concerns to what extent the analogy presents a structural isomorphism. Looking at the table, we can see that every term in each domain corresponds to exactly one and the same term in the other domain. For example, *mouse* and *patient* correspond at every point where those terms occur. Since every correspondence is one-to-one in this way, the analogy is completely consistent structurally.

Semantic similarity: This constraint concerns to what extent corresponding terms are similar in meaning, particularly at the relation and system-relation levels. For example, *look-for* and *hope-for* are similar although not identical, and *because* (mouse) and *because* (patient) are identical. Since the corresponding terms at these levels are similar or identical in meaning, the analogy is strong in this respect also.

Pragmatic coherence: This constraint concerns how well the overall story suggested by the analogy addresses the facts at hand. In this case, the analogy suggests how treatment with an antidepressant relates to subsequent behaviors in both mouse and human. Since it helps to make sense of these observations, the analogy can be considered pragmatically coherent as well.

Since these constraints are well satisfied by the mouse-human analogy, the analogy is clearly very coherent overall. In turn, its coherence implies that this mouse model of antidepressants in human beings enjoys strong construct validity: the longer safety-seeking of the mouse in the model well represents the greater hope for achievement of goals in depressed human beings.

4 Disanalogies and construct validity

I have now illustrated how construct validity can be captured by the concept of analogy developed in cognitive science research, such as the multiconstraint theory. I can now raise an objection similar to LaFollette and Shanks's argument concerning face validity. In brief, what happens to our assessment of the construct validity of animal models when there are disanalogies between model and target?

LaFollette and Shanks showed that models that are causally similar to their targets may also be causally dissimilar as well. This observation tends to undermine the face validity of animal models. Now, face validity is a matter of relatively little concern, but what if a similar problem applies to the *construct validity* of animal models? Perhaps models that are analogous to their targets in terms of the multiconstraint theory may also be disanalogous as well. If true, this situation means that the construct validity of any animal model, such as the Forced-Swim Test is not assured by the existence of even a strong analogy between model and target. If so, then the validity of animal models in general is cast into doubt.

In order to explore and assess this argument, we need to find an appropriate concept of disanalogy. The concept of disanalogy that I have developed in my recent work fits exactly this requirement. In [Shelley, 2002a] and [Shelley, 2002b], I defined a disanalogy as an extension to a given analogy that supports a conclusion incoherent with the conclusion supported by the original. In other words, a disanalogy arises when we have an analogy that leads us to one conclusion initially but that leads us to an incompatible conclusion when augmented with new information.

To clarify this concept, let us consider an example of a disanalogy to the Forced-Swim Test [Kolata, (28 March 2004)]. In a mouse, the effect of an antidepressant is most likely produced by affecting the lower areas of the brain. The frontal cortex in a mouse is small in relation to its lower brain regions in contrast with the relatively large frontal cortex of a human being. So, since the antidepressant is probably acting on a brain area *associated* with non-cognitive mental states, it is likely that it is affecting the mouse's non-cognitive mental state. Now, we can add this information to the existing analogy between mouse and human and draw the corresponding conclusion regarding the human target. That conclusion is that the mental state induced in a depressed person by treatment with antidepressants is also a non-cognitive mental state.

Here we encounter some incoherence. This conclusion does not fit with the rest of the analogy. In particular, the hope that a depressed person experiences as a result of treatment, represented by the *hope-for* relation in the analogy, is an irreducibly cognitive state. People do not merely hope, they hope for some things in particular, like publishing a paper or being promoted at work. People represent these hopes to themselves cognitively, in the form of mental representations, using their big frontal cortices. Yet, from what we were just saying, the hope for achievement of life goals in these patients should be like the safety-seeking in mice: instinctual rather than involving sophisticated mental states. It appears very much as though this one mouse model of human depression is leading us to two, incompatible conclusions. If this is so, then we should probably consider the whole comparison misleading and so reject the construct validity of the Forced-Swim Test.

This disanalogy is represented explicitly in Table 2. This table is simply a copy of the previous one with the new information added. The added correspondences are indicated with asterisks. The new causal relation in the bottom row states that the effect that an antidepressant has on a mouse's non-cognitive state is what causes the mouse to persist for a longer time when looking for safety. The contents of this causal relation are also added in as the *affect* relation and the *non-cognitive-state* attribute in the appropriate places. Equivalent representations have been placed in the human target domain. The result is a new analogy that appears to be as coherent as the original one in terms of the multiconstraint theory.

Mouse	**Human**
mouse	patient
antidepressant	antidepressant
safety	goals
longer-time	further-extent
*non-cognitive-state [ncs]	*non-cognitive-state [ncs]
look-for(mouse,safety)	hope-for(patient,goals)
persist(mouse,longer-time)	persist(patient,further-extent)
receive(mouse,antidepressant)	receive(patient,antidepressant)
*affect(antidepressant,ncs)	*affect(antidepressant,ncs)
because(persist,look-for&receive)	because(persist,hope-for&receive)
*cause(affect,persist)	*cause(affect,persist)

Table 2. The disanalogy concerning the Porsolt Forced Swim test.

It is clear that this new analogy meets the definition of a disanalogy given above. It is derived from the existing mouse-human analogy by adding new information to it. Furthermore, the new information suggests a conclusion: that the antidepressant affects a non-cognitive mental state. This conclusion does not fit with the implications of the original version of the analogy. Therein lies the problem: Essentially one-and-the-same analogy suggests two, mutually incompatible conclusions. Given this situation, it seems that we cannot rely on *anything* that this analogy suggests.

5 Defeasibility and construct validity

I have now given a reconstructed version of LaFollette and Shanks's argument aimed at construct validity. As things stand, the whole idea that we can sensibly compare mice and depressed humans is in jeopardy. When we generalize this difficulty, we get an argument threatening the construct validity of any animal model. So, it seems that the validity, and therefore morality, of the whole practice has been thrown into doubt. What I will now argue is that this argument to the general non-validity of animal models does not work either. The reason is that disanalogies, as I have described them, are *defeasible*.

In general, to say that an inference is defeasible is to say that its conclusion may be retracted in the light of further evidence. In the case of disanalogies, defeasibility means that the incompatible conclusions suggested by the disanalogy may be reconciled with each other by the introduction of new information into the context of the analogy. For example, one might point out that cognitive and non-cognitive states are not independent in humans: The frontal cortex *interacts* with lower brain areas in the realization of mental states. So, even if antidepressants do act in the first instance on the non-cognitive component of mental states in mice and humans, nothing prevents them from thereby affecting the *cognitive* states of humans as well. Indeed, such an interconnection is to be expected given the strong emotional component of depression, which is strongly associated with non-cognitive areas of the brain. So, the two conclusions supported by the disanalogy discussed earlier seem to fit together after all: It is not so odd to think that antidepressants might act on both cognitive and non-cognitive mental states at the same time.

This example illustrates why the general argument against the construct validity of animal models does not work. Whether or not a disanalogy succeeds in undermining the validity of an animal model depends upon the details of the particular model in question. In some cases, there may be no way of reconciling the incoherence produced by raising a disanalogy. In other cases, such as the one we have discussed here, there are reasonable ways of resolving that incoherence. Each case will depend upon the context of each model. So, no general conclusion about the construct validity of animal models follows from these considerations. What we can say in general is that, with the proper concepts of analogy and of disanalogy in hand, we can hope to make reasonable determinations of construct validity in each individual case.

6 Conclusions

The most immediate conclusion that we can draw from this discussion is that the argument made by LaFollette and Shanks against the validity of animal modeling in general is not effective. In its original form, it addresses face validity, which is a relatively inconsequential matter. When revised to address construct validity, the argument becomes potentially more damaging. However, because the argument relies on disanalogies and because disanalogies are defeasible, no general conclusions follow about the construct validity of animal models.

Moreover, I argue that I have greatly clarified one important answer to the question in the philosophy of science of what makes animal models valid. One vital quality is the construct validity of animal models. The best way to understand the construct validity of animal models, I claim, is to apply recent cognitive theories of analogy—and of disanalogy. I have tried, in particular, to show how consideration of disanalogies is indispensable to understanding the validity of animal models.

Finally, these theories help us not only to better understand what makes an animal model valid, but they also provide us with the means for principled evaluation of them. Having these conceptual tools is extremely important, given the continued reliance on animal models in biomedical research, and given the ethical controversy regarding them that continues today.

Acknowledgment

Previous versions of this paper were given as talks at the University of Waterloo and at Carleton University. Thanks to Lorenzo Magnani for making it possible for me to give this talk at the Model-Based Reasoning conference in Pavia in December 2004.

BIBLIOGRAPHY

[Davidson et al., 1987] M. K. Davidson, J. R. Lindsey, and J. K. Davis. Requirements and selection of an animal model. *Israel Journal of Medical Sciences*, 23:551–555, 1987.

[Gentner, 1983] D. Gentner. Structure-mapping: A theoretical framework for analogy. *Cognitive Science*, 7(2):155–170, 1983.

[Holyoak and Thagard, 1995] K. J. Holyoak and P. Thagard. *Mental leaps: Analogy in creative thought*. The MIT Press, Cambridge (Mass.), 1995.

[Kolata, (28 March 2004)] G. Kolata. Why test animals to cure human depression? *New York Times*, (28 March 2004).

[LaFollette and Shanks, 1996] H. LaFollette and N. Shanks. *Brute Science: Dilemmas of animal experimentation*. Routledge, London, 1996.

[Mill, 1872] J. S. Mill. *A system of logic*. 8th edition., Longman, London, 1872.

[Porsolt et al., 1977] R. D. Porsolt, A. Bertin, and M. Jalfre. Behavioral despair in mice: a primary screening test for antidepressants. *Archives internationales de pharmacodynamie et de therapie*, 9(2):327-36, 1977.

[Sarter and Bruno, 2002] M. Sarter and J. P. Bruno. Animal models in biological psychiartry. In H. D'Haenen, J. A. den Boer, and P. Willner, editors, *Biological Psychiatry*, pages 1–8. John Wily & Sons, 2002.

[Shelley, 2002a] C. Shelley. Analogy counterarguments and the acceptability of analogical hypotheses. *British Journal for the Philosophy of Science*, 53(4):477–496, 2002.

[Shelley, 2002b] C. Shelley. The analogy theory of disanalogy: When conclusions collide. *Metaphor and Symbol*, 17(2):81–97, 2002.

[Willner, 1991] P. Willner. Methods for assessing the validity of animal models of human psychopathology. In A. A. Boulton, G. B. Baker, and M. T. Martin-Iverson, editors, *Animal models in psychiatry*, pages 1–23. Humana Press, 1991.

Cameron Shelley
12 Clark St. W.
Guelph, ON
Canada N1H 1S9
Email: `cam_shelley@yahoo.ca`

Cognitive Fictions
GIOVANNI TUZET

ABSTRACT. *Cognitive Fictions* represent non-actually-observable states of things inferred from actually observable ones. Examples of such are simulations and artificial representations of past or future events. I distinguish three kinds of simulations: (A) *Non-ampliative simulations*, that do not add any novel information to the amount at disposal before their realization; (B) *Determinist simulations*, realized on the basis of deterministic laws; (C) *Rule-based simulations*, based on the rules of a game, or behavioral rules, or other rules governing a social interaction. Simulations of kind (A) are the less interesting. Simulations of kind (B) and (C) are informative, but rule-based simulations do not give unique solutions to the problems specified (as determinist simulations do), since they elaborate different possible scenarios. The main issue becomes the evaluation of the different scenarios. So, which criteria determine the credibility and acceptability of a cognitive fiction? The criteria of *coherence* and *inferability* are considered. The paper concludes that devices like cognitive fictions can be of great importance for our cognitive tasks but must be used with due caution and many constraints must be put on their use.

1 Some issues on cognitive fictions

The idea of cognitive fiction came to me considering the function of the legal process and some aspects of contemporary evidentiary practices. The legal process has a *cognitive* function (cf. [Ferrua, 1997] and [Ferrajoli, 2000]) and certain *fictions* have, as it seems to me, an important role to that purpose. So, if that is true, we shall call them *Cognitive Fictions*. They are fictions which have a cognitive function or purpose. The artificial reconstruction of the crime scene is the most remarkable example of that.

It is important to specify what I do *not* speak of. I do not speak of legal fictions in the technical sense of legal entities, *viz.* corporations, or in the sense of statements knowingly false but accepted in order to yield a good legal consequence (see [Fuller, 1967]). I speak of the technological artefacts whose function is to represent something which is not observable. In this sense, strictly speaking they are not hypotheses, but incorporate hypotheses or predictions.

I call *Cognitive Fictions* those fictions representing non-actually-observable states of things inferred from actually observable ones. Examples of cognitive fictions are simulations and artificial representations of past or future events. Here I won't deal with the technological aspects of such devices; rather, with some logical and epistemological issues they raise.

To begin, why call them fictions if the purpose they are created for is a cognitive one? In fact, they differ from what we ordinarily call a fiction, namely something which is knowingly false. They are fictions not for their being knowingly false but, as it seems to me, for their phenomenological content. Let me expand on this point.

In a possible classification of fictions, cognitive fictions are those most approaching a representation of reality. On the one hand, they are fictions because they do not pretend to represent reality as it appears to us. On the other they have a cognitive value, since they are supposed to inform us of certain aspects of reality. How can they do that? My answer is this: they consist in *evidence-based artificial representations of non-actually-observable states of things*, in particular of past or future events. In both cases (past or future events) they are representations of non-actually-observable states of things inferred from actually observable ones. In the case of past events they incorporate hypotheses. In the case of future events they incorporate predictions. But notice that, *qua* fictions, they are artificial representations of such hypotheses or predictions.

Now, what do I mean by "artificial representation"? Consider the difference between a picture by a camera and a simulation by a computer with a high degree of abstraction. Cognitively speaking they are both artificial devices but the former is supposed to represent reality as it appears to us, while the latter is not. So we may say the following: an artificial representation is not supposed to represent reality as it appears to us but it is nonetheless supposed to inform us about reality. Distinguishing two kinds of content, it has an informational content which is supposed to inform us of certain aspects of reality and a phenomenological content which differs from the way reality appears to us. Cognitive fictions consist in this sense in artificial representations.

Could we also say that they consist in models of reality? In a ample sense of the word "model", of course they do. In a specific sense of the word, it depends. If a model is a generalization, I would say no. If a model is an abstraction of reality, I would say yes. In that sense, cognitive fictions are models of reality. But I prefer to say that they consist in artificial representations, in order to stress their artificial nature and avoid the ambiguity between generalization and abstraction.

There are two main problems concerning cognitive fictions:

1. Which criteria are used in realizing them?

2. Which probative value do they have?

Here I won't deal with the second problem, which would require, in the context of legal cognitive fictions, a discussion on the legal theory of evidence and proof. I would say something on the first problem, that is the problem of criteria, after having classified three kinds of cognitive fictions. Then I will end by considering, as a case study, the epistemological profile of what is called, in the Italian law, Judicial Experiment. Before going into that, let me remark the inferential profile of the matter. We said that cognitive fictions can consist in artificial representations of past or future events, namely in hypotheses or predictions. In the first case they draw abductive inferences. In the second deductive inferences. Focusing on the legal process, the most interesting case is the use of cognitive fictions in representing past events, namely drawing evidence-based abductive conclusions, using for instance simulations supported by visual imagination or other cognitive devices[1]. But, as we shall see, the use of such devices in the legal context is as interesting as problematic.

2 Three kinds of cognitive fictions

A simulation, as it seems to me, is an example of cognitive fiction. Let me try to classify some kinds of cognitive fictions using the example of simulations[2]. We can distinguish three kinds of simulations.

(A) *Non-ampliative simulations*. These are the simulations that do not add any novel information to the amount at disposal before their realization. That is the case, for instance, when a graphical representation of the information at disposal is realized. Sometimes they are called *post-processing simulations*. With this kind of simulation, the information is already given and the task comes to that: the realization of another (more simple, or more useful, depending on the context) representation of it.

(B) *Determinist simulations*. These are the simulations realized on the basis of deterministic laws. On such a basis a mathematical model is constructed capable of giving, for any problem, in terms of equations, the only solution which is correct. These simulations are informative and the solution they offer is unique. Predictive uses of this kind of simulations are possible (simulating the performances of boats, planes, cars, earthquake-proof

---
[1] On abduction and legal reasoning cf. [Tuzet, 2004a], [Tuzet, 2004b], [Tuzet, 2004c]. On visual abduction [Magnani, 2001, pp. 104-115]. On imagination and simulation [Nersessian, 1999, pp. 17-18]

[2] On simulations in general [Kuipers, 1994], [Casti, 1997], [Parisi, 2001]. On simulations and social sciences cf. [Hegselmann et al., 1996], [Conte et al., 1997].

buildings) and abductive uses as well (ballistics). But the condition of their grounding on deterministic laws makes their use in the social sciences quite restricted. However they have important uses: the case of simulations based on ballistics is a remarkable use of cognitive fictions of deterministic kind in a legal process.

(C) *Rule-based simulations*. These are the simulations based on rules governing a social interaction, both normative and non-normative. Namely the rules of a game, or behavioral rules, or other rules governing a social interaction. On such a basis computer programs can be realized capable of playing a game or representing a social interaction. (Think of the programs playing chess, as the most obvious example). The simulations so performed are informative, but the difference with simulations of kind (B) is that simulations of this kind do not give unique solutions to the problems specified, but elaborate different possible scenarios. (In particular, in case of simulations based on the rules of a game, we cannot predict how the program will play).

When simulations of this kind are used retrospectively, *viz.* abductively, they have this interesting character: on the basis of the evidence at disposal, of stochastic laws, behavioral regularities and normative rules, they represent different possible versions of the events at issue. The unpredictability of the solutions offered by the program increases when it consists in a multi-agent simulation, namely when it has to retrospectively simulate the conduct of two or more interacting subjects: the program capable of playing the game defined by the rules imagines the possible moves of the subjects and the possible replies to those moves, attempting to represent different possible ways of arriving at the known results (the evidence at disposal). Plainly the main issue becomes the evaluation of the different scenarios realized by the program: the *best explanation* of what happened has to be selected from the possible scenarios represented by the program. It is clearly an abductive problem, or the problem of the inference to the best explanation[3].

Notice that in simulations of kind (B) and (C) there is a fundamental *relevance* issue[4]. These simulations exclusively represent the information which is relevant to the problem at hand. Think of the abstract character of simulations by computers: human figures, for instance, are not represented in their natural aspect but, more often, in some specific features that distin-

---

[3] On the notion of *Inference to the Best Explanation* cf. [Harman, 1965], [Lipton, 1991], [Josephson and Josephson, 1994]. Notice that one of the examples made by [Harman, 1965, p. 89] is quite a legal one: "When a detective puts the evidence together and decides that it *must* have been the butler, he is reasoning that no other explanation which accounts for all the facts is plausible enough or simple enough to be accepted."

[4] On relevance [Sperber and Wilson, 1988].

guish one figure from another relatively to the relevant information and the nature of the issue.

There is something more to be noted. Even if the outcomes of simulations of kind (B) seem to be the most certain – or at least the most reliable – in a context like the legal one many problems can hinder the relevant inference. If the initial set of data is poor or inaccurate the scenarios compatible with it might be too numerous to provide a unique solution or at least to permit the selection of an explanation more plausible than others. In other words, one might exclude some hypotheses without being capable of determining the best explanation of the case in hand.

On the contrary, a simulation of kind (C) may be less complex than it might thought to be if the events took place in a quite restricted space: then more precise data may be found in virtue of the place being delimited. To put it differently: the possible worlds compatible with the data are (quite paradoxically) less than the ones compatible with a case of kind (B) when the data are poor or inaccurate.

3 Some criteria for credible cognitive fictions

Remember the problem I raised in the first section: Which criteria are used in realizing cognitive fictions? There is also, of course, a normative version of the problem: Which criteria are to be used in realizing cognitive fictions? An answer to the normative version of the problem needs an epistemological evaluation of the criteria that are used or can be used in realizing cognitive fictions. When is a cognitive fiction reliable? When is it credible? When is it acceptable?

In this presentation I shall confine myself to the criteria of *coherence* and *inferability*. As to the domain of application, I shall focus on legal reasoning.

The first criterion, *coherence*, is, as it seems to me, a necessary condition for the credibility of a cognitive fiction. If coherence means logical consistency, how could a logically inconsistent fiction be credible? How could it be a credible representation of past or future events? Given that there are no logical inconsistencies between events themselves, a logically inconsistent fiction would be deprived of credibility. Anyway, this is not the only meaning that coherence can be given. Paul Thagard [1988; 1992; 2000] has claimed that coherence is a fundamental criterion not only for assessing scientific hypotheses but also for evaluating thought and action, and distinguished different kinds of coherence. Such a criterion constitutes a well-known issue in legal theory and theory of legal reasoning as well[5]. Consider in par-

---

[5]On *narrative coherence* in evidentiary discourses [Pastore, 1996, pp. 197-200]. For a narrativist conception of the legal process cf. [Jackson, 1988]; for a critique of narrativist conceptions see [Taruffo, 1992].

ticular the distinction made by Neil MacCormick [1995] between *narrative coherence* as a test for factual claims and *normative coherence* as a criterion for normative claims (cf. [Pastore, 1996, pp. 206-211]). Coherence in MacCormick's sense is something more than logical consistency (absence of contradictions): it is a sort of "making sense", that concerns not only the logical relations between some sentences but also the relations of those sentences to the world and in particular to the facts and events at issue. But, as it seems to me, this way of assessing coherence relies on the very vague notion of "making sense". So, even though I agree with the purpose lying behind this way of assessing coherence, I prefer to use the criterion of coherence as logical consistency[6]. But obviously the purpose must be met: logical consistency is not a relation with the world.

To respect the criterion of coherence as logical consistency is necessary, but not sufficient, for the credibility of a cognitive fiction. A credible cognitive fiction must not be a piece of pure fiction or imagination (which may be coherent as well). It must embody a plausible hypothesis inferred from real elements[7]. It must be based on actually observable things in order to represent a non-actually-observable state of things. The relevant actually observable things are those from which the unobservable state represented in the fiction can be inferred. They are – to put it in legal terms – the evidence at disposal.

Thus, to be credible, cognitive fiction must be not only *coherent* but also *inferable from real elements* (actually observable things). In short there are *real constraints* on which its credibility depend.

This concerns the credibility of a cognitive fiction. But what about the case in which, as there are competing hypotheses, there are competing cognitive fictions that are all credible? Suppose that the parties in a legal process provide different cognitive fictions (artificial reconstructions of the events) compatible with the evidence at disposal. It seems we need something more: we need criteria for assessing rival credible cognitive fictions. So, a credible cognitive fiction is not necessarily a fiction worth accepting. Its acceptability is determined by the testing of the hypothesis elaborated in the fiction. In this sense, predictive cognitive fictions have an important role in the testing of abductive cognitive fictions. But many specifications are needed, in particular (for a predictive simulation):

- which hypotheses it is supposed to test;

---

[6]But see [Thagard, 2000] for a classification of six main kinds of coherence: explanatory, deductive, conceptual, analogical, perceptual, and deliberative.

[7]I cannot discuss here the issues of plausibility, likelihood and probability. But I think that no hypothesis may be plausible without being inferable from real elements.

- how the hypotheses are embodied in the artefact;

- which level of abstraction is used;

- how to evaluate the outcomes of the simulation.

So the acceptance of an abductive cognitive fiction can be determined by successful predictions, *viz.* deductions, from the previous hypothesis; but, for the test being reliable, many constrains must be put on the use of predictive cognitive fictions.

Besides these logical and technical questions there are more general epistemological questions concerning the credibility and acceptability of cognitive fictions. Notably this, to put it very roughly: Do the points of view influence the outcome of a cognitive fiction?

Philosophically we must distinguish the *evaluative* points of view consisting in a choice of value or preference, from the *epistemic* points of view consisting in the peculiar perspective from which something is observed.

Let me start with the evaluative points of view. The thesis that our choices of value determine in general the contents of our cognition is quite disputable. I cannot enter into such a dispute here; I just note that the strongest value conflicts occur when there is a basic agreement on facts and a disagreement on their value. However, what about the cases in which a certain choice of value is relevant to the representation of what is at stake? Think of the case in which someone attributes an intention to someone else on the basis of the observed behavior. For instance: "I saw him behaving in a menacing way". What does mean "menacing"? Is there a fact of the matter about it? As Hilary Putnam has recently remarked in many cases a strong dichotomy of facts and values collapses [Putnam, 2002].

Coming now to the epistemic points of view, I claim they do not influence the outcomes of cognitive fictions. For instance, once a geometrical and mathematical model is constructed it can be used in different ways, observed from different sides, zooming, focusing on details, and so on. The properties of the models do not change. A different question is the presentation of the model. In deciding how to present a model, we can decide from which side, with which details, we prefer to present it. Obviously such a choice does not prevent the possibility of representing it otherwise. What is at stake are the different rhetorical effects of different presentations, despite the fact they are presentations of the same model. What is important is to distinguish the realization of the model from its presentation. The first is evidence-based, the second rhetorically-oriented. The properties of the model shall not be identified with the properties of its presentations.

4 The case of judicial experiment

Many authors working on simulations claim they can be of great help in developing our social sciences (see [Parisi, 2001]). It is probably true. But what is the case for such social sciences like the legal sciences? Again, those authors claim that simulations could help the historical sciences, providing causal explanations of historical events. But what is the case for the kind of inquiry involved in a legal process? It is basically historical, at least in the sense that its methods are more akin to the methods of historical sciences than they are to the methods of empirical sciences like physics or biology. Now, do simulations, either of kind (B) or (C), provide historical causal explanations capable of resolving a legal case?

I am skeptical about that, because simulations basically concern kinds of events, while a legal process basically concerns singular events[8].

Let me consider the case of Judicial Experiment. In our legal domain, what we call cognitive fictions fall under the notion of judicial experiment. It is a broad notion indeed. If we look at the legal doctrine, we learn that it covers artificial reproductions, artificial experiences, artificial representations, simulations, of the events at issue. In principle it can consist in a determinist simulation as well, but in most cases in consists in rule-based simulations.

Articles 218 and 219 of the Italian Code of Criminal Procedure concern the so-called Judicial Experiment (*Esperimento giudiziale*). Article 218 states its presuppositions. The first paragraph runs as follows (my translation): "The judicial experiment is admitted when it is needed to ascertain whether a fact has or might have happened in a certain way." The second paragraph says: "The experiment consists in the reproduction, as far as possible, of the situation in which the fact is said or believed to have happened, and in the repetition of the modalities in which the fact itself occurred."

Article 219 states the modalities of the judicial experiment itself. It is the business of the judge to fix them. In particular, a specification of the day, time and place of the experiment is required. But of course with computer simulations these are not the relevant specifications. What is important – I would say better, what is essential – is the specification of the hypotheses, laws, rules or regularities embodied in the artefact. And even in that case, namely in the case in which hypotheses, laws, rules or regularities are made explicit, the simulation does not necessarily provide conclusive evidence.

Suppose that subject S was killed and subject T is supposed to have

---

[8]Simulations "do not deliver any explanations of a single real world event, but they do lead to understanding how certain types of events are possible in principle." ([Hegselmann et al., 1996, vii])

killed him. Then suppose that the killing occurred in place A and T declares he was in place B ten minutes before the killing. Suppose also that some testimonies state that subject T was indeed in place B ten minutes before the killing. Now, a judicial experiment could be set in order to ascertain whether it is possible to reach place A from place B in ten minutes, given certain conditions of traffic, weather and the like. Provided we know which were those relevant conditions at the time of the killing, a computer simulation could do the job much better than an old-style reconstruction (with real cars, people, sunny or rainy weather, and a subject going from B to A). What are the possible outcomes then? In case the simulation shows that it is not possible to reach A from B in ten minutes, it does count as evidence in favor of subject T. But in case it shows that is it possible, it does not count as evidence against him. It merely states the possibility of reaching place A from place B in ten minutes. In any event it says nothing about the killing of subject S.

Briefly, the point is that simulations are of great help in studying *kinds* of events and their relations. They are of minor help in studying *singular* events. Our simulation says whether or not it is possible to reach A from B in ten minutes given certain conditions. It does not say whether subject T killed subject S, even though it can exclude an event of the like.

To resume. A cognitive fiction like a simulation can provide conclusive evidence against a hypothesis on a singular event in case it falsifies the hypothesis. It cannot provide conclusive evidence in favor of such a hypothesis. (At least, we might suppose that it could provide supporting evidence in favor of it, but not conclusive evidence). Nevertheless, given its rhetorical impact, when used without due caution and epistemological scrutiny (which is often the case in courtroom and in particular in legal systems with juries[9]) it may be taken as a reliable device giving conclusive evidence also in favor of a hypothesis on a singular event. And that would be a completely groundless claim from an epistemological point of view and a step towards injustice from a legal point of view, since it might lead to the conviction of an innocent person.

So, all things considered, devices like cognitive fictions can be of great importance for our cognitive tasks but must be used with due caution. Moreover, a specific discipline of their forensic use and their probative value is needed in order to avoid error and injustice based on epistemological confusions about the outcome of a cognitive fiction.

---

[9]Cf. [Thagard, 2003] on O.J. Simpson's case: emotions have a strong influence on juror's inferences (but emotional coherence, Thagard claims, is not arbitrariness nor wishful thinking: it has a cognitive component).

BIBLIOGRAPHY

[Casti, 1997] J. Casti. *Would be Worlds: How Simulation is Changing the Frontiers of Science*. John Wiley, New York, 1997.

[Conte et al., 1997] R. Conte, R. Hegselmann, and P. Terna. *Simulating Social Phenomena*. Springer, Berlin, 1997.

[Ferrajoli, 2000] L. Ferrajoli. Il giudizio penale. In S. Nicosia, editor, *Il giudizio*. Carocci, Roma, 2000.

[Ferrua, 1997] P. Ferrua. *Studi sul processo penale*, volume III. Giappichelli, Torino, 1997.

[Fuller, 1967] L.L. Fuller. *Legal Fictions*. Stanford University Press, Stanford, 1967.

[Harman, 1965] G. Harman. The inference to the best explanation. *The Philosophical Review*, LXXIV:88–95, 1965.

[Hegselmann et al., 1996] R. Hegselmann, U. Mueller, and K.G. Troitzsch, editors. *Modelling and Simulation in the Social Sciences from the Philosophy of Science Point of View*. Kluwer Academic Publishers, New York, 1996.

[Jackson, 1988] B. Jackson. *Law, Fact and Narrative Coherence*. Deborah Charles, Chicago, 1988.

[Josephson and Josephson, 1994] J.R. Josephson and S.G. Josephson, editors. *Abductive Inference*. Cambridge University Press, Cambridge, 1994.

[Kuipers, 1994] B. Kuipers. *Qualitative Reasoning: Modeling and Simulation with Incomplete Knowledge*. The MIT Press, Cambridge (Mass.), 1994.

[Lipton, 1991] P. Lipton. *Inference to the Best Explanation*. Routledge, London, 1991.

[MacCormick, 1995] N. MacCormick. *Legal Reasoning and Legal Theory*. Clarendon Press, Oxford, 1995.

[Magnani, 2001] L. Magnani. *Abduction, Reason, and Science*. Kluwer Academic/Plenum Publishers, New York, 2001.

[Nersessian, 1999] N.J. Nersessian. Model-based reasoning in conceptual change. In L. Magnani, N.J. Nersessian, and P. Thagard, editors, *Model-Based Reasoning in Scientific Discovery*. Kluwer Academic/Plenum Publishers, New York, 1999.

[Parisi, 2001] D. Parisi. *Simulazioni*. Il Mulino, Bologna, 2001.

[Pastore, 1996] B. Pastore. *Giudizio, prova, ragion pratica*. Giuffrè, Milano, 1996.

[Putnam, 2002] H. Putnam. *The Collapse of the Fact/Value Dichotomy and Other Essays*. Harvard University Press, Cambridge (Mass.), 2002.

[Sperber and Wilson, 1988] D. Sperber and D. Wilson. *Relevance. Communication and Cognition*. Harvard University Press, Cambridge (Mass.), 1988.

[Taruffo, 1992] M. Taruffo. *La prova dei fatti giuridici*. Giuffrè, Milano, 1992.

[Thagard, 1988] P.R. Thagard. *Computational Philosophy of Science*. The MIT Press, Cambridge (Mass.), 1988.

[Thagard, 1992] P.R. Thagard. *Conceptual Revolutions*. Princeton University Press, Princeton, 1992.

[Thagard, 2000] P.R. Thagard. *Coherence in Thought and Action*. The MIT Press, Cambridge (Mass.), 2000.

[Thagard, 2003] P.R. Thagard. Why wasn't o.j. convicted? Emotional coherence in legal inference. *Emotion and Cognition*, 17:361–383, 2003.

[Tuzet, 2004a] G. Tuzet. Abduction in legal reasoning. *Semiotiche*, 2:79–90, 2004.

[Tuzet, 2004b] G. Tuzet. Abduzione: quattro usi sociologico-giuridici. *Sociologia del diritto*, 31:117–131, 2004.

[Tuzet, 2004c] G. Tuzet. Le prove dell'abduzione. *Diritto e Questioni Pubbliche*, 4:275–295, 2004.

Giovanni Tuzet
University of Ferrara
Ferrara (Italy)
Email: `giovanni.tuzet@unife.it`

Cognitive Design Principles: from Cognitive Models to Computer Models

BARBARA TVERSKY, MANEESH AGRAWALA, JULIE HEISER,
PAUL LEE, PAT HANRAHAN, DOANTAM PHAN,
CHRIS STOLTE, AND MARIE-PAULE DANIEL

ABSTRACT. Visualizations are everywhere, on signs and billboards, in newspapers and texts, informative or instructive. Designing effective visualizations is a challenge. We have developed a collaboration of cognitive and computer science for uncovering and instantiating cognitive design principles to generate visualizations automatically. The program is iterative, using psychological research to uncover mental representations of the content to be communicated as well as interpretable visual devices for conveying that content and computer graphic techniques and knowledge to produce visualizations. We illustrate the program by describing projects developing effective route maps and assembly instructions.

1 Foreword

Model is a word replete with meaning. There are models in the mind, coherent and complete ones, and sketchy ones. There are models in the world, some coherent and complete, some sketchy. Whether in the mind or in the world, models are meant to check, capture, explore, and communicate ideas for one's self and for others. Here we review a project in which uncovering models in the mind informs creating effective models in the world.

Before there were written languages, there were visualizations, painted on caves, inscribed in stone, carved on wood. Visualizations of things that are actually visible, such as maps, building plans, people, and wildlife, are ancient and widespread. Visualizations of things that are metaphorically visual, like graphs of economic data or charts of organizations, are a relatively new invention, first appearing in Europe in the late 18th c [Beniger and Robyn, 1978]. Like written language, visualizations are cognitive tools, designed to augment the capacity of the human mind (see [Tversky, 2001] and [Tversky, 2005] for further discussion). Then, as now, they serve myriad purposes, to record information, to lighten the burden of working memory, to convey information to others, to promote discovery, inference, and insight, to facilitate collaboration. Primary among their roles is communication. Some visualizations, notably maps, have undergone years of informal user testing as communication tools. The consequent refinements provide suggestions for good design, and the process of refinement provides suggestions for uncovering design principles. Other visualizations are the latest products of the latest computer tools, often in need of refinement.

Now, visualizations seem to be everywhere, in automobiles, newspapers, airports, textbooks, and instructions. As users know all too well, many of these visualizations are frustrating; they are complex and cluttered, often with extraneous and distracting decorative details. Like all communication, to be effective, visualizations should schematize effectively, that is, they should extract, emphasize, and even distort the information that is important to the task and eliminate the information that is not. Take maps as an example. An aerial photograph, despite its' photorealism, is not an effective road map. It portrays detail that is not only irrelevant to the road structure but hides it. An effective map for driving emphasizes roads, intersections, and other features that aid navigation. In fact, road maps exaggerate the size of roads, so that they can be seen in the maps. Maps serve purposes other than driving. A good map for hiking will display other features such as trails and topography. A tourist map may deliberately mix perspective, to show an overview of the streets to allow people to find their ways and frontal views of the tourist sites, to allow people to recognize them. In order to facilitate perception and comprehension, maps may violate certain spatial relations, for example, exact distance, size, angle, and perspective.

In the best cases, visualizations have developed in a community of users who produce and comprehend them, refining them and fine-tuning them to suit the circumstances through casual user testing in the wild. This informal process can be systematized in the laboratory. Participants can be asked to produce visualizations; characteristics of those visualizations can be extracted. These characteristics can be systematically varied, and tested for comprehension in other participants. What is important and what is effective in visualizations depends on cognition, on how humans represent the task at hand and how they perceive and represent the information presented in visualizations.

Creating effective visualizations, as found in instructions, textbooks and other media, requires collaboration between graphic designers and domain experts, as well as testing the target audience. The domain experts know the content to be communicated; the designers know techniques for conveying content. In practice however, tightly intertwined collaboration between designers and domain experts is not always possible. Even when such collaboration is possible, it is not ideal, as designers may be educated in design, but not in the discipline and domain experts may be educated in the discipline, but not in general design principles. In addition, designing is a time-consuming, labor-intensive process that cannot keep up with the increasing demand. Try to imagine, for example, the number of maps that are downloaded daily from websites. Currently computers save labor, but primarily by replacing only low-level tools such as pens, brushes, ink, paint, and paper. To keep up with demand, it is essential that computers provide higher-level design tools that can make it easier to quickly produce effective visualizations

We have undertaken a novel use for computers, to instantiate cognitive design principles in algorithms that automate the process of generating effective visualizations. Our approach combines research in cognitive and computer science in three iterating steps:

1. Revealing the mental representations people have for a given domain and the visual devices they use to convey it, yielding domain cognitive design principles

2. Development of algorithms that create effective visualizations based on the domain cognitive design principles.

3. Testing the visualizations to insure that they adequately convey the desired information.

Below, we consider the issues involved this three-step approach and we illustrate the approach in two domains we have worked on, route maps and assembly instructions.

2 Approach

General cognitive design principles

In summarizing a large number of studies comparing static and animated visualizations for teaching a broad range of concepts, we suggested that effective visualizations conform to two general principles [Tversky *et al.*, 2002]. According to the *Principle of Congruence*, the structure and content of a visualization should correspond to the structure and content of the desired mental representation. According to the *Principle of Apprehension*, the structure and content of a visualization should be readily and accurately perceived and comprehended. To illustrate the depth of these principles, consider animations, increasingly popular as tools for creating them become available. Many animated diagrams, such as those showing how the heart works or how to operate equipment (see http://www.howstuffworks.com/at-0.htm for examples) at first appear to fulfill the congruence requirement in that they use change over time to convey change over time. Dozens of experiments, however, have failed to show benefits of animations over equivalent still diagrams in conveying information, from animations illustrating how a bicycle pump works to those illustrating how computer algorithms work. This is surprising, given the premises that animations use change in time to convey change in time, and the conclusion arouses controversy. Some of the failures of animation may be because the animations violate the Apprehension Principle. They are too complex or too rapid to be accurately perceived and conceived and, unlike static graphics, they cannot be re-inspected at the viewers' own pace. However, animations also seem to violate the Congruence Principle. Although events in the world are continuous, they are typically understood as a sequence of discrete steps (e.g., [Zacks et al., 2001]). The steps are the joints, the transitions from action to action. The joints of events performed by hands are often objects or object parts; those performed by feet are turns at landmarks [Tversky et al., 2004]. The joints of events do not come at regular temporal intervals. If people conceive of animated events as sequences of discrete steps, then it may be more effective to visualize events in steps rather than requiring the user to do the segmentation. A sequence of stills for example, may actually provide a more compatible cognitive match than an animation [Tversky, 2005]. A sequence of stills allows viewers to directly compare the state of the system at each important step.

Individualizing visualizations

The utility of many visualizations depends on their adaptability. For example, with maps different schematizations of the same environment are desirable depending on the task and the user. Hikers, bikers, drivers, travelers, and surveyors all need different information just as those unfamiliar with an environment need different information from those familiar with it. Hand-designed maps can take such needs into account. But creating effective individualized maps by hand for every user in every situation would be too expensive and time-consuming to be practical. Even within the domain of route maps for drivers, the number of possible routes that drivers may want is inconceivably large.

Our vision is to create individualized visualizations automatically, using computer algorithms that instantiate cognitive design principles. To do this we require methods for uncovering cognitive design principles and methods for incorporating these principles into computer algorithms. We describe both, using examples from two domains representative of large classes of visualizations, maps and assembly instructions. Maps are one of the most ancient and pervasive of visualizations, whether drawn on paper or carved in wood or incised in stone (e.g. [Brown, 1979]). Assembly instructions are representative of a large class of visualizations that includes instructions on how to put something together, how to operate a complex system as well as how complex systems, from hearts to corporate structures function. Underlying such visualizations are the parts of the system and their spatial, temporal, or functional relations. Visualizations of systems, too, are ancient; frescoes in Egyptian tombs show how crops are grown and harvested. Within each domain, we have chosen to explore and develop examples likely to be familiar to the general public, for maps, route maps and for systems, assembly instructions.

Revealing cognitive design principles

To make visualizations that are congruent with the desired mental representations requires techniques that reveal those internal mental representations. Cognitive psychology has a large bag of tricks for externalizing the internal. Reaction times are commonly used to this end. The reasoning typically goes: if the mental representation has a certain format, then responses for retrieval tasks congruent with that format should be faster and more accurate than responses for formats that are not congruent with the format, as they require more onerous and time-consuming mental transformations. For example, maps that are not oriented in the direction of travel must be mentally transformed to correspond with the navigator's current orientation. Therefore people are generally slower and less accurate when using such maps. Similarly, certain patterns of error, of grouping, and of description follow from certain types of mental representations, elucidating them. Returning to maps, people draw streets that are not parallel as parallel, suggesting that that is how they think about them (e.g. [Tversky, 1981]).

For complex mental representations, more open-ended techniques may be more revealing. One that we have adopted and will describe here is to ask participants to construct descriptions and depictions for a given domain. This captures the natural way that communication occurs, and provides us with a rich set of data. Analyzing the structure that is common to both descriptions and depictions provides the commonalities and differences between them. The common structure reveals the underlying representation. The features particular to descriptions and depictions provide insights into the design of visualizations, as visualizations are typically accompanied by language. These insights do not fully determine the visualizations. Guidelines must be drawn from other sources as well, and then tested by users in order to qualify as design principles. Comprehension usually goes beyond production, from babies learning to speak to adults using diagrams, so production sets a lower limit. Comprehension tasks, then, complement and supplement production tasks.

Creating computer algorithms

Creating effective visualizations entails numerous design decisions. Here are a few. For a visualization of a route, what landmarks should be included, and where should they be included? How much can angle and distance be distorted without confusing the user? For a visualization of a process that extends in time, how should the process be segmented into steps? How should the steps be ordered? For each step, what is the best view or perspective to show? What detail should be included, and what omitted? What details need to be distorted, or shown as insets? What extra-pictorial features need to be added, features such as lines, arrows, and highlighting? The cognitive research will inform these decisions, but they do not completely determine the algorithm. The cognitive principles may be in the form of trade-offs, or simply insufficient, so some aspects of algorithm generation may still rely on the educated sensibilities of the designer. User testing can help overcome these shortcomings, but it may never be possible to test all the possible design variants. Our approach is to build tools that are based on the cognitive design principles, but also allow users to override the automated decisions as necessary. Many of the design issues that arise are not unique to the particular examples we have chosen, so that their solutions will have generality to other domains.

3 Route maps

Revealing cognitive design principles for route maps

The map someone sketches to show a friend how to get to a party doesn't usually resemble a map from the USGS or Rand McNally. Nonetheless, such sketch maps have been used for hundreds of years, presumably with success. Sketch maps differ from the efforts of geographers in several ways. To find out how, [Tversky and Lee, 1998] approached students near a campus dorm, asking them if they knew the way to a popular fast food place. If they did, they were asked to either sketch a map or write down directions to the restaurant. Typical instructions appear in Table 1, and a typical sketch map in Figure 1. The maps shared a number of characteristics. They had an infrastructure of lines formed into paths and turns. The paths included, in this case, roads, were primarily those that comprised the route. Other paths intersecting the route were included when they were useful for keeping the traveler on track, for example, intersections just before or after a turn. Paths were simplified, primarily to straight and curved lines. Turns were also simplified, to approximately 90 degrees. Distances were altered; long distances with little action were shortened and short distances with many turns were enlarged. Note that these simplifications and alterations are actually distortions that violate the metric properties of the environment. However the topology of the sketched routes – the points of intersection between the roads – corresponded to the true topology of the routes in the real environment. Landmarks were included when they were important for turns or for keeping on track; they were typically expressed as names of streets or blobs representing structures and usually labeled. The text route descriptions shared many of the same properties. Exact angles of turns and distances were typically absent. The shape of paths was dichotomized just as in the depictions; for straight paths, informants wrote, "go down" and for curved ones, they wrote, "follow around". Despite these simplifications, omissions, and distortions, such schematized depictions and descriptions are usually sufficient to arrive at the destination because the environment provides the missing information about angles of turns, shapes of roads, and distances. What is essential is the sequence of paths and turns. Visualizations are used in contexts, and the contexts can provide missing information and disambiguate. Thus, the sequence of paths and nodes is the skeletal mental representation underlying both depictions and verbal descriptions of routes.

Of course, a highway map could be used for the same purpose. But a highway map, even one with the route marked on it, has disadvantages. It is cluttered with irrelevant information that makes finding the relevant more difficult. It has a single scale, so large portions of the map convey no information, and at the same time, discerning many small turns may be difficult. Because a highway map doesn't extract the information needed for a mental representation of a route, it demands considerable time-consuming processing before it can be useful.

4 Applying cognitive design principles to computer algorithms for route maps

Two basic design principles follow from the cognitive analysis:

1. People think of routes as sequences of paths and turns at landmarks and therefore the topology of the route (i.e. the turning points) must be depicted accurately.

2. People don't accurately apprehend or represent distances or angles, and therefore such geometric information can be simplified to increase emphasis on the turning points.

DW 9
From Roble parking lot
R onto Santa Theresa
L onto Lagunita (the first stop sign)
L onto Mayfield
L onto Campus drive East
R onto Bowdoin
L onto Stanford Ave.
R onto El Camino
Go down few miles. It's on the right.

BD 10
Go down street toward main campus (where most of the buildings are as opposed to where the fields are) make a right on the first real street (not an entrance to a dorm or anything else). Then make a left on the 2nd street you come to. There should be some buildings on your right (Flo Mo) and parking lot on your left. The street will make a sharp right. Stay on it. That puts you on Mayfield road. The first intersection after the turn will be at Campus drive. Turn left and stay on campus drive until you come to Galvez Street. Turn Right. Go down until you get to El Camino. Turn right (south) and Taco Bell is a few miles down on the right.

BD 3
Go out St. Theresa
Turn Rt.
Follow Campus Dr. way around to Galvez
Turn left on Galvez.
Turn right on El camino.
Go till you see Taco Bell on your Right

Table 1. Examples of route directions (from [Tversky and Lee, 1998]).

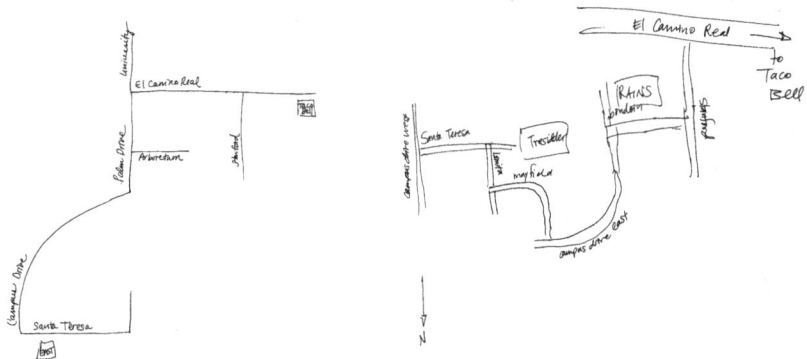

Figure 1. Two example sketch maps (from [Tversky and Lee, 1998]).

We have developed a system called LineDrive that automatically designs route maps for any given origin and destination based on these principles [Agrawala and Stolte, 2001]. LineDrive is responsible for choosing graphic attributes, such as, position, orientation, size, etc. for each of the graphic elements in the map, including roads, labels, cross-streets and landmarks. The space of possible map designs encompasses all possible choices of graphic attributes for each of the graphic elements. LineDrive uses search-based optimization over this large multi-dimensional space to find the map that best adheres to the design principles. The design principles are instantiated as layout constraints within the search-based optimization framework.

Because LineDrive maps are based on a cognitive model of how people think about routes they are far easier to follow while driving than standard highway maps (see Figure 2). LineDrive has been commercially deployed on large internet map service sites (see www.mapblast.com) and the maps been received enthusiastically by a large community of users.

5 Assembly instructions

Maps have been used to communicate spatial information within communities over the millennia, much like spoken language. As for spoken and written language, this provides a natural user-testing laboratory, where some construct maps and others use them, with greater or lesser success, an ongoing process that refines and improves. Instructions for assembly or operation have not undergone that refinement, and indeed are the frequent recipient of groans, complaints, and slogans. An informal, but wide-reaching survey of instructions for assembling or operating the sorts of things people

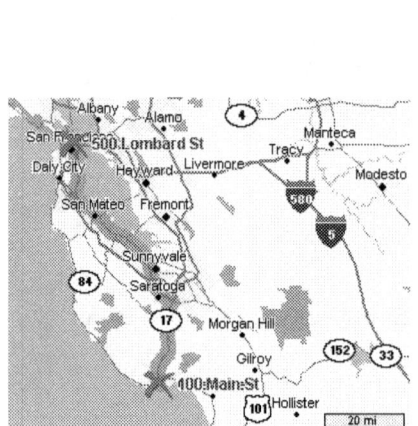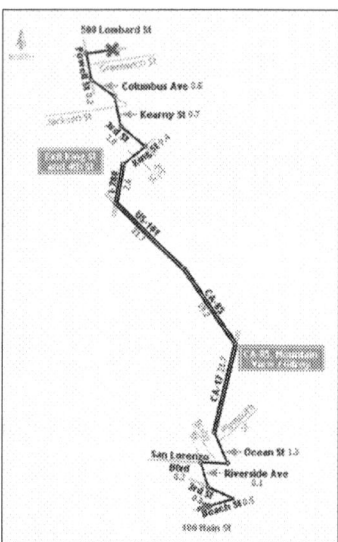

Figure 2. Highway map and LineDrive map for the same route (from [Agrawala, 1992]).

bring into their houses, furniture, cameras, cell phones, computers, and the now proverbial VCR, reveals a number of common difficulties (some of this collection appears in [Mijksenaar and Westendorp, 1999]). The typical visualization is an exploded view of the parts with lines and arrows. Such diagrams are usually at one scale and from one perspective, both of which may not be appropriate for all steps of assembly. They are frequently cluttered with so many parts and connections that it is difficult to discern particular components. They all too frequently show the entire assembly or operation at once, so that the sequence of assembly is not given. They often use extra-pictorial features such as arrows in multiple ways, with insufficient context to disambiguate meaning. As we shall see, even the instructions produced by student novices correct many of these problems. One notable exception to these common shortcomings of instruction is the widely admired instructions for Lego. Lego instructions are step-by-step; they also change perspective and scale when needed to show how to attach components.

Revealing cognitive design principles for assembly of objects

The cognitive structures underlying assembly are multiple: a mental model of the object to be assembled, a mental model of the actions required for assembly, and a model for ordering the actions. People think of objects as a hierarchy of parts [Tversky and Hemenway, 1984]. The parts segmented are those that are perceptually salient and functionally significant. In most cases, perceptual salience, that is, contour distinctiveness, and functional significance correlate, as in the wheels of an automobile or the handle of a pump or the legs of a chair. The correspondence between perceptual salience and functional significance promotes inferences from structure to function, especially when the form of the part suggests function, as in wheels, handles, and legs. Thus, objects though sometimes seamless, are nevertheless perceived as consisting of distinct parts, segmented by appearance and by function or behavior. The same holds for actions, such as making a bed or assembling an object. Though typically continuous in time, actions are thought of as a sequence as of discrete steps, distinct in both perceived action and conceived function. Goal-directed action sequences such as assembly are also conceived of hierarchically, with the higher level segmented by actions on separate objects or significant object parts and the lower level segmented by finely articulated actions on the same object or object part [Zacks *et al.*, 2001].

To reveal the mental representations underlying assembly and to reveal graphic preferences at the same time, we followed the same general strategy for uncovering cognitive design principles as we had for route maps. Heiser, Daniel-Ginet, and Tversky (in preparation) asked students to assemble a TV cart using the picture of the assembled cart on the package as a guide (see Figure 3).

After assembling the cart, the students were asked to construct instructions for assembling it under one of four conditions: 1) Use sketches and language to create instructions so someone else can easily assemble the TV cart; 2) Use sketches and language, but confine yourself to short, concise instructions; 3) Use only language; 4) Use only sketches. This is the Instruction Production Study. As expected, the steps corresponded to the major object parts, yielding 5 steps. Of the many possible sequences, participants primarily used two, corresponding to mechanical ease of assembly. The students had been divided by a median split into high and low spatial ability on the basis of spatial tests of mental rotation [Vandenberg and Kuse, 1978] and perspective-taking [Money and Alexander, 1966]. There were vast differences in the sketches produced by high and low ability participants, which we describe below. In a second study, the Instruction Rating Study, a subset of instructions from the production study were selected to span a range of sketch manners and techniques. These were given to a new group of participants, as before, split into high and low ability, who first assembled the TV cart using one of the sets of instructions and then rated the instructions

Figure 3. Participants used the photo of the fully assembled TV cart on the box (to the left) to assemble the parts (to the right) (from Heiser, Daniel-Ginet, and Tversky, in preparation).

for quality. As noted, there were large differences in the sketches produced by high and low spatial ability participants. Interestingly, the more sophisticated techniques used in the sketches produced by those with high ability were exactly those preferred by participants of all ability levels in the rating study.

What were the diagrammatic techniques that the high ability participants used and that received high ratings?

Action sketches

Those high in ability produced more *action* sketches than *structural sketches* whereas low ability participants produced relatively more structural sketches (see Figure 4). As the labels imply, an action sketch shows the parts as well as how they are assembled whereas a structural sketch shows only the spatial relations among the parts. For assembly of objects, the action sketches are generally superior to structural sketches because they contain all the information in structural sketches and in addition show the actions required to assemble, without introducing to much additional visual.

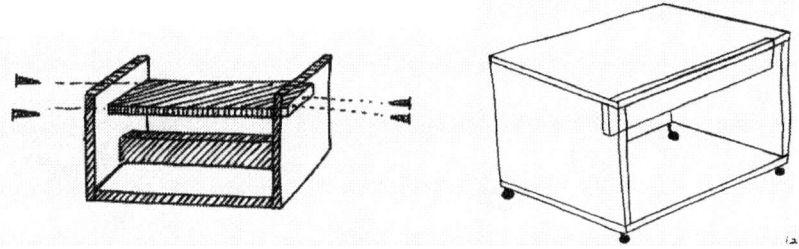

Figure 4. Examples of an action sketch and a structural sketch. The action sketch is easier to follow because it shows the assembly action that must be performed. In this case the assembly action is screwing in the shelf.

Step-by-step sketches

The high ability participants produced more step-by-step sketches, reflecting the hierarchical organization of assembly.

3-D perspective

The sketches of high ability participants showed the assembly from a 3-D perspective that made the assembly operations visible.

Morphograms

Highly rated sketches produced by high ability participants used extrapictorial devices that we have termed *morphograms* [Tversky et al., 2002], notably, guide lines and arrows, to show points, direction, and mode of attachment and sequence of steps. Morphograms are a class of simple geometric elements used in diagrams, such as lines, arrows, crosses, and boxes, that are readily interpreted partly through their Gestalt or mathematical characteristics and partly through context [Tversky et al., 2000].

Effective integration of text with diagrams

In highly rated sketches, the text referred to the diagrams, elucidating each assembly step. They also supplied caveats and other information not readily available from the diagrams.

Figure 5 shows the instructions produced by one high ability participant, with some of the features characterizing good instructions highlighted.

Just as in our study of route instructions, there were striking parallels between the depictions and the text descriptions. Nearly half of the statements described actions, such as "Attach other side panel to the open side". Another quarter described the components to be assembled, for example,

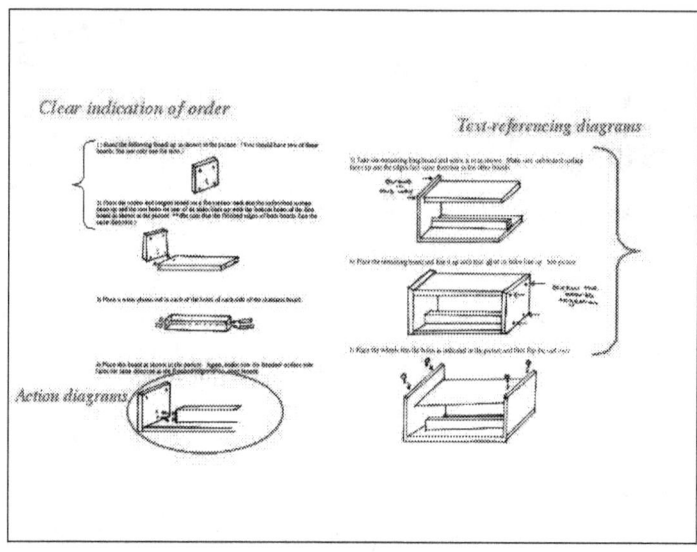

Figure 5. Instructions produced by one high ability participant, with some of the features characterizing good instructions highlighted.

"The top piece is rectangular". The remaining statements were commentaries and caveats, statements that began "remember that" or "make sure that". Under the concise condition, the proportion of action statements increased and the proportion of descriptions of components and commentaries decreased, suggesting that users believed that the assembly actions, the sequence of steps, were the critical information to be conveyed. In addition, the descriptions were hierarchical; that is, the higher-level statements were retained in the concise descriptions at the expense of the lower-level statements.

A third study, the Instruction Evaluation Study, tested whether the instructions that received high ratings were more effective. A new group of participants, again split by spatial ability, used instructions varying in rated quality to assemble the TV stand. Those high in ability assembled the stand faster and with fewer errors. For them, the quality of instructions made no difference. For those low in spatial ability the instructions rated highly led to better performance than those with low ratings, indicating that people's intuitions about this type of instruction are calibrated. Recall that the participants in these studies were students in a highly selective university, so that the low ability students are probably more representative of the population at large than the high ability students. These studies are ongoing, the analyses are by no means complete, and we expect further insights to emerge as we continue.

These experiments provide cognitive guidelines for the construction of computer algorithms for assembly. They inform how we can segment a set of assembly actions into steps and sub-steps, based on the structure of the objects and actions. They also inform how to produce diagrammatic visualizations: we should use step-by-step action diagrams using 3-D views that show how to make the connections, enriched by lines and arrows. But they do not yet answer all of the design issues.

6 Applying cognitive principles to computer algorithms for assembly of objects

From the results of these experiments some design principles are already clear:

1. People prefer step-by-step instructions in which each step shows how one major part is attached rather than a single diagram showing all the assembly steps at once.

2. Each new part added to the assembly must be clearly visible in each step of the instructions.

3. The instructions must show how the new parts attach to the other parts in the assembly.

4. Action sketches separate the new parts from the old parts and use arrows and guidelines to show how the new parts should attach. Thus action sketches explicitly show the operations required to perform each attachment and are preferable to structural sketches which only show the parts included in each step.

5. In each step the object may be oriented in either its most common real-world orientation (i.e. legs of table should point down and table top should appear horizontal) or in the orientation optimal for that assembly step. Some assemblies may become unstable or unbalanced as parts are attached. In such cases the instructions may orient the object so that it would remain stable against a ground plane

We have developed an automated assembly instruction design system based on these principles [Agrawala et al., 2003]. As input our system requires a geometric model of the assembly with each part in its final assembled configuration. Users can optionally specify additional semantic information about the model, by labeling the parts. In this manner users can specify symmetries between parts and divide parts into fasteners and fastened parts.

Our system divides the task of generating assembly instructions into two phases; *assembly planning* and *diagram production*. In the assembly planning phase the system analyzes the geometry of the model to compute the sequence of steps required to build the object. Assembly planning algorithms have been well-studied in robotics [Romney *et al.*, 1995] [Wilson, 1992]. These algorithms first compute all possible disassembly sequences for the model by analyzing which parts block other parts This analysis yields a directed acyclic graph encoding the removability constraints on each part and every valid topological sort of this removability graph produces a geometrically valid assembly sequence for the object.

The goal of the prior robotics algorithms was to plan a sequence of instructions that a machine-tool could use to build the object. Each valid assembly sequence is evaluated for the particular machine-tool it will be implemented on and the "best" sequence is chosen as the final assembly plan. Since these robotics techniques are designed to produce an assembly plan for a robotic machine-tool, they never need to produce a visual representation of the instructions and therefore do not consider the requirements of a human builder. As a result the assembly sequence produced by such systems can be difficult for humans to follow. Our system computes the set of geometrically valid assembly sequences using the removability analysis approach. However, unlike the prior systems, we evaluate the resulting sequences based on the design principles outlined above to produce step-by-step assembly sequences that are well-designed for humans. For example, visibility in the instructions affects decisions about assembly order.

Once we have computed the assembly plan we must generate a diagram showing each step in the plan. We can generate structural sketches by simply showing the set of parts added in each step in their final assembled positions. Our system ensures that all of the parts added in each step and their points of attachment will be visible in each step. This approach yields instructions that are similar to those included with Lego and we are able to produce assembly sequences that are easy for humans to use and follow. The assembly planning algorithm also determines the set of directions in which the new parts can be moved in to attach to the previous parts. We have recently extended our system to use this information to produce action sketches (see Figure 6).

The new parts are placed a short distance away from the previous parts and guidelines and arrows are added to indicate the motions of the parts as well as the points of attachment.

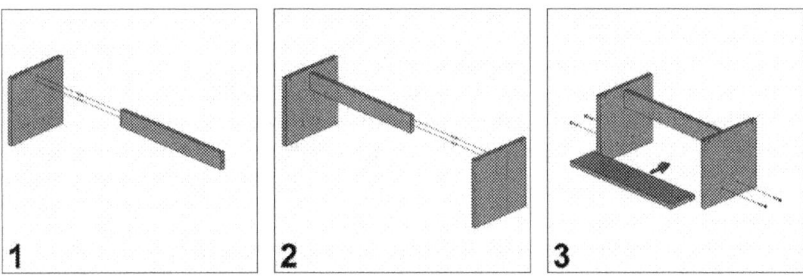

Figure 6. Instructions for assembling the TV stand as generated by our system. Only the first three steps are shown.

Testing the computer-generated instructions

We compared the instructions generated by computer to those that came in the box, with the TV cart, and to the best hand-drawn instructions [Heiser et al., 2004]. The instructions in the box consisted of a menu of parts, an exploded diagram of the TV cart with guidelines and arrows indicating attachment, and an enlargement showing attachment specifics. Because the TV cart has relatively few parts and attachment operations, these are not bad instructions. Nevertheless, the computer-generated instructions won hands-down. Ten participants assembled the TV cart using each set of instructions. Those using the computer-generated instructions made significantly fewer errors than those using either of the other sets of instructions while assembling 30% faster. They also reported greater satisfaction with the computer-generated instructions on a number of measures.

7 Structure of visualizations

Route maps and assembly instructions belong to a larger class of visualizations that are used in many contexts and for many purposes, practical, educational, and aesthetic. The design principles developed for route maps and assembly instructions have broader application to other visualizations as well as providing guidelines for establishing design principles in other visualization domains. In particular, many biological, physical, and conceptual systems have parts in a spatial, temporal, or abstract structure. The design principles developed for routes and assembly suggest that visualizations of these systems should clearly delineate the parts and their relations. Many of the systems have action or change; the design principles developed for assembly suggest that the changes can be conveyed through the use of extra-pictorial features, especially arrows and guidelines.

The program for generating more specific design principles generalizes to other domains. The program entails first revealing the cognitive structures underlying understanding. Asking knowledgeable participants to produce descriptions and visualizations of the systems is one way to reveal cognitive structures. The structure common to both descriptions and depictions represents the cognitive structure underlying understanding. The descriptions and depictions also suggest verbal and pictorial devices for conveying the concepts. Together, these can be used to develop cognitive design principles. The second step is to instantiate the design principles into algorithms that generate visualizations for a large class of instances. Then the effectiveness of the design principles is tested by comparing learning with the computer-generated visualizations to learning with standard visualizations. This program is applicable to myriad visualizations.

Visualizations use space to convey meanings that are spatial or metaphorically spatial. For the most part, proximity in graphic space is used to convey proximity in real or metaphoric space, preserving information at the level of category, order, interval, or ratio.

Icons and figures of depiction

Visualizations use elements as well as space to convey meaning. The simplest and most direct kind of element is an icon, where the element bears resemblance to the thing it represents. These are as old as ideographic languages, where schematic animals and edibles represented their real-world counterparts, and as new as the latest computer or Olympics icons. But many useful concepts cannot be readily depicted. Figures of depiction have been spontaneously adopted, again since ancient times. Synechdoche, where a part represents a whole, is common, as in the horns or head of a sheep to stand for sheep. Similarly, metonomy, where an entity associated with a concept stands for the concept, as in a crown for a king or scales for justice or scissors for delete. The same devices, of course, appear in figures of speech. Icons that are related to the things they represent by figures of depiction also appeared in ancient scripts and appear in contemporary machinery. The advantage of icons and figures of depiction is that their meanings are readily understood and remembered.

Morphograms

There is another kind of element that is prevalent across a wide range of graphics and that is readily understood in context [Tversky et al., 2002]. Lines, crosses, arrows, and blobs are simple, schematic geometric figures that are an integral component of many kinds of graphics, maps, graphs, and mechanical diagrams for examples. Their meanings are related to the their geometric or Gestalt properties, but are context-dependent. Lines, for example, connect, they serve as paths from one point to another, suggesting a relationship between the points. Crosses are intersections of lines. And arrows are asymmetric lines, suggesting an asymmetric relationship. Blobs are two-dimensional, suggesting an area. Their amorphous shape suggests that shape is irrelevant. Like words in language, morphograms can be combined in various ways to create varying meanings. Like words in language, there are constraints on how they can be combined. Finally, like words such as *relation, intersection,* and *field*—words corresponding to lines, crosses, and blobs, morphograms are rich in possible senses, which context specifies.

Visualizations as cognitive tools

Visualizations are a cognitive tool, designed to augment human cognitive capacities [Tversky, 2001]. They aid memory by off-loading it from limited working memory or fallible long-term memory. By off-loading, visualizations relieve working memory of its' memory functions, leaving more capacity for processing. Visualizations, for example, doing calculations on paper, can also augment information processing. Visualizations can be explored and altered, promoting inference and insight. They are public, and can be examined and revised collectively, insuring common ground.

Acknowledgements

Portions of the research were supported by Office of Naval Research Grants NOOO14-PP-1-O649, N00014011071, and N000140210534 to Stanford University. The paper has originally been published in G. Allen, ed., *Applied Spatial Cognition,* Erlbaum, Mahwah, NJ, 2006. Reprinted by permission.

BIBLIOGRAPHY

[Agrawala and Stolte, 2001] M. Agrawala and C. Stolte. Rendering effective route maps: Improving usability through generalization. In *Proceedings of SIGGRAPH '01*, pages 241–250, 2001.

[Agrawala et al., 2003] M. Agrawala, D. Phan, J. Heiser, J. Haymaker, J. Klingner, P. Hanrahan, and B. Tversky. Designing effective step-by-step assembly instructions. In *Proc. Siggraph 2003, ACM Transactions on Graphics*, pages 828–837, 2003.

[Agrawala, 1992] M. Agrawala. *Visualizing route maps.* PhD thesis, Department of Computer Science, Stanford University, 1992.

[Beniger and Robyn, 1978] J.R. Beniger and D.L. Robyn. Quantitative graphics in statistics. *The American Statistician*, 32:1–11, 1978.

[Brown, 1979] L. Brown. *The story of maps.* Dover, NY, 1979.

[Heiser et al., 2004] J. Heiser, D. Phan, M. Agrawala, B. Tversky, and P. Hanrahan. Identification and validation of cognitive design principles for automated generation of assembly instructions. In *Proceedings of Advanced Visual Interfaces '04, ACM*, 2004.

[Mijksenaar and Westendorp, 1999] P. Mijksenaar and P. Westendorp. *Open here: the art of instructional design.* Thames and Hudson, London, 1999.

[Money and Alexander, 1966] J. Money and D. Alexander. Turner's syndrome: Further demonstrations of the presence of specific cognitional deficiencies. *Journal of Medical Genetics*, 3:47–48, 1966.

[Romney et al., 1995] B. Romney, C. Godard, M. Goldwasser, and G. Ramkumar. An efficient system for geometric assembly sequence generation and evaluation. In *Proc. ASME International Computers in Engineering Conference*, pages 699–712, Boston, 1995.

[Tversky and Hemenway, 1984] B. Tversky and K. Hemenway. Objects, parts, and categories. *Journal of Experimental Psychology: General*, 113:169–193, 1984.

[Tversky and Lee, 1998] B. Tversky and P.U. Lee. How space structures language. In C. Freksa, C. Habel, and K.F. Wender, editors, *Spatial cognition: An interdisciplinary approach to representation and processing of spatial knowledge*, pages 157–175. Springer-Verlag, 1998.

[Tversky et al., 2000] B. Tversky, J. Zacks, P.U. Lee, and J. Heiser. Lines, blobs, crosses, and arrows: Diagrammatic communication with schematic figures. In M. Anderson, P. Cheng, and V. Haarslev, editors, *Theory and application of diagrams*, pages 221–230. Springer-Verlag, 2000.

[Tversky et al., 2002] B. Tversky, J.B. Morrison, and M. Betrancourt. Animation: Can it facilitate? *International Journal of Human Computer Systems*, 57:247–262, 2002.

[Tversky et al., 2004] B. Tversky, J.M. Zacks, and P. Lee. Events by hand and feet. *Spatial Cognition and Computation*, 4:5–14, 2004.

[Tversky, 1981] B. Tversky. Distortions in memory for maps. *Cognitive Psychology*, 13:407–433, 1981.

[Tversky, 2001] B. Tversky. Spatial schemas in depictions. In M. Gattis, editor, *Spatial schemas and abstract thought*, pages 79–111. MIT Press, 2001.

[Tversky, 2005] B. Tversky. Functional significance of visuospatial representations. In P. Shah and A. Miyake, editors, *Handbook of higher-level visuospatial thinking*. Cambridge University Press, 2005.

[Vandenberg and Kuse, 1978] S.G. Vandenberg and A.R. Kuse. Mental rotations. a group test of three-dimensional spatial visualization. *Perceptual Motor Skills*, 47:599–604, 1978.

[Wilson, 1992] R. Wilson. *On Geometric Assembly Planning*. PhD thesis, Department of Computer Science, Stanford University, 1992.

[Zacks et al., 2001] J. Zacks, B. Tversky, and G. Iyer. Perceiving, remembering and communicating structure in events. *Journal of Experimental Psychology: General*, 136:29–58, 2001.

Barbara Tversky
Stanford University, Stanford
Email: `bt@psych.stanford.edu`

Maneesh Agrawala
University of California, Berkeley
Email: `maneesh@cs.berkeley.edu`

Julie Heiser
Adobe Systems
Email: `jheiser@gmail.com`

Paul Lee
NASA-Ames
Email: plee@mail.arc.nasa.gov

Pat Hanrahan
Stanford University, Stanford
Email: hanrahan@cs.stanford.edu

Doantam Phan
Stanford University, Stanford
Email: dphan@stanford.edu

Chris Stolte
Tableau Software, Seattle
Email: chris@tableausoftware.com

Marie-Paule Daniel
Laboratoire d'Informatique Pour
La Mecanique et Les Sciences d L'Ingenieur (LIMSI)
Email: mpd@limsi.fr

Cognitive Complexity and the "Supports" of Modeling

Zhikang Wang

ABSTRACT. This paper advances the idea of how to understand the cognitive complexity by a theory of hierarchical structure of thinking systems. From three typical cases results indicate that scientists use many supports to solve the problems of rationality during the process of model-based reasoning. The supports of reasoning come from different consciousness-layers and support each other. The paper concludes with the view that scientific discovery is a complicated event. The complexity caused by interacting between different layers in thinking systems is the source of creation of new ideas and thought.

1 Introduction

Most scientific research is model-based. In the process of model-based reasoning, scientists have to solve the problems of rationality about their models, such as, "what is the basis of a model?" "Why select model A and not model B?" "How to choose a model?" "Where to get original forms of models?" The keys to these problems are called the "supports" of modeling. We would like to know how the problems are solved.

The philosophers have been asking the question "What is the foundation of knowledge?" and empiricism and rationalism have been arguing with each other about the rationality of knowledge for several centuries. Kant was conscious that the bases of empiricism or rationalism can't be replaced by one another; they are in different thinking levels. Because he was using subjectivism, Kant failed in his attempt to account for the contradiction between empiricism and rationalism. Later, phenomenologist E. Husserl, dialectics expert W.F. Hegel, as well as experts of the analytical school of thought, attempted to explain the bases of knowledge, from their own points of view. The characteristics of a whole system of cognition, as well as the pure process of acquiring knowledge tend to be ignored. It seems that the epistemology that we have now is one-sided and far from scientific practice. Reductionism is not suited for our analysis of model-based reasoning, and cognitive science needs a new theoretical framework.

This paper will adopt a new hypothesis with new concepts that come mainly from modern complexity science and its recent development, and based on examining typical cases from the history of the science.

2 A new theoretical framework for cognitive science

There are ten reasons that have caused us to consider thought as a complexity system: (1) until now the law of thought in general has not been explained by any theory, (2) the mechanism of thought can not be simulated in practice up to now, (3) we have a long way to go to reach our expected aim in artificial intelligence research, (4) the unanimous principle between matter structure and its function has met with serious setbacks in the study of brain science, (5) it seems the creative ability of a psychopath differs little from that of the scientist, (6) it is difficult for science and religion to have a conversation, (7) research of psychology deviates from empirical science, (8) to make philosophy into a science seems to be getting to be more difficult, (9) physiology plays a greater role in the research of thinking, (10) evidence indicates that the spatial-temporal location of thought is a false question.

The difficulty to understand and simulate the mechanism of thought suggests that thought is a system; hence, it is better to analyze it in terms of a theory of complexity, rather than that of simplicity. Under the view of traditional theory of simplicity, thought is considered as "thinker is thinking". The foundation of thought is just logic and image, or sometimes logic, sometimes image, and the thinking process is linear, plane and closed. By the theory of simplicity one can't explain scientific discovery unless there is the mechanism with which the basis of rationality could be provided from outside. The key word here is "base of rationality". No matter what situation we are in, logic reasoning or image reasoning, we have to have one or more non-logic and non-image supporting points beforehand. In other words, as K. Gödel showed in the "incompleteness theorem", no set of logical relations can be established that does not also imply the existence of still other relations with which the set itself cannot cope [Richards, 1983]. Gödels findings give us reasons for clearly considering thought as complexity system, and for understanding the inappropriate and limitations in analyzing thinking by traditional theory of simplicity.

Obviously, it is impossible to find the supporting point of rationality about thought in same administrative level if thought is defined as a simple system, in which there are no hierarchies. It seems that if the research is expected to conform to cognitive practice, a new theoretical framework has to be established. Human cognition is a complicated event and that cognitive science must absorb new ideas from modern complexity science.

2.1 Complexity

Kauffman advances a very useful idea, which is to descriptively analyze the complexity of systems, that a system can be viewed from a number of different perspectives, and that these perspectives may severally yield different non-isomorphic decompositions of the system into parts. Here is an application of his point: systems for which these different perspectives yield decomposition of the system into parts, whose boundaries are not spatially coincident, are properly regarded as more descriptively complex than systems whose decompositions under a set of perspectives are spatially coincident.

Another important method to analyze the complexity of systems was developed by Herbert Simon and others [Wimsatt, 1974], which is called the judgment of the interaction complexity of systems. Their main point is that many systems can be decomposed into subsystems for which the intra-systemic causal interactions are all stronger than the extra-systemic ones, and that under the concept of "near-complete decomposability", we can make sure whether the subsystems decomposed by S-decomposition and denoted by $\{S^i \in c\}$ cross boundaries between the different K-decompositions of a system. A system is interactively simple if none of the subsystems in $\{S^i \in c\}$ cross boundaries between the different K-decompositions of a system, and interactively complex in proportion to the extent to which they do.

A new idea is to understand the complexity of the world base on the idea of administrative levels, which originated from Kant and Engels and then penetrated into or was hidden in the creative work by L.V. Bertalanffy, C. Shannor, N. Wiener, W.R. Ashby, I. Prigogive, R. Thom, H. Haken, J.V. Neumann, H.A. Simon, and S.A. Kauffman. "What is complexity?" – Complexity is the leap of different layers (or levels) of system, the expression of the interrelation of striding cross layers, and the existing form of the objects which have more than two layers [Wang, 1990; Wang, 1991].

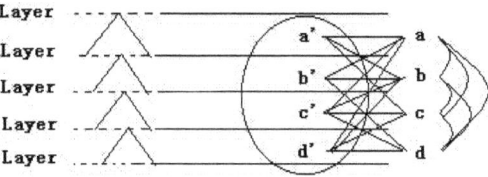

Figure 1. Layers and complexity.

2.2 Simplicity system and complexity system

Complexity rises from the understanding of the relation of "one and multiple", a pair of old conceptions as categories in science and philosophy. There are two basic forms of unity of diversity in the universe: one is that there is relatively independent relation without containing and being contained in things contacting one another, another is that there is dependent relation containing and being contained materially in these things. The unity of diversity of the world is a limitless layer structure. Kant was the first one who named such a universal layer structure as world system [Kant, 1972]. Engels asserted that, "there are no leaps in nature, precisely because nature is composed entirely of leaps" [Engels, 1954], and he affirmed the objectivity of chance, qualitative change and layers. The world and the relations of things can be observed and studied from the view of simplicity or complexity, both based on objective reality. Complexity is used to represent the unity quality that cannot be reduced among different layers, while simplicity is used to emphasize the unity quality of the same layer. Hence, complexity can be defined as the striding-across-layers interrelation of the world, which cannot be reduced with the theories and laws of the same layer. At first, the relations of the things in the same layer are recognized, and different subjects are formed according to different objects in different layers. It is simplicity science based on the simplicity system. And then, the researcher will go deep into the relations of things striding across layers or of different layers, and develop a interdisciplinary-science, comprehensive-science. It is complexity science based on the complexity system.

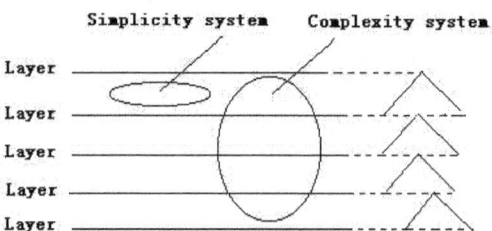

Figure 2. Simplicity system and complexity system.

If the qualities are decided by the relationship of the same layer, such a system is a simple system possessing simple relations of unity of diversity, while if the qualities are decided by the interrelation of different layers, it belongs to a complexity system possessing complex relations of the unity of diversity.

2.3 Self-organization and evolution

Self-organization and evolution are the basic characters of the complexity system. These characters caused by the interactions of all factors of different layers in a system drive the system, engendering harmony and from one form to another, by the mechanisms of mutation, restriction, fluctuation and coding. So the conflict between the 2nd law of thermodynamics and Darwinism can be settled in the framework of complexity system theory. The out-of-order changes of same layers are the order changes of different layers, and the latter changes restrict the former changes. All changes are input, stored and managed in different layers, and operate striding-across-layers. The relations of unity of diversity, inside and outside system, keep on renovating and adapt the changes of the whole environment by the mechanisms of feed back, sorting and magnifying, layer by layer. Obviously, the new ideas of evolutionism are established by the theory of administrative levels, which is the first principle of the evolution of all things, although it hasn't been researched deeply since it was put forward and affirmed by Kant and Engels. Moreover, up to now, people haven't found out the mechanisms that can illustrate "chance consequence" and "causality consequence" to be unified in a simple system. It is obvious that the organization of simplicity system can only be achieved through outside intervention therefore it is not Self-organization [Wang, 1993]. The trend restricting the increase of entropy rises only when three conditions listed below are met: (1) the cause is an essential one and produces a certain result; (2) at the same time, the cause is a casual one and produces casual result; (3) the essential result and the causal result are harmonized together and construct a new cause and effect relationship. It is only complexity systems that have the conditions by which the casual consequence and the essential consequence can be harmonized and become a unity that spans different layers and produces mutual function, and enables the system inside to produce the unifying relationship among the new qualities and parts.

3 Thinking System

The framework in which to understand the complexity of the world can be a tool to analyze human thinking.

Human thinking is a complex hierarchical system composed by the consciousness in different administrative-levels (consciousness-layers). The same as natural hierarchy, the consciousness hierarchy cannot be reduced unless the thinking is not functioning. Each consciousness-layer has its own substance with special form comes from abducting of higher level and by abstracting of lower level and by influence from outside the system. All consciousness-layers work together supporting each other, none of them work alone. During thinking they all play a supporting role. The cognition is a complex event worked on by the whole brain (consciousness). It is unnecessary to rely on the basis of rationality that is supposed beforehand. The rationality of knowledge can only be found if the thinking is considered as multi-layered complexity system, and if the content of one layer is supposed as the foundation of others, and if the relationship of striding across layers is defined as the organizer of a new idea.

3.1 The administrative level structure of thinking

We can build the administrative-levels mode of the thinking system. It is organized by consciousness with many layers – direct perception / indirect perception / rational faculty / world outlook / consciousness / subconscious / top-consciousness. The existence of top-consciousness is especially emphasized, which is the consciousness related with the physical mechanism, and which is still the unclear and unconfirmed part of consciousness [Wang, 2003]. The illustration about the layer structure of the thinking system is absolutely necessary to understand the integer, creativity, and process or mechanism of cognition. It shows that the understanding or cognition of human beings is not a linear course (layer can't be reduced) and it is not the leaping simply only through perception and rational faculty. During the process of actual thinking or cognition, many layers work together. They are independent comparatively, inter-containing, inter-restraining and inter-contact, forming complex interacting relationships. This process is a creative course and its creativity comes from the interacting relation, which possesses qualities that contain and are contained by each other, and which is irreducible, striding across layers and non-linear among layers. The thinking systems have standards of values and psychology besides the characteristics of general complexity system, such as mutation, restriction, coding and organization.

An interesting idealized experiment about brain changing can show the most essential difference between human's thinking system and the present computer. It is a standard to judge whether or not an object is a thinking system, which is similar to that given by A.M. Turing for judging machines and human beings. Suppose there are two men (A and B) who want to be the other ($A \rightarrow B$ and $B \rightarrow A$) by exchanging brains. The process begins with changing the hardware of brains from outside to inside, first the organ, then the cortex, by separating and transplanting each part to the corresponding position. The brain change can be completed using modern surgical procedures. A problem arises: "How much change do we expect if A is B and B is A?" This problem can only be solved successfully if the consciousness is composed by parts of the brain, and the consciousness

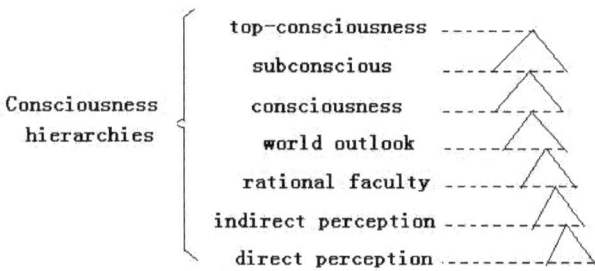

Figure 3. The administrative level structure of thinking.

is decided by the statistical law of the physical and chemical characters of brain matter. But, we know, all of them are not possible. The question about degree of change has no answer. Consider the change of software and ask two men to forget all their own knowledge and accept that of the other. The following steps are the changes of rational knowledge, their world view including religion, and the sub-consciousness. The same problem comes up: "what degree should the changing reach if A is B and B is A?" It is obvious that the question is false according to modern neurology and brain science. From the research of cybernetic and information science, during the course of the change of information, every import and export leads to the change of the whole content inside. Though the experiment of changing brains has not happened, the analysis concludes that thinking is a complex system with layers of structure, which is composed of matter and of consciousness. If the separate parts (including hardware and software) are replaced and re-organized into a new integer step by step, only inter-changing the space between two objects, such is not a thinking system. If the reorganizing integer is not only changed in space, but also in its function, mechanism and the content of consciousness, it can be regarded as a thinking system.

3.2 The complex relations of consciousness-layers

Various thinking modes or layers, in essence, construct a series of hierarchical relationships that contain and are contained, restrict and are restricted by one another. Objects in any layer are conditioned on various striding-across-layers relationships. The relative independence among layers is also obvious, because different layers of thinking interact with each relevant objects and subjects. The variation in a layer can change not only the condition of its layer, but also the condition of other layers, and even the unified relationship of the whole layer structure. Experience and ideas imported from various sources in different layers will build up the unified relationship across layers. Hence, our thinking structure selects a certain mode, which is the result produced by experience in our daily life, communication and education, and within our thinking.

Our thinking would not stop for a moment and it is always brisk. So it would not be in a single primary mode. For this reason, the relations of changing information in every layer can be coordinated. The things in every layer in our thinking system may often be replaced, may even be in chaos. These cases happen probably because our physiology has changed, we are delusional, we have become more sensitive, or because we have accepted a new logic. We can rectify the things in every layer at any time to make them be in order and to reflect the objective world rationally. No prearranged procedure can enable creative thinking and foresight except the mechanism processed by the mutual relationship of strides-across-layers.

Processing information coming from different layers must be independent of each other. Although we have known that our thinking is autonomous, the mechanism of independence has not been illustrated till now. The example of splitting apart one's brain shows that there is more than one conscious self working independently during one's thinking. And this point enlightens the fact that we can understand the independence of thinking and the working principle of creativity and foresight from the angle of value. Every layer and every part in our thinking system may be the conscious subject and the conscious object of other layers at the same time. Among conscious subjects of different layers a relationship of value may be built up. The distinction of every part does not depend on previous logic or procedures, but on value. And the concept "value" can be understood as the rationality satisfying the situation of every layer, as the satisfaction of the object layers to the subject layers. Value here is the crux of the independence of thinking, which is decided by the nature of matter and the layer-structure of the thinking system itself. The layers of the brain and the informational layers of the thinking system both keep in touch with outside information directly or indirectly. And the contact enables the rationality of layers to have standards. There is no other standard for rationality because of the leaps or the discontinuous relationship of layers until multi-subjects are satisfied. Only the value standard satisfying its rationality can connect or communicate the thinking situation and the physical situation. Based on this, we can boldly draw a conclusion that the brain can construct its spiritual ego with such relation of value, and the independence of thinking is realized. The difference of creativity and adaptation emerged and vanished through the coordination and the conflict of values among subjects of layers. This is a rapid procedure of acquiring, selecting and eliminating information. From this, we may conclude that if the excitement of one's thinking concentrates and stays in a certain layer and cannot transfer, that is to say, there is no standard of value and the new ideas lose its rationality, one will become a psychopath. Surely, it is difficult to judge whether the inventor is a psychopath or a genius only with a single new idea or a group of odd ideas. However, the problem can be settled easily if these ideas are considered with the theory of relation of layers. The thinking of a psychopath is on a single level and he regards the rationality of a certain layer as its standard. While the thinking of a scientist is three-dimensional with many layers, a scientist can acquire a rational standard from the value relations of different layers. Though many thoughts of scientists are put forward as assumptions or hypotheses, and though many of them are proved to be wrong later, it is different from the fantasy of a psychopath. A psychopath can only think within a simple mode, while a scientist within a complex one.

It is hard for a normal person to imagine the pain of a psychopath suffering from failure to find rationality for their fantastic ideas in a layer. From observation and analysis of psychopaths, we know that their ability to think in any layer has not been destroyed and their thinking has not ceased. Their disease results from no coordination of certain layers in their thinking structure. And the direct cause of this is that the importing port of changing information is blocked, and the mechanism of information reflection does not work properly. The changing information stays in a certain layer and can not move across layers and can not acquire value standards. The loss of value standards leads to suffering and mental disorder. The spirit of a psychopath has many egos because of the blocking of layers. This case is obvious in religious psychopaths. The medicine given to them can not relieve their pain. The unique way is to open the importing port of information relating with layers, try to transfer the thinking excitement to different layers, evoke the intact ego, enable the thinking to acquire the value relation that creativity needs again, and change suffering into happiness.

A normal person always avoids suffering and searches for happiness, so he always finds rationality among changing information in layers, and creates new ideas to get more satisfaction on value. Based on this, if a person can consciously realize the complexity of thinking layers when he is considering a problem, he may connect with the results in different layers positively, and put the freedom of a layer into more layers positively. This person will not suffer the feeling of a psychopath but will experience happiness.

Humans are not endowed with layer structure of thinking. It is formed through maturity and social activity. In conclusion, when we are born, we may only have two thinking layers: "top-consciousness" and "direct perception". These contact directly or indirectly with layers inside or outside our bodies. Hence, the judgments of facts and value mainly depend on feelings. With the continuity of the passing of time and metabolism, only two layers of thinking structure can not satisfy examination of the continuous changing information from outside and reflect information effectively. A series of middle layers appear between the two basic layers. Layers of thinking increase constantly and the structure improves. We acquire rational faculty gradually and have a world outlook and personality develops. Basically, the rationality among layers changes from the field of feeling to the field of ideas. Hence, we have the spirit of independence.

The thought that can be realized and controlled consciously is only a small part of the structure of our thinking system – a layer or several layers. During the process of thinking, many layers work together independently and then build up the complex mutual relationships inside and outside our thinking. We can't locate the complexity of the thinking system and can't formulate it. Its development depends on itself. What we can discuss are the relations of limited layers. That is the reason why a group of psychopaths can not be treated with a prescription, the reason why many priests fail to preach, and the reason why it is hard for intelligent machines to simulate the whole function of human's thinking system.

4   The "supports" of model-based reasoning

From the discussion above the answer to the question of how scientists solve the problems of rationality of their scientific discovery is obvious. Following are some typical cases in the history of science to be examined.

4.1   The process of model application

During research, in order to gain a complete theory rather than superficial descriptions of the research object, scientists build a model in thought instead of the object. The research object then becomes a model, which is composed of known ideas and laws, and indispensable hypothesis, because thinking can only process models. Based on results of examining the model, which is a portrayal of the essential connection of object, scientists establish theory, then lets the theory approach the object, and gets exact knowledge about the object.

In order to build a model, researchers have to, first understand the protoplast of the object, generally by observation and experiment, and judge which aspect of the protoplast will be researched, and second have the indices needed to research abstracted from the protoplast, and form an abstract object, and thirdly analyze the relations among factors within the abstract object, and seek a new abstract object of thinking by the results of the analysis. The new abstract object of thinking is a model of original object. In short, the process of model application can be included as "abstraction-transfer-concreteness".

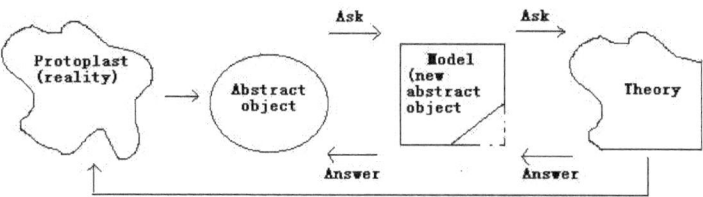

Figure 4. The Process of model application.

The scientific model has three characteristics, which are: simplification, direct observation and hypothesis, and all of them need supporting points in thinking. To get a model suited with the object of research scientists have to possess a kind of ability by which all supporting points of the rationality about the model can be synthesized together. The crux here is "How does the scientist seek substitute of the protoplast during the process of model application?" from which a series of questions emerge, they are: "Does the scientist get the new abstract object of thinking by logic reasoning or imagining or belief or instinct, or by all of them, standing on introspective experience or feeling of practice, or both of then?" "What is the basis of each support?" "Is the rationality of bases given in advance or in the time of 'thinker is thinking'?" "Is there common foundation of the rationality in based-model reasoning?" and more, "What make the scientists to firmly believe the models true?"

4.2 Typical cases of model application in biology

In order to get more distinct answers for the questions above than those given by theorization, examining some successful cases in history of science are needed. There are three famous models in biology: the target-model of radiation-biology, by which Lea and other bio-scientists succeed in explaining the mechanisms of radiogenic death; the double-helix-model of DNA, by which Watson and Crick succeeded in founding the theory of molecular biology; and the model of unfolded coefficient of $(a+b)^n$, by which Mendel succeeded in establishing the second law of genetics.

Lea says that biologists found the experiential curve of radiogenic death by observation, experiment and statistics after atomic bombs exploded in Hiroshima and Nagasaki, but they didn't know the mechanism of radiogenic death and what reason caused the curve. Lea, a theory-biologist, used the method of model to solve these problems. Following are Lea's steps:

(A) To give basic quantities according to the relative theories.

(B) To have a comparison with thinking in images, and have the model set up on the process resembled with direct observation. Lea studied "shoot-target" early on and built the "target-model" for theory analysis. Lea could see the process of radiogenic death with target-model in his mind. He could not see it with his eyes, so he had the research work transformed to the model that he set up.

(C) To analyze the factors of the process on the basis of the model which is a new abstract object of thinking, and re-settled basic quantities.

(D) To make one or two hypothesis for the model. Lea concluded that the ray as the tommy gun must be shooting around, and that the target is small and has a small probability to be hit.

(E) To establish the theory for the process of radiogenic death based on the model. Lea and other theoretical biologists accomplished their task very facilely, to build the theory so-called "target-theory" by means of statistics and theory of classical probability. The calculation of the theory is identical with experiential curve.

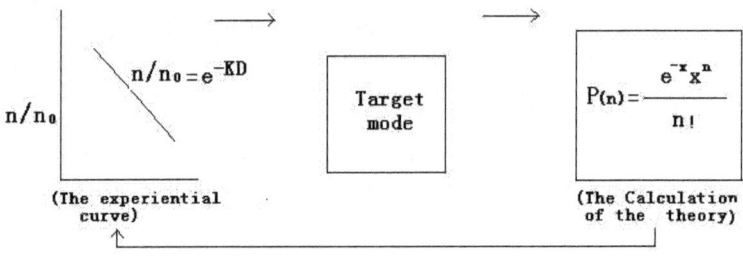

Figure 5. The target model of radiation-biology.

In every step Lea sought the bases of rationality, which may be from his feeing (direct perception), or imagination (indirect perception), or logic (rational faculty), or faith (world outlook), or sense (consciousness), or personality (subconscious), or instinct (top-consciousness). He compared the bases and linked them together and when he did this work he had no guidelines or prior knowledge because he could not take part in the actual process of radiogenic death. It is the creative mechanism of the complexity of thinking system. In (B) which is the key to using model-based reasoning, there are many abstract objects of thinking for Lea to choose as a model by which the theory could be constructed. Having been proved in other scientific cases, the same theory can be built from different models, and a model that plays a part of tool is often rejected when the theory have been constructed. The brain (the thinking system) is new every day, so, it is possible that a scientist chooses different things as models in different time for same research object, and that different scientists use different models for same research object. In (C), (D) and (E) Lea understood and dealt with the model and decided the tools of analysis to achieve theory from model according to the supporting points in different consciousness. He had no plan but he did have ideas about the interdependent relationships of consciousness-layers. Nothing is prepared for "thinker is thinking". In a certain sense Lea was successful because of the special state of his thinking system in that time. The complexity of thinking system, as F.T. Arecchi pointed out, is the relationship about the state of our knowledge.

While to get the double helix model of DNA was not as smooth as to get target model of radiogenic death, scientists experienced repeated failures. They had been checking and composing unceasingly various modes of the models that they could think of, such as, the model of single helix, of tri-helix, of tetra-helix, and perhaps of having no helix, before they confirmed the rationality of double helix for DNA.

The above two cases have a common ground, that is, both rely on seeking new abstract objects of thinking and on judging the things if they fit the research object. To be successful there must be many supporting points on different consciousness layers of scientists' thinking system. For example, Lea must have had the situation of "shooting the target with tommy gun" or have heard of it. Watson and Crick must have used the ladder and seen the helix. Knowledge that is non-logical, acquired by different modes and formed at different times and places becomes long term memory information stored in different thinking layers. They become supporting points for choosing and judging a new model.

Mendel used a mathematical formula as a substitute of prototype of his research object. Do not think of it as another kind of model, called "mathematical model", because the model of unfolded coefficient of $(a + b)^n$ is not abstraction of the prototype. The mathematical formula here is a new abstract object of thinking for setting up a scientific model, which goes step further for establishing the theory. Like Lea, Watson and Crick, Mendel correlated all objects in different consciousness for the supports for the rationality of his model.

5 Short conclusion

During the process of model application, logic (which is usually not the main factor), imagination, belief, instinct, introspection, feeling, all factors in the thinking system supporting each other, formed a condition which I call "comfort" and "satisfaction" of thinking system, being stronger causal interactions of different consciousness-layers. "Comfort" and "satisfaction" that are main characteristics of the complexity of thinking system, are the bases of rationality of scientific discovery and creation of thinking.

BIBLIOGRAPHY

[Asbby, 1960] W.R. Asbby. *An Introduction to Cybernetics.* W.J. Wiley, New York, 1960.

[Edmonds, 1999] B. Edmonds. *What is Complexity? The Philosophy of Complexity.* Brussels Free University, Brussels, 1999.

[Engels, 1954] F. Engels. *Dialectics of Nature.* Foreign languages Publishing House, Moscow, 1954.

[Helighen, 2000] F.H. Helighen. The grouth of structural and functional complexity during evolution. In Yaneer Bar-Yan, editor, *Unifying Themes in Complex Systems*, pages 257–272. ABP Perseus Books, Cambridge, MA, 2000.

[Kanffman, 1971] S.A. Kanffman. Articulation of parts explanations in biology. In R.C. Buck and R.S. Cohen, editors, *Boston Studies in the Philosophy of Science*, pages 257–272. Kluwer, 1971.

[Kant, 1972] I. Kant. *Allgemeine Naturgeschicht und Theorie des Himmels, oder Versuch Von der Verfassung und dem mechanischen Ursprungedes ganzen Veltgebaudes nach Newton'schen Grundsatzen abgehandelt.* People Publishing House, Shanghi, 1972. Chinese edition.

[Laszlo, 1998] E. Laszlo. *The Grand Synthesis.* Social Science Publishing House, Bejing, 1998.

[Popper, 1994] K. Popper. *The Myth of the Framework. In Defence of Science and Rationality*, chapter VIII. Routledge, London, 1994.

[Prigogine, 1986] I. Prigogine. *Exploring Complexity.* W.H. Freeman & Company, New York, 1986.

[Quine, 1953] W.V.O. Quine. *From a Logical Point of View.* Harvard University Press, Cambridge, MA, 1953.

[Richards, 1983] S. Richards. *Philosophy and Sociology of Science.* Blackwell, Oxford, 1983.

[Rivadulla, 2004] A. Rivadulla. The newtonian limit of relativity theory and the rationality of theory change. *Synthese*, 141:417–429, 2004.

[Simon, 1999] H.A. Simon. Can there be a science of complex system? In Yaneer Bar-Yan, editor, *Unifying Themes in Complex Systems*, pages 257–272. ABP Perseus Books, Cambridge (Mass.), 1999.

[Wang, 1990] Z. Wang. The concept of complexity: its source, definition, characteristic and function. *Philosophical Research*, 3, 1990.

[Wang, 1991] Z. Wang. The meaning of complexity. In *LMPS91*. Uppsala University Press, Uppsala, 1991.

[Wang, 1993] Z. Wang. *Mutation and Evolution.* Guangdong Higher Education Publishing House, Guangzou, 1993.

[Wang, 1996] Z. Wang. Complexity and evolution of social system. *Studies in Dialectics of Nature*, 8, 1996.

[Wang, 2003] Z. Wang. On the administrative level structure and complexity of thinking system. *Studies in Dialectics of Nature*, 10, 2003.

[Wimsatt, 1974] W.C. Wimsatt. Complexity and organization. In R.C. Buck and R.S. Cohen, editors, *Boston Studies in the Philosophy of Science*, pages 67–86. Kluwer, 1974.

Zhikang Wang
Department of Social Science
Zhongshan University, Guangzhou, P.R. China
Email: zdwangzk@tom.com

A Diagrammatic Proof Search Procedure as Part of a Formal Approach to Problem Solving

DIDERIK BATENS

ABSTRACT. This paper aims at describing a goal-directed and diagrammatic method for proof search. The method (and one of the logics obtained by it) is particularly interesting in the context of formal problem solving. A typical property is that it consists of attempts to justify so-called bottom boxes by means of premise elements (diagrammatic elements obtained from premises) and logical elements. Premises are not preprocessed, whence most premises lead to a variety of premise elements.

The method is simple and insightful in three respects: (i) diagrams are constructed by drawing the goal node and superimposing the top node of a new diagrammatic element on a bottom box of an element that occurs in the diagram; (ii) diagrammatic elements are built up from binary and ternary relations that connect nodes (comprising one or two boxes) to boxes (entities containing a single formula); (iii) diagrammatic elements are obtained in view of existing bottom boxes by a unified approach. At the propositional level, the method is an algorithm for derivability (but leaves choices to the user). Extended to the predicative level, it provides a criterion for derivability and one for non-derivability.

The method is demonstrably more efficient than tableau methods and has certain advantages over linear methods and certain other goal-directed methods. Apart from making certain properties of search paths more visible, the method also led to a simplification of the metatheoretic proofs.

1 Aim of this paper

In [Batens, 2003b] and [Batens, 2003a], I spelled out the basics of a formal approach to problem solving. An important part of it consists of a proof search method that is largely pushed into the proofs.[1] Some first results

---
[1] One objection against a strict distinction between rules of inference ("definitory rules") and heuristic rules ("the strategy") – see for example [Hintikka, 1999] – is that

were meanwhile published – see for example [Batens and Provijn, 2001], [Batens, 2002] and [Batens, in print].

The approach followed in those papers is prooflike, and there are a number of arguments in favour of this. However, both for finding a correct formulation and for finding the proofs of the metatheorems, it turned out extremely helpful to think about the proof search in terms of a tree-like structure. Moreover, I often had the impression that much insight could be gained by explicitly spelling out a diagrammatic version of the proof search procedure. The present paper contains a (semi-formal) formulation of this version as well as some first results that derive from it.

2 Two examples

Suppose that one tries to derive $s \vee q$ from the premises

$$\sim t,\ p \supset q,\ \sim r \vee p,\ p \supset \sim s,\ (p \vee t) \wedge \sim s.$$

The search process may have been the following. Clearly, $s \vee q$ cannot be obtained directly from any premise. So one tries to derive either s or q. While s cannot be obtained from a premise, the second premise tells us that q can be obtained if p can be obtained, whence one looks for p. In view of the third premise, p can be obtained if r can be obtained. But r clearly cannot be obtained from any premise. So one looks for another way to obtain p. The fifth premise entails $p \vee t$; in view of this, p can be obtained if $\sim t$ can be obtained. And $\sim t$ is the first premise. So one is home.

The previous paragraph *describes* the search process, but does not make it fully explicit. Thus the claim that $s \vee q$ cannot be obtained directly from any premise implicitly states that all five premises have been considered. So, although the fourth premise is never explicitly mentioned in the description of the search process, it was considered repeatedly. However, the fourth premise failed to be useful at any point, and hence did not lead to any search path. The third premise led to a search path, which however was unsuccessful.

In view of what was said in section 1, I now present the proof search in a diagrammatic way.

---

heuristic reasoning is itself a form of reasoning.

A Diagrammatic Proof Search Procedure

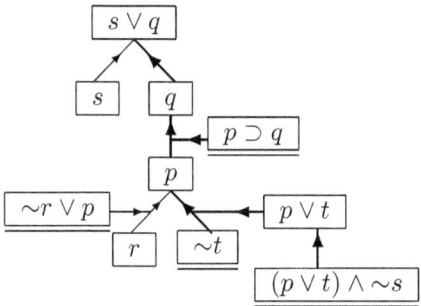

The diagram can best be seen as composed of seven (overlapping) diagrammatic elements. The top half of the diagram is constructed from the following elements:

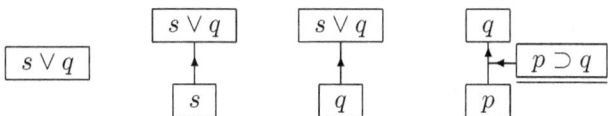

The leftmost element is the *goal element*. The diagram is started by drawing it. The next two elements are analysing elements or *logical elements*, expressing respectively that $s \vee q$ is obtained from s and that $s \vee q$ is obtained from q. The fourth element is a *premise element* – the premise occurs in the underlined node. The element represents that $p \supset q$ is a premise and that it justifies the transition from p to q. Here are the other elements from which the diagram is constructed:

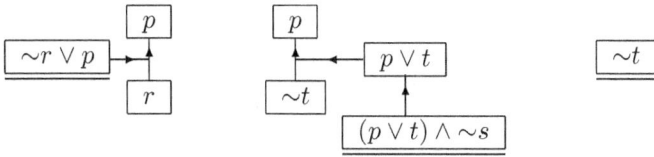

All of these are premise elements – obtained from premises in a way that is spelled out in section 3.

Sometimes there is a slight complication, as is illustrated by the diagram for $s, p, s \supset q \vdash (r \vee p) \wedge q$ (the diagram obtained in trying to derive $(r \vee p) \wedge q$ from the premise set $\{s, p, s \supset q\}$).

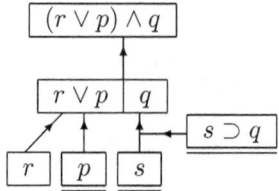

No premise element has $(r \vee p) \wedge q$ as its top node (see section 3), and hence the diagram cannot be extended downward from the goal node by means of such an element. So the goal node has to be analysed by means of a logical element. This analysis gives us a two-box node: $(r \vee p) \wedge q$ is obtained if both $r \vee p$ and q are obtained. Remark that two diagrammatic elements are 'attached' to the box containing $r \vee p$, whereas only one element is 'attached' to the box containing q. In general, all nodes contain either one or two *boxes*, and every box contains a single formula.

We have seen that there are three kinds of diagrammatic elements: the goal element, logical elements, and premise elements (a premise node or a set of connected nodes obtained from a premise). It is useful to distinguish within the elements *top nodes* (nodes from which no arrow departs) and *bottom nodes* (nodes in which no arrow arrives). In single node elements, the top node and bottom node coincide. For the other elements, there is exactly one top node and there are one or more bottom nodes.

A diagram is constructed starting from the top, by drawing the goal node, and is extended downward by superimposing the top node of a new element on a *box* of a bottom node that already occurs. It is important to stress this: a diagram is extended downward from a bottom box in an attempt to justify the formula that occurs in the box. If a bottom node comprises two boxes, both formulas have to be justified.

There are two kinds of *arrows* in the diagram. A simple arrow, for example the one from s to $s \vee q$, is a *logical arrow*. It corresponds to $s \vdash s \vee q$ (or, where Γ denotes the premise set, to "if $\Gamma \vdash s$, then $\Gamma \vdash s \vee q$"). A combined arrow (an arrow at which another arrow arrives) is a *contingent arrow* – the combined arrow represents a ternary relation. Thus the contingent arrow between $p \supset q$, p and q expresses that the transition from the minor p to the conclusion q is warranted by the major $p \supset q$.[2] It corresponds to $p \supset q, p \vdash q$ (or to "if $\Gamma \vdash p \supset q$, then if $\Gamma \vdash p$, then $\Gamma \vdash q$"). Incidentally, in the diagrams presented in this paper, there is at most one arrow to an arrow, Contingent arrows occur only in premise elements, and the arrow of a logical element is always a logical arrow.

---

[2]Nothing prevents that $s \supset (s \vee q)$ is a premise, justifying the transition from s to $s \vee q$.

Several paths may be distinguished on a diagram. For the time being, a path can be seen as a chain of diagrammatic elements. A path is successful if (again, for the time being) all its bottom boxes (boxes in which no arrow arrives) are premises, which is seen on the diagrams by the fact that they are underlined – more precise definitions follow in section 3.

Once a successful path on the diagram is identified, there is an algorithm for transforming it to, for example, a Fitch-style proof – see [Provijn, 2002] for some first results. While the search three was obtained by moving down from the goal node (see the paragraph on the search process), the proof is stepwise obtained by moving up along the only successful search path in the first diagram.

1 $(p \vee t) \wedge \sim s$ Premise
2 $p \vee t$ 1; Simplification
3 $\sim t$ Premise
4 p 2, 3; Disjunctive Syllogism
5 $p \supset q$ Premise
6 q 4, 5; Modus Ponens
7 $s \vee q$ 6; Addition

The following proof is derived from the successful path of the second diagram.

1 p Premise
2 $r \vee p$ 1; Addition
3 s Premise
4 $s \supset q$ Premise
5 q 3, 4; Modus Ponens
6 $(r \vee p) \wedge q$ 2, 5; Adjunction

If one compares the first proof to the heuristic reasoning, the most striking feature is that several search paths were tried but do not occur in the proof. Proofs are supposed to demonstrate that a conclusion can be obtained from the premises, and they are supposed to do so in an elegant way. This is the reason why the third premise (from the list in the first paragraph of this section) is not even introduced in the proof. More importantly, the proof contains no trace of the unsuccessful search for r that was induced by the third premise.

3   Diagrammatic elements, paths, successful paths, and descendants

In order to systematically and concisely describe the way in which logical elements and premise elements are obtained, I distinguish between a-formulas and b-formulas (varying on a theme from [Smullyan, 1995]). Let $*A$ denote the 'complement' of A, viz. B if A is $\sim B$ and $\sim A$ otherwise. To each 'complex' formula two other formulas are assigned according to the following table:

a	a_1	a_2		b	b_1	b_2
$A \wedge B$	A	B		$\sim(A \wedge B)$	$*A$	$*B$
$A \equiv B$	$A \supset B$	$B \supset A$		$\sim(A \equiv B)$	$\sim(A \supset B)$	$\sim(B \supset A)$
$\sim(A \vee B)$	$*A$	$*B$		$A \vee B$	A	B
$\sim(A \supset B)$	A	$*B$		$A \supset B$	$*A$	B
$\sim\sim A$	A	A				

Most premises generate several diagrammatic elements. First of all, each premise gives us a one-node premise element. Next, a premise element may be extended by a so-called upward move to its top node. The *upward moves* for a-formulas and b-formulas are respectively as follows:

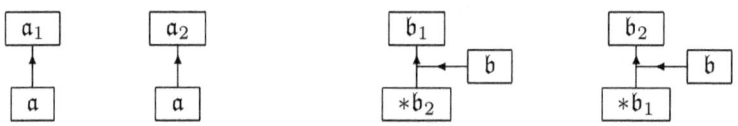

So both a-formulas and b-formulas may be extended in two ways, depending on whether one wants to obtain the first or the second associated formula. An important difference is that an a-formula justifies both its associated formulas, whereas a b-formula justifies the transition from the complement of one of its associated formulas to the other associated formula. Extending a-formulas upward leads to logical arrows. Extending b-formulas upward leads to contingent arrows.

If the top node of the thus obtained element is itself an a-formula or a b-formula, two new diagrammatic elements may be obtained by a further upward move, and so on. It is instructive to list the diagrammatic elements that are obtained from a premise, for example $(p \wedge q) \supset r$, which delivers the following five elements:

A Diagrammatic Proof Search Procedure

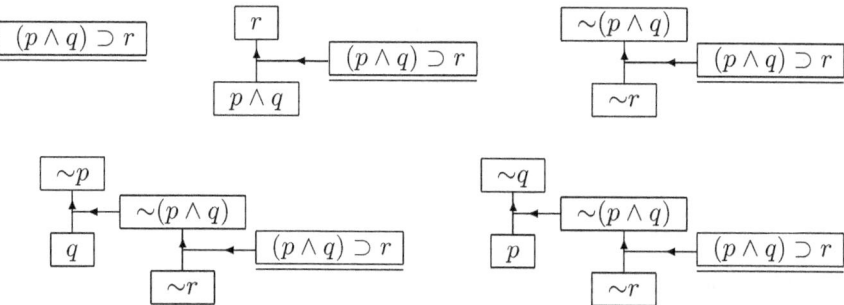

Let us now turn to the way in which logical elements are obtained. We have seen that α-formulas are justified by their associated formulas *taken together*. *Downward move for α-formulas*:

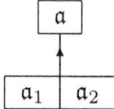

For b-formulas, the following downward extensions are certainly correct:

However, this approach does not provide a complete search system for Classical Logic (henceforth **CL**). One of the reasons for this is that Excluded Middle is absent from it. This can be seen from the diagram for $q \vdash p \vee \sim p$, but it is instructive to consider an example that illustrates the problem in a more general way, viz. the diagram for $s \supset (p \supset q), q \supset r \vdash s \supset (p \supset r)$.

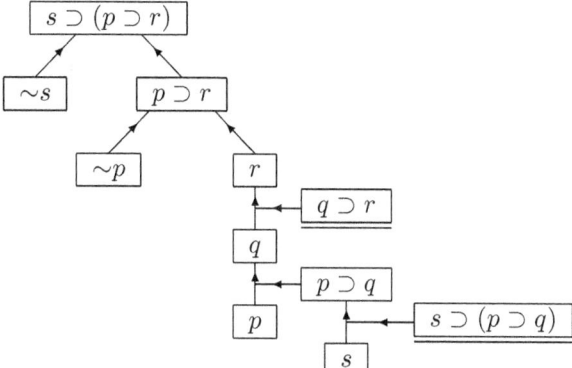

On the present criteria, there is no successful path (no path of which all nodes are justified). Yet the conclusion is derivable from the premises by **CL**. And indeed, this can be seen from the diagram as follows. Either p or $\sim p$ is true, and either s or $\sim s$ is true. If $\sim s$ is true, $s \supset (p \supset r)$ is true by the leftmost path of the diagram. If $\sim p$ is true, $s \supset (p \supset r)$ is true by the middle path of the diagram. And if both p and s are true, then $s \supset (p \supset r)$ is true by the rightmost path of the diagram.

So, it seems that we need a supplementary criterion for justified nodes or paths. The criterion that is implicitly used in the previous paragraph – let us call it the EM (excluded middle) criterion – seems to offer just the required way out. And yet, there is a more attractive approach.

Consider the following fragment (of the rightmost path) of the diagram:

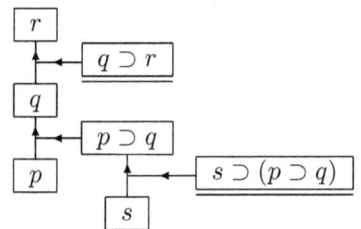

This fragment shows that, if s and p were premises, then r would be derivable: $s \supset (p \supset q), q \supset r, s, p \vdash r$. By applying the deduction theorem twice to this, one obtains $s \supset (p \supset q), q \supset r \vdash s \supset (p \supset r)$ as desired. This criterion is more appealing and more intuitive than the EM criterion, but it sometimes fails, as may be seen from the diagram for $p \vee q \vdash (p \vee r) \vee (s \supset q)$ – it is part of the next diagram. However, the criterion shows us the way to a better approach.

When precisely does one want to apply a supplementary criterion? This happens in cases in which neither b_1 nor b_2 is derivable from the premises, whereas b is. Thus neither $\sim s$ nor $p \supset r$ is derivable from $\{s \supset (p \supset q), q \supset r\}$, but $s \supset (p \supset r)$ is. However, we know that $\Gamma \vdash b$ holds true iff $\Gamma \cup \{*b_2\} \vdash b_1$, and also iff $\Gamma \cup \{*b_1\} \vdash b_2$ holds true. So the introduction of a new criterion for justified nodes can be avoided if one allows that $*b_2$ is used as a supplementary premise on the paths on which one tries to justify b_1, and that $*b_1$ is used as a supplementary premise on the paths on which one tries to justify b_2. This leads to the following *downward moves for b-formulas*:

in which a formula in an 'oval', attached to a node, indicates that this formula may be used as a supplementary premise on the paths to which this node belongs – it will only be used 'below' this node, whence no ambiguity arises.

Let us see what becomes of the diagram for $p \vee q \vdash (p \vee r) \vee (s \supset q)$ in view of the downward move for b-formulas – I do not draw the additional premises on the third level from the top as they are useless anyway.

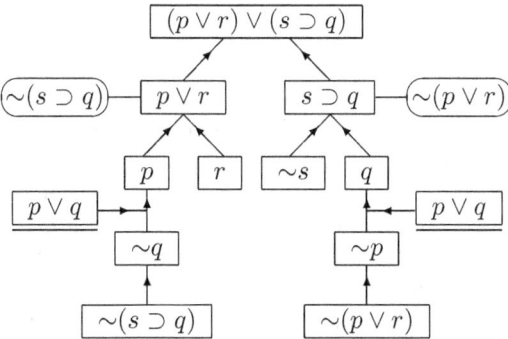

Both outer paths (the first and the fourth one) are successful. Of course there is no need to even start the second path once one found that the first path was successful, but I drew all possible paths in order to show that the previous criterion (the one without the new premises) does not work.[3]

This ends the description of the way in which logical elements and premise elements are obtained. It is important to realize that these elements are restricted in such a way that they lead only to the analysis of targets (formulas one tries to justify) and to the analysis of premises. Thus, if A is the formula of a bottom box of a diagrammatic element, no move enables one to introduce, for example, the following 'logical' element:

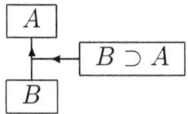

It obviously holds that, where Γ denotes the premise set, "if $\Gamma \vdash B \supset A$, then, if $\Gamma \vdash B$, then $\Gamma \vdash A$". However, introducing boxes containing B and $B \supset A$ in order to justify A is an arbitrary and inefficient move, not a goal-directed move. For this reason, the aforementioned restriction on logical

---

[3] A successful path on which a supplementary premise is used, is most naturally turned into a Fitch-style proof by starting a subproof with the supplementary premise as its hypothesis, applying Conditional Proof once the desired formula is obtained, and next transforming the result as required.

elements and premise elements is essential. In section 7, I shall mention a further (optional) restriction on premise elements, which is directed towards the efficiency of the search process.

Before turning to to the definition of a path on an diagram and of a successful path on a diagram, I need to specify the relations represented by the arrows. I shall do this in an informal way, leaving the obvious precise mathematical description to the reader.

A logical arrow connects a node to a box; a contingent arrow connects two nodes to a box. Apart from what is obvious, the essential further convention is that, while many arrows may end up in the same box, it is never the case that two arrows (whether logical or contingent) depart from a node. Of course, nothing prevents one to use the same diagrammatic element at different points in the same diagram.

A *path* on a diagram is a smallest set of connected diagrammatic elements defined by: (i) the goal node belongs to every path, (ii) if a box is the top node of n different diagrammatic elements, then each of these elements belongs to a different path (leading from the element to the goal node). There are three paths in the first diagram of section 2. It may be less obvious that there are two paths in the second diagram of that section, viz.:

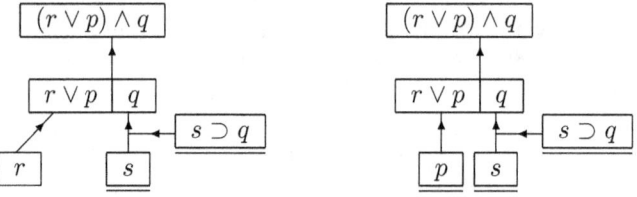

The node that comprises two boxes belongs to both paths. On each path, each of these boxes is the top node of a diagrammatic element.

A path is *justified* iff (if and only if) all its nodes are justified. A node is justified iff all its boxes are justified. That a box is justified (in view of the premises) is recursively defined by the following clauses:

1. A node is justified if all its boxes are justified.

2. Premise boxes (underlined boxes) are justified.

3. A box that is reached by a logical arrow from a justified node or by a contingent arrow from two justified nodes is justified.

There is one justified path in each diagram of section 2 – in the first diagram, the path is marked by bold arrows, in the second diagram the justified path is the rightmost one in the figure that displays them separately (in this section).

A *bottom box of a diagram* is a box in which no arrow arrives; a *bottom box of a path* is a bottom box of the diagram that belongs to the path. These should not be confused with the bottom boxes of a diagrammatic element.

In the sequel, I shall also need the notion of a descendant of a box. That a box is a *descendant* of another box is recursively defined by (i) box b is a descendant of box b' if there is a logical arrow from the node containing b to the box b', (ii) box b is a descendant of box b' if there is a contingent arrow from the node containing b (and from another node) to the box b', and (iii) if box b is a descendant of box b' and box b' is a descendant of box b'', then box b is a descendant of box b''.

If two boxes occur on the same path, it is possible that neither is a descendant of the other. This is due to the occurrence of contingent arrows and to the downward move for α formulas. Thus, in the first diagram of section 2, the box containing $(p \vee t) \wedge \sim s$ is not a descendant of the box containing $\sim t$, nor is the latter box a descendant of the former. In the second diagram of section 2, the descendant relation does not obtain in either direction between, for example, the box containing p and the box containing s.

4 Reasoning *ex absurdo*

The method described up to now still characterizes a logic that is weaker than Classical Logic (henceforth **CL**). One of the reasons for this can be seen from the diagram for $\sim p \supset q, q \supset p, p \supset r \vdash r$.

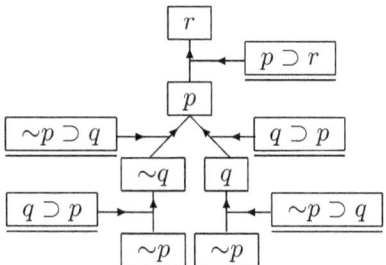

On the present criteria, there is no successful path in this diagram, whereas r is **CL**-derivable from $\{\sim p \supset q, q \supset p, p \supset r\}$. Indeed, this may be seen from the diagram. Actually, both paths show that $\sim p$ is sufficient to justify p. But if $\Gamma \cup \{\sim p\} \vdash_{\mathbf{CL}} p$, then $\Gamma \vdash_{\mathbf{CL}} p$.

In order to build this in into the search method, the definition of a justified box needs to be extended with: *if box b is a descendant of box b', and box b contains the complement of the formula of box b', then box b is justified.*

Let us return to the last diagram. In view of the change, the two bottom boxes containing $\sim p$ are justified. That the box containing $\sim p$ is justified has the effect that the box containing p is also justified. In other diagrams, there may be unjustified boxes containing other formulas; an example is the diagram for $\sim p \supset q, q \supset p, s \supset (p \supset r) \vdash r$, which, even on the modified definition, contains no justified path, as required.

5 Where went Explosion?

The Explosion rule, also known as *ex falso quodlibet*, states that every formula can be derived from an inconsistency $(A, \sim A/B)$. This rule warrants that, in **CL** and other systems that share the rule, $p, \sim p \vdash q$. In its present state, the diagrammatic method does not validate Explosion. Thus the proof search diagram for $p, \sim p \vdash q$ consists only of the goal node, containing q; no diagrammatic element obtained from the premises has q as its top node.

It is not difficult to add a further (and final) move to handle Explosion. The idea is that, if A is a premise $(A \in \Gamma)$, and $\Gamma \vdash \sim A$, then anything can be derived from the premise set $(\Gamma \vdash B$ for every $B)$. This is expressed in diagrammatic terms by the following move, which may be seen as an upward move generated by a premise $A \in \Gamma$:

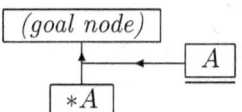

As the display suggests, I restrict this move by requiring that the top node of the element is the goal node. Although this element could in principle be attached to any box, the only sensible way of proceeding consists in attaching it to the goal node.

The procedure then looks as follows: first one tries the method described in the previous sections. If this fails, one applies the Explosion move to the first premise, next to the second premise, and so on. A very simple example is the diagram for $p, \sim p \vdash q$:

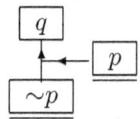

This completes the description of the proof search diagrams. The following theorems can be proved about the method.

THEOREM 1 *If Γ is finite, then so is the diagram for $\Gamma \vdash A$.*

THEOREM 2 $\Gamma \vdash A$ *iff the diagram for* $\Gamma \vdash A$ *contains a successful path.*

If the procedure is upgraded from propositional **CL** to full **CL**, Theorem 1 obviously fails (full **CL** is undecidable). Still, Theorem 2 may be adapted to the following: if the diagram is finite and contains a successful path, then $\Gamma \vdash A$; if the diagram is finite and contains no successful path, then $\Gamma \nvdash A$.

6 The Logic CL$^-$

The Explosion move is isolated from the others. What happens if one does not add it to the procedure?

Not to apply the Explosion move is often sensible. Indeed, in many cases one does not want Explosion. Thus, if one is trying to derive a prediction from a theory together with a set of data, one is not interested in any prediction that is derivable by Explosion – if A is so derivable, so is $\sim A$. No one will act on A in this case, and A will not be considered as part of the predictive power of the theory – similarly for the explanatory power of the theory.

Let **pCL**$^-$ be the diagrammatic method without the move handling Explosion, and let $\Gamma \vdash_{\mathbf{pCL}^-} A$ iff the method without the Explosion move leads to a diagram for $\Gamma \vdash A$ in which at least one path is successful. One can then define the logic **CL**$^-$ as follows:

DEFINITION 3 $\Gamma \vdash_{\mathbf{CL}^-} A$ *iff* $\Gamma \vdash_{\mathbf{pCL}^-} A$.

The following theorems can be proved – see a forthcoming paper.

THEOREM 4 *If* Γ *is finite, then so is the diagram for* $\Gamma \vdash_{\mathbf{CL}^-} A$.

Incidentally, means to handle infinite premise sets are described in [Batens and Provijn, 2001].

THEOREM 5 *If* Γ *is consistent, then, for all* A, $\Gamma \vdash_{\mathbf{CL}^-} A$ *iff* $\Gamma \vdash_{\mathbf{CL}} A$.

THEOREM 6 *If* Γ *is inconsistent, then there is an* A *such that* $\Gamma \vdash_{\mathbf{CL}^-} A \wedge \sim A$.

In other words, **CL**$^-$ leads to exactly the same consequences as **CL** if the premise set is consistent (this is the intended domain of application of **CL**), whereas every inconsistent premise set has an explicit contradiction as a **CL**$^-$-consequence, as desired.

As was shown in [Batens, to appear], **CL**$^-$ is sound and complete with respect to a semantics. In that paper some further properties of **CL**$^-$ are studied and more motivation for the system is presented.

For years, logicians have claimed that Explosion cannot be isolated in **CL**, because if one wants to avoid Explosion, one needs to give up, for example, either Disjunctive Syllogism or Addition (as well as one of many other plausible inference rules). The procedural approach shows that these claims are wrong. If a logic is defined by a procedure – roughly by the results of a search process – it is possible to isolate Explosion.

7 Devising the elements one needs

If the premise set is large, it seems a drawback of the procedure that all diagrammatic elements of all premises have to be prepared before the construction of the diagram can start. Actually, this can easily be avoided, viz. as follows. One starts the diagram with the goal node as before. Next, one considers a non-justified box of a bottom node. Let the box contain the formula A. One then considers a premise B, starting with the first one, and checks whether A is a positive part of B, $pp(A, B)$, which is recursively defined as follows:

1. $pp(A, A)$.

2. $pp(A, \mathfrak{a})$ if $pp(A, \mathfrak{a}_1)$ or $pp(A, \mathfrak{a}_2)$.

3. $pp(A, \mathfrak{b})$ if $pp(A, \mathfrak{b}_1)$ or $pp(A, \mathfrak{b}_2)$.

4. If $pp(A, B)$ and $pp(B, C)$, then $pp(A, C)$.

If A, the contents of the box one tries to justify, is a positive part of the premise B, one applies upward moves until the single box of the top node of the diagrammatic element contains A. There is a simple algorithm for obtaining this diagrammatic element, which is then used to extend the diagram downward. If A is not a positive part of any premise, a downward move is applied to A – see section 3.

8 A matter of elegance

The application of the downward move for $\mathfrak{b}$-formulas is somewhat inefficient in that, more often than not, the left move (the one introducing $\mathfrak{b}_1$ with $*\mathfrak{b}_2$ as a supplementary premise) leads to a justified path iff the right move does so. So trying out both paths seems a loss if time.

The trouble is with the "more often than not". Consider the top of the following diagram for $p \vdash q \supset p$.

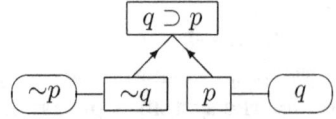

The right path will obviously be justified (without the supplementary premise), but the left path will not, unless one were to apply the Explosion move to the bottom node containing $\sim q$, which would complicate what was said in section 5, would not work in $\mathbf{CL}^-$, and would result in an ugly path (and an ugly proof) while nice alternatives are possible. A compromise is obtained by recasting the downward moves for b-formulas as follows:

The change is justified by the following consideration. Let Γ be the premise set as before and let a b-formula be analysed according to the modified moves, introducing b_1 on the left path and b_2 with the supplementary premise $*b_1$ on the right path. If the right path is not successful[4] because $\Gamma \cup \{*b_1\}$ is inconsistent, then $\Gamma \vdash b_1$, whence the left path is bound to be successful. The choice to add the new premise to the right, rather than to the left, is obviously arbitrary, except that the choice for the right path leads to a nice Fitch-style reconstruction in terms of Conditional Proof.

9 Improving the efficiency of the procedure

Not much attention was paid until now to the efficiency of the procedure. My main aim was to spell out a *sensible* procedure that leads to sensible diagrams. But even with respect to sensibility the procedure may be made more (but not completely) deterministic and efficient. The improvement offered in this section concerns the marking of paths.

Dead ends At every stage of the diagram, a bottom box containing A will be marked as a *dead end* iff A is not a positive part of any premise and is neither an a-formula nor a b-formula (i.e. cannot be analysed by means of a logical element). If this obtains, no logical element and no premise element has a top node containing A.

If a box of a path is marked as a dead end, the path is a *dead end path*. Even if all other bottom nodes on the path can be justified, the marked bottom box will forever remain unjustified. So dead end paths are not continued.

Some care is required here. If a path contains a dead end node n, it is still possible that an higher node of the path may be extended downward, resulting in a new path that does not contain n. The new path may obviously be a justified one. This shows the importance of defining a path in terms

---
[4] More precisely: if no path to which the node containing b_1 belongs is successful. Similarly below in the text.

of diagrammatic elements. If a node is marked, justifying another bottom node of the same diagrammatic *element* is useless, but justifying a bottom node of an element that is located higher in the path may be useful because it starts a new path.

A simple example of a dead end path occurs if one tries to derive $q \vee p$ from a premise set containing $r \supset (s \supset q)$ and r, but not containing any premise of which either $q \vee p$ or s is a positive part. The diagram might start as follows:

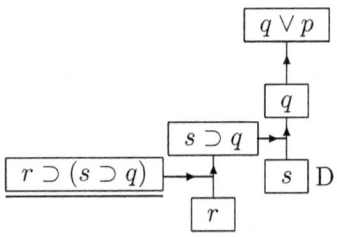

As s is a dead end, one has either to find an alternative justification of q or one has to apply the second downward move for b-formulas to the goal node.

Inconsistent paths A path is inconsistent iff, for some A, two unjustified boxes that belong to the path contain respectively A and $\sim A$.[5] If this is the case, the lowest box of the two is *marked as inconsistent*. Inconsistent paths are not continued.

This requires two comments. First, even if the premises are inconsistent, it is still possible that neither A nor $\sim A$ can be justified. So it seems advisable to give up inconsistent paths and, if necessary, to apply Explosion moves. Next, inconsistent paths should *not* be marked in diagrams for the logic $\mathbf{CL}^-$, where they are required for demonstrating the inconsistency of the premise set.

Here is a simple example of an inconsistent search path in a diagram for $p \supset (q \supset r), s \supset p, \sim s \supset q, s \wedge t, t \supset q \vdash r$. The boxes containing respectively p and $\sim p$ are I-marked. This indicates that at least *one* of them cannot belong to a successful path. As $p \supset (q \supset r)$ is the only premise of which r is a positive part, we have to find a path that does not contain $\sim p$. As $s \supset p$ is the only premise of which $\sim s$ is a positive part, we have to find an alternative justification for q, and q is a positive part of $t \supset q$.

---

[5] The qualification 'unjustified' is required in view of the *Ex Absurdo* justification from section 4.

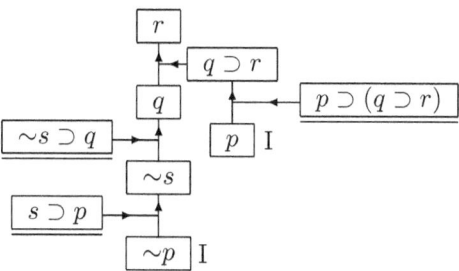

So we have to find an alternative justification for q along this road, which indeed leads to a justified path, viz. the right one.

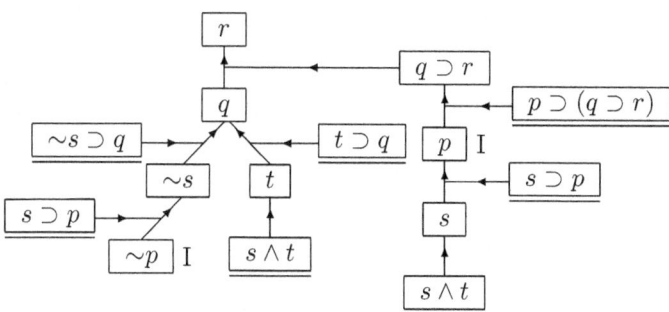

Redundant paths Where p is a path on a diagram, let B(p) be the set of unjustified bottom boxes of p. A path p is *marked as redundant at a stage of the diagram* iff, at that stage, there is a path p' such that B(p') ⊂ B(p) and the premises that are available on path p (including the ones available from downward moves to b-formulas) are also available on path p'. The underlying idea is obvious: if the members of B(p) can be justified by the premises, then so can the members of B(p'), but *not* vice versa. A path is marked as redundant by marking (with a R) an unjustified bottom node from which departs the 'lowest' arrow of the path. In terms of the previous paragraph, the obvious choice is to mark a node in p − p'.

By way of an example, suppose one is trying to derive $p \supset q$ from a premise set containing $p \supset r$ and $q \supset (r \supset s)$. If both logical moves are tried on the goal node, the top of the diagram may look as follows:

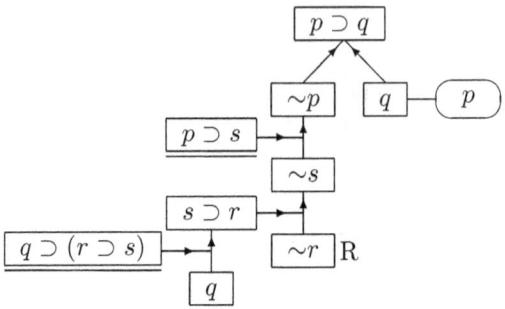

Clearly the right path will be justified if q is justified, whereas the left path is only justified if both q and $\sim r$ are justified. Two roads are open once the left path is R-marked. Either one continues the right path or one looks for an alternative justification for $\sim s$ or for $\sim p$.

Once a path p is marked as redundant at a stage of the construction, there can be no reason to remove the mark at a later stage. The example nicely illustrates this. Even if the justification of the box containing q on the right path would require the justification of $\sim r$, the right path would still be more efficient than the left one in that $\sim r$ has to be justified only once, viz. as part of the justification of q. Indeed, $\sim r$ would have to be justified twice on the left path, once because it is the formula of the R-marked bottom box and once for the justification of the box containing q.

Circular paths A path is *marked as circular* if both box b and box b′ belong to the path, b is a descendant of b′, and b and b′ contain the same formula. The underlying idea is that, if the lower box b can be justified by a chain of diagrammatic elements, then this chain can used to justify the higher box b′, which results in a shorter path. A path is marked as circular by marking (say with a C) the descendant box b.

By way of an example, suppose one is trying to derive p from a premise set containing $q \supset p$ and $r \equiv q$. The diagram might start as follows:

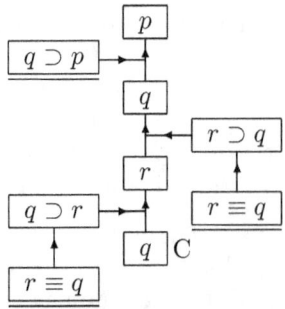

The only node to be justified is the (single box) bottom node containing q. If this node can be justified, the justification may just as well be attached to the higher box containing q. So one either looks for an alternative justification of r, or for an alternative justification of q, or for an alternative justification of p.

10 In conclusion

The advantages of the diagrammatic approach show if one compares it to the prospective dynamic proofs from [Batens and Provijn, 2001].

First, the diagrammatic approach has the advantage to clearly identify the paths, and to make the application of the marking definitions extremely transparent. This mainly results from the fact that a prospective element (a formula-plus-condition) may belong to several paths. So defining paths in terms of prospective proofs is tiresome and paths that have to be marked are easily overlooked. A further difficulty is that it is not always clear which prospective elements should be marked in order to indicate that a path has to be marked.

Next, the diagrammatic approach has advantages for the metatheory, for example for the completeness proof – this comes basically to showing that $\Gamma \nvDash A$, viz. that a model of the premises falsifies the conclusion, iff a model falsifies a non-justified bottom node of every path on a completed diagram.

It can be shown that the goal-directedness of the search method makes it more efficient than tableau methods. The method also stands out if compared to the goal directed methods from [Gabbay and Olivetti, 2000].

In view of what was said in section 7, the premises need not be preprocessed (for example turned into Horn clauses). An advantage is that if $\Gamma \vdash A$, and Γ' and A' are obtained by substitutions of letters occurring in Γ and A, then (if a decent strategy is followed) the proof for $\Gamma' \vdash A'$ has the same structure as the proof for $\Gamma \vdash A$.

Acknowledgement

Research for this paper was supported by subventions from Ghent University and from the Fund for Scientific Research – Flanders, and indirectly by the Flemish Minister responsible for Science and Technology (contract BIL01/80). I am indebted to Joke Meheus for useful comments on a former draft.

BIBLIOGRAPHY

[Batens and Provijn, 2001] D. Batens and D. Provijn. Pushing the search paths in the proofs. A study in proof heuristics. *Logique et Analyse*, pages 173–175:113–134, 2001. Appeared 2003.

[Batens, 2002] D. Batens. On a partial decision method for dynamic proofs. In H. Decker, J. Villadsen, and T. Waragai, editors, *PCL 2002. Paraconsistent Computational Logic*, pages 91–108, 2002. (= *Datalogiske Skrifter* vol. 95). Also available as cs.LO/0207090 at http://arxiv.org/archive/cs/intro.html.

[Batens, 2003a] D. Batens. A formal approach to problem solving. In C. Delrieux and J. Legris, editors, *Computer Modeling of Scientific Reasoning*, pages 15–26, Universidad Nacional del Sur, Bahia Blanca, Argentina, 2003.

[Batens, 2003b] D. Batens. Notes on problem solving, 2003. Technical report.

[Batens, in print] D. Batens. A procedural criterion for final derivability in inconsistency-adaptive logics. *Journal of Applied Logic*, in print.

[Batens, to appear] D. Batens. It might have been Classical Logic, to appear.

[Gabbay and Olivetti, 2000] D.M. Gabbay and N. Olivetti. *Goal-Directed Proof Theory*. Kluwer, Dordrecht, 2000.

[Hintikka, 1999] J. Hintikka. Is logic the key to all good reasoning? In *Inquiry as Inquiry: A Logic of Scientific Discovery*, pages 1–24, Dordrecht, 1999. Kluwer.

[Provijn, 2002] D. Provijn. How to obtain elegant Fitch-style proofs from goal directed ones. In H. Blockeel and M. Denecker, editors, *Proceedings of the Fourteenth Belgium-Netherlands Conference on Artificial Intelligence*, pages 243–250, Leuven, Belgium, 2002.

[Smullyan, 1995] R.M. Smullyan. *First Order Logic*. Dover, New York, 1995. Original edition: Springer, 1968.

Diderik Batens
Centre for Logic and Philosophy of Science
Universiteit Gent, Belgium
Email: Diderik.Batens@ugent.be

Diagrams as Physical Models to Assist in Reasoning

BALAKRISHNAN CHANDRASEKARAN

ABSTRACT. Diagrams serve a variety of roles in applied reasoning. At times, they are best viewed as "sentences" in a 2-D language, with specialized rules of inference that provide new diagrams. Other times, however, they are best understood as providing a physical model, much like an architectural model of a building or a 3-D molecular model of a chemical compound, for a state of affairs. In this paper, I discuss the notion of a physical model for a logical sentence, and the role played by the causal structure of the physical medium in making the given sentence as well as a set of implied sentences true. When the physical model is *prototypical*, it supports the inference of certain other sentences for which it provides a model as well. I also informally discuss a proposal that diagrams and similar physical models help to explicate a certain sense of *relevance* in inference, an intuition that so-called Relevance Logics attempt to capture.

1 Introduction

In this paper[1] I wish to consider some issues related to the use of diagrams as aids in reasoning. A particular use of diagrams corresponds to treating the diagram as a physical model for the premises. The problem solver sees that the representation is also a model for another assertion that is not explicitly part of the premises, and concludes that the assertion follows from the premises. The phenomenon of interest is not specific to diagrams – other examples of external representations as part of problem solving include 3-dimensional architectural models during design and physical 3-dimensional molecular models in problem solving in chemical engineering– but diagrams are ubiquitous, so we will largely focus on them in the paper. Regarding a main concern of logic – accounting for justifiable inferences – this style of reasoning based on a physical model needs to be part of any account of

---

[1] An earlier version of the paper was presented under the title, "Diagrams as Physical Models", at Diagrams 2006 – Fourth International Conference on the Theory and Application of Diagrams, June 28-30, 2006, Stanford University, CA, USA. The Proceedings are to be published by Springer as part of their Lecture Notes in Computer Science series.

Lorenzo Magnani, editor, *Model–Based Reasoning in Science and Engineering*, pp. 285–300 © 2006, B. Chandrasekaran

natural reasoning. It puzzles me that more has not been said in logic about the use of such physical models as aids to reasoning, given how prevalent diagrams are in everyday as well as professional reasoning, and the role played by architectural and molecular models in their respective disciplines. So this paper's goal is to raise the profile of physical models in logic. I raise a set of issues for deeper consideration by logicians.

Before I get to the main issues of concern, I take a brief detour on the multiple, and at times contradictory, senses of the term "model" in science, logic, and cognitive science. This seems appropriate in a conference devoted to model-based reasoning.

2 Various senses of "model"

The term "model-based reasoning" is in a sense redundant – all thinking and reasoning is model-based in the sense that these processes involve making use of knowledge (or beliefs) about the domain of discourse, i.e., a model of the domain. Craik [1983] specifically characterized cognitive activity in general as model–based in this sense, though he left unspecified the form in which the model was represented.

The term has been used in multiple, sometimes opposite ways, leading to much confusion. In one sense, a domain is a model of a description; in another usage, a description is offered as a model of a domain. For example, in logic, a domain provides a model of a set of axioms, e.g., arithmetic is a model of Peano's Axioms and plane geometry is a model of Euclidean Axioms. If the axioms are a description, the domain that fits the description is a model of the description. A different but related usage is a model for a sentence that is constructed by assigning truth values to the elements of the Herbrand Universe.

In philosophy and practice of science, the direction is from domain to description. A description – a set of equations, e.g. – is a model of a domain if the description can be used to predict phenomena in the domain. Thus, Maxwell's Equations model electro-magnetic phenomena and physicists speak of the Newtonian model versus the Einsteinian model.

In cognitive science the term "mental model" [Johnson-Laird, 1983] is used to describe a postulated internal diagram-like representation during syllogistic reasoning, containing a model of the situation. This is close, but not identical, to the sense of model in logic. There are additional usages, such as the phrase "model-based reasoning" as it was used, in the 1970-80's, in the literature on diagnostic reasoning in AI. One type of representation was a collection of heuristic relations between observations and diagnostic hypotheses. This was called *rule-based*, and was contrasted with a *model-based* representation that made use of knowledge about the functions and the structure, i.e., the components and their connectivity, of the device being diagnosed [Chandrasekaran and Mittal, 1983; Chandrasekaran et al., 1989]. Perhaps the term "structural model-based" might have been more descriptive, since even the set of diagnostic rules in the "rule-based" approach still constitutes a model of the device.

As mentioned, in this paper, I am concerned with the use of diagrams to assist problem solving and reasoning, where the diagram is a model in the sense of logic, a "situation" whose particulars make the sentence true. However, I wish to look at the diagram as a *physical* model.

3 Roles of diagrams in reasoning

Diagrams perform many different types of assistance during problem solving. We identify four roles here: helping extend short term memory, helping organize problem solving by spatial organization of related variables, providing a model of the premises so that plausible subtasks in theorem proving may be hypothesized, and providing a model of the premises from which consequents can be inferred and asserted.

First, they extend short term memory, by providing a spatially organized external location in which to note down information. Second, they help *organize* problem solving. Larkin and Simon [Larkin and Simon, 1987] use the example of analyzing a pulley system – they show how the diagram of the pulley system helps the problem solver organize the sequence of equations to solve, or variables to assign values to. The problem solver can use his visual perception to locate the pulley that a strip of rope goes over, and thus to choose which tension variable to consider next.

A diagram might provide a model of the set of premises, i.e., it depicts a situation that satisfies the description. In the third role for diagrams, and the first role for diagrams as models, the model suggests hypotheses. This is exemplified by diagrams as used in proving theorems in Euclid. In this kind of use, a diagram is a model of the premises.

The problem solver knows that it is not a general model, i.e., while what is asserted in the premises is true in the diagram, all that is true in the diagram doesn't necessarily follow from the premises. In any case, nothing can be asserted on the basis that it is true in the diagram. Nevertheless, it can provide information that can be used to set up and select subtasks. For example, the fact that two angles are adjacent, and the theorem involves one of the two adjacent angles might suggest to the theorem prover that perhaps stored theorems involving adjacent angles may be useful in advancing the proof. Lindsay [1988] provides a review of the issues in the use of diagrams in geometry theorem proving. It has been estimated that the use of the diagram in this way reduces the search space by an order of 300 or more. In the traditional use of Venn or Euler Diagrams in proving theorems in Set Theory, the diagrams play a similar role. It is important to emphasize that the information from the model is not *asserted* as conclusion, but only used to find other strategies for arriving at the general conclusion. It may be pointed out in passing that the heuristic use of Venn and Euler diagrams in proving propositions in Set Theory stimulated a direction of research by Barwise and associates [Allwein and Barwise, 1996] on new proof procedures in which certain sequences of diagrams can be considered proofs in themselves. As I understand it, the diagrams are viewed as "diagrammatic sentences", i.e., as two-dimensional syntax–controlled compositions of diagrammatic symbols[2].

The fourth role for diagrams is that they directly support inference, when they are models of the premises. The information obtained from the model is asserted as conclusion in the general case, though exactly when and how much to generalize are issues for which answers differ from one diagrammatic application to another. This role of diagrams is my focus in this paper.

Let us consider two very simple examples. Given a simple addition problem in arithmetic, say to show $1 + 3 = 2 + 2$, suppose we draw four points (or arrange four stones on the ground) as below:

Figure 1.

Under the appropriate mappings, the situation is a model of $1+3$. But it is also a model of $2+2$. We can demonstrate to a child that $1+3 = 2+2$ by using the above diagram. Here the generalization issue is trivial: the child could, but typically wouldn't, say, "Maybe this is true when we add 1 star to 3 stars, but is it true when we add 1 slice of pizza to 3 slices of pizza?". Our intuitions about numbers seem sufficiently robust that this issue doesn't arise in a child or an adult. "Individuals that keep their distinct identity"

---

[2]The Stanford Encyclopedia of Philosophy entry on Model Theory (http://plato.stanford.edu/entries/model-theory/) says, "[...] the overwhelming tendency of this work is to see pictures and diagrams as a form of language rather than as a form of structure. For example Eric Hammer and Norman Danner (in the book edited by Allwein and Barwise) describe a 'model theory of Venn diagrams'; the Venn diagrams themselves are the syntax, and the model theory is a set-theoretical explanation of their meaning". It is worth noting, however, that the Hyperproof work by Barwise and associates is not in the "diagrams as sentences" framework. It is closer to the diagrams as models view of the current paper.

seems to be the background intuition that is operational here, and using that we generalize from star marks on paper or stones on the ground to numbers in general.

Let us consider another example, one that we will use often in the rest of the paper. Given "If A is to the left of B, B is to the left of C, is A to the left of C?", people often draw a diagram as in Figure 2:

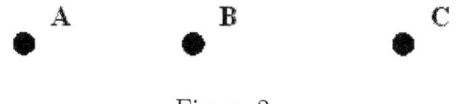

Figure 2.

There is a natural sense in which the physical diagram is a model of the problem situation[3]. The problem solver notices that indeed A is to the left of C, and declares that the inference is true. Of course, the diagram only represents one specific way in which the points can be located to provide a model. Yet the problem solver makes bold to assert that the conclusion is true for all the specific ways in which the points could be located. Let us first consider how to relate the notion of physical models to the sense of model in predicate logic.

4 Physical fragments providing models for logical sentences

Let us start by restating the standard definition of a model for a sentence in Logic. An *interpretation* for a sentence S consists of:

- A non-empty, possibly infinite, domain D of individuals

- Assignment of specific individuals in D to constant symbols in S

- Assignment to each n-ary function symbol in S of an n-ary function that maps from D^n to D.

- Assignment to each n-ary predicate symbol in S of an n-ary function that maps from D^n to $\{\text{True}, \text{False}\}$.

An interpretation for S is a *model* for it if S evaluates to True under the interpretation.

---

[3]More precisely, it a model of the conjunct of the given premise with axioms that capture the structure of space in terms of which the predicate Left is defined. For someone for whom the semantics of Left is that of spatially left in ordinary language, the axioms are implicit, and the Figure provides a model for the premise.

4.1 Modeling physical things

Our goal here is to show how a physical entity may be used to provide an interpretation of a sentence.

Domain of Individuals. Let Π be a fragment of physical world. Let $\Delta_\Pi : \{\pi_1, \pi_2, \ldots\}$ be a (possibly infinite) set of entities, each π_i a *part* – a subfragment – of Π. The entities need not be physically disjoint – one entity may be a physical part of another entity; nor it is necessary for Δ_Π to exhaust Π, i.e., the totality of physical fragments represented by the elements of Δ_Π to be equal to the matter represented by Π.

We will use two concrete examples to illustrate the ideas: Π_1, the set of points constituting a finite physical horizontal straight line, say drawn on a piece of paper; and Π_2, a physical object intended to be an architectural "model" of a house.

Example Π_1: The entire finite straight line is Π. Each point in it is a π, thus Π has an infinite number of parts in this model. Another model for the same physical object might subdivide the line into various segments, each providing a π.

Example Π_2: The physical entity (the architectural "model") as a whole is Π, and the physical matter corresponding to various rooms, walls, doors, etc., are the π's.

Functions and Predicates. Let $\{\phi_i | 0 \leq i \leq k\}$ be a finite set of functions of various arities, such that if ϕ_i is n-ary, it is a function from Δ_Π^n to Δ_Π. Similarly, let $\{\rho_i | 0 \leq i \leq l\}$ be a set of functions of various arities, such that if ρ_l is n-ary, it is a function from Δ_Π^n to $\{T, F\}$. The ρ's are predicates defined on the physical variables, and thus the values that they take for their various arguments is determined by the causal structure of Π.

Example Π_1: The function, $\text{right}_1(\pi_i)$, defined as "the point that is exactly 2 inches to the right; if there is no such point, the right end point" is a unary function. Example of a binary predicate ρ is: $\text{Left}(\pi_i, \pi_j)$, with the obvious interpretation.

Example Π_2: Unary function $\text{Entrance-to}(\text{room}_i)$, which takes values from the subset of parts of type "door". Thus, e.g., $\text{Entrance-to}(\text{room}_5) = \text{door}_6$.

Example of ρ: $\text{Bigger-than}(\text{room}_i, \text{room}_j)$ is a binary function which evaluates to True if the area of room_i is larger than that of room_j, and False otherwise.

Properties and Causal Structure of Π. For the purpose at hand, the physical structure is modeled in terms of a set of variables, selected *attributes* of the physical system. A specific physical instance will have specific values for these variables. Let $\Theta_i : \{\theta_{i1}, \theta_{i2}, \ldots, \theta_{ik}, \ldots\}$ be a set of variables in terms of which entity π_i is modeled, and let $\Theta = \bigcup_{i=1}^{n} \Theta_i$. The *causal structure* of Π, which constrains the values of the variables in Θ, determines the truth values of the various predicates for various values for their arguments, and thus the truth values of sentences composed out of these predicates.

Thus, part of modeling a physical fragment for the purpose of providing an interpretation for a sentence involves identifying a physical system with the right properties to provide an interpretation, and then setting its parameters to where the fragments provides a model for the sentence.

Example Π_1: Let part π_i be modeled in terms of a single variable x_i, the x-coordinate of π_i from some origin. Left(π_i, π_j) is defined by the values of x_i and x_j. Additionally, the constraints of the physical line result in constraints between predicates: if Left(π_i, π_j) and Left(π_j, π_k) are both True, then Left(π_i, π_k) is constrained to be True.

Example Π_2: The parts of the house may be modeled in terms of their length, width, height, area, etc. Color and material out of which a part is made may also be in the set of variables. Whether room$_5$ is larger than room$_3$ is fully determined by the physical dimensions of Π; there is no additional freedom to assign T or F. Additionally, if in a physical architectural model room$_1$ is larger than room$_2$ which in turn is larger than room$_3$, the model will necessarily satisfy the predicate, larger-than(room$_1$, room$_3$).

In order to avoid confusion between different usages in science and engineering on one hand and in logic on the other, we use the term *p-model* to refer to a description of a physical entity as in the next definition.

Definition. A *p-model* of a physical fragment Π consists of the following specifications:

- Δ_Π, a set of individuals consisting of parts of Π.

- A set $\{\phi\}$ of functions of various arities, such that an n-ary function is a mapping from Δ_Π^n to Δ_Π.

- A set of functions $\{\rho_l\}$ of various arities, such that an n-ary function is a mapping from Δ_Π^n to $\{T, F\}$.

- a set of variables Θ in terms of which Π and elements of Δ_Π are modeled; a causal structure Π_{ax}, that defines the causal constraints between the variables in Θ.

Remark. There is an infinity of p-models for a given physical entity.

4.2 A physical entity supporting a logical model

Let a p-model M_Π of a physical fragment provide an interpretation for a sentence S, i.e., each constant symbol in S is mapped to a specified element in Δ_Π, each n-ary function symbol in S is mapped to a function in $\{\phi_k\}$, mapping from Δ_Π^n to Δ_Π, and each n-ary predicate symbol in S is mapped to a function in $\{\rho_l\}$, mapping from Δ_Π^n to $\{\text{True}, \text{False}\}$.

Definition. If a sentence S evaluates to True under the interpretation provided by a p-model M_Π of a physical fragment Π, we say that Π provides a *physical model* for S.

Remark. What makes a predicate true or false in a physical model is that the variables take specific values in the physical fragment, and the causal structure Π_{ax} constrains values between variables.

Examples

Consider the following sentence S:

$$\forall x \forall y \forall z (L(x,y) \ \& \ L(y,z) \rightarrow L(x,z)) \tag{1}$$

Let Π be a physical 1-D spatial line fragment, and let the following be a p-model M_Π for Π.

- Δ_Π : the (infinite) set of points in the line fragment, $\{x_i\}$.

- Θ: a single attribute, the co-ordinate of a point x_i with respect to some origin.

- $\{\phi\}$: null set.

- $\{\rho_l\}$: a single function, Less-than(x_i, x_j) = True, if the co-ordinate of x_i is less than that of x_j; False, otherwise. $\tag{2}$

Under the interpretation M_Π, S True in Π. A physical 1-dimensional line fragment is thus a physical model for S.

As more complex example, consider S':

$[(\forall x \forall y \forall z (L(x,y) \ \& \ L(y,z) \rightarrow L(x,z)) \ \&$

$(\forall x \forall y (L(x,y) \ \& \ L(y,x) \rightarrow \text{Eq}(x,y))] \ \&$

$L(A,B) \ \& \ L(B,C) \tag{3}$

Consider the physical diagram in Figure 2, with the following M'_Π.

M'_Π: M_Π as defined in (2) plus the following assignments: Constant A, B and C assigned to the points in the 1-dimensional line fragment corresponding to the coordinates as in the Figure. Eq(x,y) assigned to function "Equal(x_i, x_j) = True iff x_i is the same as x_j, and False otherwise". (4)

Under the interpretation M'_Π, S' evaluates to True, so the diagram in Figure 2 is a physical model for S'. Readers will recognize (3) as a simple axiomatization of left-ness plus the premises of the problem we stated at the beginning. M'_Π is also a model for the following:

$[(\forall x \forall y \forall z (L(x,y) \& L(y,z) \rightarrow L(x,z)) \&$
$[(\forall x \forall y (L(x,y) \& L(y,x) \rightarrow \text{Eq}(x,y))] \&$
$L(A,C)$ (5)

Remark. In applied reasoning, the agent is reasoning in some domain of interest, D, and he is interested in making a model of a sentence, say S. Let D_{ax} be the set of axioms that describe the relevant aspects of the domain of interest. Thus, the agent is looking for a physical model of $D_{ax} \& S$. When we say that Figure 2 is a model of Left(A, B) & Left(B, C), it is because we interpret Left in the spatial meaning of the terms. This interpretation assumes D_{ax}. If instead S were Goo(A, B) & Goo(B, C), we wouldn't see Figure 2 as its physical model. Successfully making a physical model of S when the agent is reasoning in D involves finding a physical medium such that its causal structure Π_{ax} has the right kind of homomorphism relation with D_{ax}.

There is no requirement that an arbitrary Π have a p-model that provides an interpretation for an arbitrary sentence S. In fact, it is a special situation where a physical model can be constructed so as to provide an interpretation for a sentence. In the next section, we discuss how such physical models are often used.

Warrant for generalization

Figure 2 is a model for S', but it is just one model. There are infinitely many configurations of points for which the corresponding physical diagrams will provide a model for S'. Nevertheless, we generalize the inference to a class of situations. Figure 2 also provides a model for "A is farther left of B than B is of C", but we know that this inference cannot be generalized.

This is quite common in applied reasoning. A chemist, who is considering whether $S_1 \rightarrow S_2$, where S_1 and S_2 are sentences in his domain, might construct a chemical reaction which is a model of S_1 (really a model of his domain axioms and S_1), see if it is also provides a model for S_2, and, though the specific chemicals in interaction model only instances of S_1, generalize to the larger class. Of course, a good chemist would know just want sort of model to construct that would bear the generalization.

This style of proof might be called *physical-model-based proof*. The *Model-Based Rule of Inference* may be stated as follows:

Given an inference problem, $S_1 \to S_2$, where S_2 is not a logical truth, in domain D with domain theory D_{ax}, and given a physical fragment Π such that it provides a p-model M_Π that satisfies $D_{ax} \,\&\, S_1$, if M_Π also satisfies $D_{ax} \,\&\, S_2$, and if M_Π *has warrant for generalization with respect to the inference* S_2, conclude $S_1 \to S_2$ in the general case in D.

One might use the term *prototypical* to describe a model that provides such a warrant for generalization.

Remark. The rule of inference blocks asserting a logical truth based on the physical model. This is because we wish the physical model to play a role in the assertion. This is related to our remarks on Relevance Logics in the next section.

In many cases, the applied reasoner has limited or no access to D_{ax} is an explicit form. However, he has a body of intuitions and practices that help him construct prototypical models for classes of S's that provide warrant for generalization and help him scope them.

Let S be a sentence in a domain D characterized by axioms D_{ax}. Let Closure($D_{ax} \,\&\, S$) be the set of all inferences that are deducible from $D_{ax} \,\&\, S$. If Π provides a model for S, it will also provide a model for all elements of Closure($D_{ax} \,\&\, S$). However, it will also provide a model for many other inferences that are not in Closure($D_{ax} \,\&\, S$). The reasoning agent needs to know how not to make the inferences that are not in Closure($D_{ax} \,\&\, S$), even though the model supports it, e.g., not to infer "A is farther left of B than B is of C" from Figure 2.

5 Prototypical models

What makes a prototypical model? The specifics depend on the domain and the predicates of interest, but some general intuitions may be useful. The following ideas might help in the development of a more formal account.

The first idea is *minimality*. Let S be Left$(A, B)\,\&\,$Left(B, C) (we are implicitly in the domain of 1-d space with a directional axis). Just as Figure 2, Figure 3 also provides a model for S. However, it provides a model for unrelated things such as Inside(D, E). Clearly, any inference based on this model, such as Left$(A, B)\,\&\,$Left$(B, C) \to$ Inside(D, E), would be a mistake. Figure 2 is in some sense minimal compared to Figure 3 for Left(A, B) & Left(B, C).

Figure 3.

The next idea is that of multiple prototype models. Let S' be Left(A, B) & Left(A, C). While Figure 2 provides a model for S', it doesn't seem prototypical for another reason: it only accounts for a subset of instances. Figure 4 provides another model.

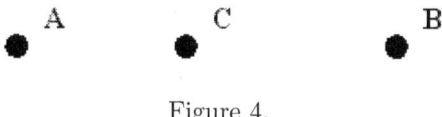

Figure 4.

Figure 2 provides a model for Left(B, C) and Figure 4 provides a model for Left(C, B), neither of which follows from S', thus neither of these inferences have a warrant for generalization. Applied reasoning in this case requires that two models be set up, each of which allows certain inferences, say Right(B, A), but not Left(B, C) or Left(C, B).

The third idea is a revisit of what we mentioned earlier, that Figure 2 doesn't support the generalization of "A is farther left of B than B is of C", though the figure provides a model for it. Suppose a new predicate boogoo(x, y, z) is defined as "x is farther left of y than y is of z". Consider S'': Left(A, B) & Left(B, C) & Left(C, D) & Left(D, E), modeled by Figure 5. Figure 5 also provides a model for boogoo(A, C, D), which has a warrant for generalization, unlike boogoo(A, B, C) in Figure 2.

Figure 5.

It can be seen, from a metatheory of Left(x, y), that it provides an order between x and y. Thus any conjunctions of Left(x,y)'s would specify an order or alternate possible orders. Only information that follows from the order information has warrant for generalization. In every domain, in principle such metatheories may be constructed, but in practice, an agent performing applied reasoning in some domain usually has no access to such metatheories. Even when he has access to the axioms for his domain, such as in the sciences, they are provisional and potentially revisable. So, reasoning in the practical world is aided by models constructed and interpreted with the aid of intuitions based on experience, training and partial theories. Such models play a large role in commonsense reasoning as well, as evidenced by the research of [Johnson-Laird, 1983] on how people solve syllogistic problems by constructing mental models that have many points of contact with the models that we describe here. Because people lack access to fully worked–out metatheories, some of the reasoning errors that occur in practical reasoning are due to mistakes in the application of generalization and in the construction of prototypical models.

Our everyday reasoning as well as reasoning in professional disciplines is full of implicit and explicit guidelines about how to construct diagrams and physical models that give the desired information perceptually and how to generalize. The ubiquity of such diagrams in our reasoning should not keep us from appreciating the hard-won nature of discoveries of appropriate diagrams for classes of problems. Such discoveries are prized – transmitted culturally for everyday reasoning, and made part of training in professional disciplines, with discoverers often honored with awards.

6 Relevance logics and physical models

Relevance Logics are the subject of a substantial body of work, a summary of which is available in Stanford Encyclopedia of Philosophy [Encyclopedia, 2005]. My knowledge of the area is limited and thus my goal will be correspondingly modest – to consider, informally, possible connections between the idea behind Relevance Logics and physical models so as to invite further attention to this connection.

Relevance logics are a response to what some people take to be paradoxes of traditional implication. The paradox arises in that in some of the inferences authorized by the semantics of traditional implication the antecedent doesn't seem *relevant* to the consequent. Examples of paradoxes mentioned in the Stanford Encyclopedia page are:

Material implication paradoxes:

- $p \to (q \to p)$
- $\sim p \to (p \to q)$

- $(p \to q) \vee (q \to r)$

Strict implication:

- $(p \,\&\, \sim p) \to q$
- $p \to (q \to q)$
- $p \to (q \vee \sim q)$

Now suppose a reasoning agent performing domain-specific reasoning uses prototypical physical models to assert consequences, as described in earlier sections of the paper.

The consequences he asserts based on physical models, at least for the examples we will consider, will not fail the test of relevance. That is, given p, he constructs a prototypical model for it in his domain W. If the model is also a model for S, and if S has a warrant for generalization, $p \to S$ will be a relevant inference *in* W. Technically, any model is also a model for logical truths, so we add a rule forbidding asserting a logical truth as a consequence of any sentence.

Consider as an example:

$$p \to (q \to p) \qquad (6)$$

In this case p doesn't seem to play a relevant role in making $(q \to p)$ true. However, suppose there is a domain D in which q is a possible cause for p, and that q would definitely cause p. In that domain, it wouldn't be a surprise to assert that if p is true, then if q is known to be present, q caused p, and consequently that the truth of q would imply the truth of p. If D_{ax} denotes the axioms characterizing D, the following would not fail the test of relevance:

$$(D_{ax} \,\&\, p) \to (q \to p) \qquad (7)$$

Suppose D describes a physical domain in which we construct a physical fragment Π that provides a p-model for q. Π, under the same p-model mappings, would also provide a model for p. That is, there is no way to construct a model for q in D without it being a model for p as well. In this case, there is no failure of relevance to assert (6). Similarly, consider an example of strict implication,

$$(p \,\&\, \sim p) \to q \qquad (8)$$

There would be no way to make a physical fragment in any domain that provides a model for $p \,\&\, \sim p$. No one will be able to assert (8) based on a physical model.

A physical model constructed for $\sim p$ will by definition not provide a model for $p \to q$, so based on the Model-Based Rule of Inference, we won't be able to assert $\sim p \to (p \to q)$.

Suppose in some domain D it is the case that $(p \to q) \vee (q \to r)$, i.e., $D_{ax} \to ((p \to q) \vee (q \to r))$. In this domain, any model that supports p will be a model for q, and any model that supports q will be a model for r. The inference calls for two prototypical models, one created to support p and the other for q. The agent will correctly assert $(p \to q) \vee (q \to r)$, and the implications are relevant.

Consider the strict implication:

$$p \to (q \to q) \qquad (9)$$

A prototypical model constructed for p would not be a model for q, but it would trivially be a model for $q \to q$. However, because of the rule that logical truths cannot be asserted as consequences of anything from a physical model, the agent will not be able assert (9). If p is Left(A, B) and q, Inside(C, D), with the obvious interpretations for the predicates in the spatial domain, the prototypical model for p will simply be two objects with labels A and B, with A to the left of B. This diagram will not provide a model for Inside(C, D) and thus the applied reasoner will not assert (9) based on the physical models. The same rule will also block the agent from asserting: $p \to (q \vee \sim q)$.

Our discussion above leads to the conclusion that judgments of relevance with respect to implications are with respect to specific domains, whose structures (causal structures if domains are physical) then can be judged to play or not play a role in the antecedent making the consequent true. That is, the issue of relevance in the various implications that we considered arises in reasoning in specific domains. Further, making such judgments is facilitated in the specific domains by constructing physical models when possible.

7 Concluding remarks

My interest in the issue of models in reasoning arose in the context of my interest in the use of diagrams. Diagrams are ubiquitous reasoning aids in many situations. I have been interested in the idea that diagrams are just the most prominent example of larger class of reasoning aids that provide physical models to premises in some domain. My goal in this paper has been to explore the logic of such physical models. Applied reasoning, where reasoning agents are concerned with inferences in specific domains rather than abstract notions of validity, has not drawn as much attention from logicians as it should. I think, however, that the use of such physical models in applied reasoning raises important issues in logic. I have attempted to formalize the notion of some piece of physical matter providing a model for a sentence. I have identified a proof technique called physical model based inference in which the prototypical models in specific domains are constructed that support useful generalizations. Developing reasoning skills in various domains includes learning or developing intuitions about how to construct prototypical models for specific reasoning situations. I also related such physical model-based reasoning to issues in Relevance Logics, where the goal is to identify when and how antecedents can be said to have a role in the consequent being true. Physical models, by incorporating the underlying causality of the domain, make it possible, under many conditions, to see whether or not the antecedents play a role in an implication being valid. My goal in this paper is to invite the attention of logicians more expert than I to look into what I think are important problems.

Model-based reasoning of the kind I have discussed is not merely an issue in logic, but in artificial intelligence. AI has focused almost exclusively on what might be called linguistic representations, mirroring the logical form of natural language sentences. However, real reasoning in humans is multi-modal, with perceptual and kinesthetic modalities often contributing to problem solving. Diagrams provide an important window into such multi–modal representations. In [Chandrasekaran et al., 2004], we describe a diagrammatic representation and reasoning architecture that integrates traditional symbolic reasoning with diagrammatic reasoning.

Acknowledgments

This paper was prepared through participation in the Advanced Decision Architectures Collaborative Technology Alliance sponsored by the U.S. Army Research Laboratory under Cooperative Agreement DAAD19-01-2-0009, and by federal flow-through by the Department of Defense under contract FA8652-03-3-0005 (as a subcontract from Wright State University and Wright Brothers Institute). I am indebted to Peter Schroeder-Heister for his comments on an earlier draft, specifically his suggestions on prototypical models and Relevance Logics. My comments here are elaborations on his basic points. I thank Gerard Allwein, Neil Tennant, Stewart Shapiro, and the anonymous reviewer that Lorenzo Magnani asked to read the paper, for helping in various ways to make this paper have fewer errors than would be the case otherwise. Finally, I thank Lorenzo for inviting me to the conference and for his organizational leadership in this area of research.

BIBLIOGRAPHY

[Allwein and Barwise, 1996] G. Allwein and J. Barwise, editors. *Logical Reasoning with Diagrams.* Oxford University Press, Oxford and New York, 1996.

[Chandrasekaran and Mittal, 1983] B. Chandrasekaran and S. Mittal. On deep versus compiled knowledge approaches to medical diagnosis. *International Journal of Man-Machine Studies*, 77:425–436, 1983. Special Issue on Expert Systems.

[Chandrasekaran et al., 1989] B. Chandrasekaran, J.W. Smith (Jr.), and J. Sticklen. Deep models and their relation to diagnosis. *Artificial Intelligence in Medecine*, 1(1):29–40, 1989.

[Chandrasekaran et al., 2004] B. Chandrasekaran, U. Kurup, B. Banerjee, J. Josephson, and R. Wilkler. An architecture for problem solving with diagrams. In A. Blackwell, K. Marriott, and A. Shimojima, editors, *Diagrammatic Reasoning and Inference*, pages 151–165, Berlin, 2004. Springer.

[Craik, 1983] K. Craik, editor. *The Nature of Explanation.* Cambridge University Press, Cambridge, 1983.

[Encyclopedia, 2005] Encyclopedia. Stanford encyclopedia of philosophy, 2005. http://plato.stanford.edu/entries/logic-relevance/.

[Johnson-Laird, 1983] P. Johnson-Laird, editor. *Mental Models: Towards a Cognitive Science of Language, Inference, and Consciousness.* Cambridge University Press, Cambridge, 1983.

[Larkin and Simon, 1987] J. Larkin and H. Simon. Why a diagram is (sometimes) worth ten thousand words. *Cognitive Science*, 11:65–99, 1987.

[Lindsay, 1988] R.K. Lindsay. Using diagrams to understand geometry. *Computational Intelligence*, 14:238–272, 1988.

Balakrishnan Chandrasekaran
Laboratory for AI Research
Department of Computer Science and Engineering
The Ohio State University
Columbus, OH, USA
Email: chandra@cse.ohio-state.edu

A Formal Model of Abduction

DOV GABBAY AND JOHN WOODS

ABSTRACT. This paper examines the basic components needed to set up a formal model of abduction. It is meant to be read in conjunction with our coompanion paper "Advice on abductive logic".

The conclusion we reach is that we need features such as structured data in databases. Procedures for input/output into databases and that abductive algorithms are dependent on the particular proof theoretic formulation of a logic and not just on its consequence relation.

1 Introduction

This note is a companion to our paper "Advice on Abductive Logic", [Gabbay and Woods, to appear]. Its aim is to give the reader an idea as to what sorts of issues formal models of *consequentialist* abduction we could reasonably be expected to address and what such models might look like. We imagine a certain cognitive agenda T, and a sentence V such that if V is obtained in a certain way, T would be realised. V is therefore a *pay-off* proposition for T. Given its role in relation to T, V may also be considered informally as a *goal*. Agendas can also be likened to targets. The structure of agendas is examined in detail in [Gabbay and Woods, 2003]. For the present it suffices to use the notion informally.

Our starting point is the AKM schema for consequentialist abduction[1]. In due course, we shall propose an alternative to the *AKM* approach, but given its popularity, it is a good place to start. In the *AKM*-model we have the following:

AKM 1 V (is the pay-off for T).

AKM 2 $\sim (K \looparrowright V)$, ($V$ does not follow from our knowledge base K).

AKM 3 $\sim (H \looparrowright V)$, ($V$ does not follow from our proposed hypothesis H).

---

[1]Among its supporters are [Aliseda-LLera, 1997; Kuipers, 1999; Magnani, 2001; Meheus *et al.*, forthcoming]. Hence the acronym "AKM".

AKM 4 $K(H)$ is consistent, (adding H to K to obtain $K(H)$ gives a consistent result).

AKM 5 $K(H)$ is minimal, (otherwise why not always take $K(V)$).

AKM 6 $K(H) \hookrightarrow V$, (the new theory $K(H)$ does yield V).

AKM 7 Therefore we are provisionally justified in assuming H.

Let us examine what kind of mechanisms are required in our logic to be able to model (AKM 1)–(AKM 7) above.

1. First we need to specify the structures that our knowledge bases K can take and explain what it means for us to move from K and H to $K(H)$, i.e. explain the process of insertion of H into K.

2. We also need a notion of the consistency of a knowledge base.

3. We need a notion of $K(H)$ being minimal relative to adding H to K. This can be part of the process of constructing $K(H)$ or of the definition of $K(H)$.

The above is not enough. Suppose we have as a goal V_1 and we add the hypothesis H_1. Suppose the process continues and we consider $V_2, V_3, \ldots$ and add $H_2, H_3, \ldots$ We thus get a sequence of knowledge bases $K(H_1)$, $K(H_1)(H_2)$. We need to put forward some theory about how to deal with such a sequence i.e. we need a theory of iterated abduction.

Note that if the agenda cannot be closed from our knowledge base K, then we abduce H and form $K(H)$. If later on we chance upon additional knowledge k_1 which, together with K can close the agenda, then we abandon H and move to $K \cup \{k_1\}$. But now note what has happened here. We formally have a database $K(H)$ and when we add k_1 to it, we decide to revise it (because of some internal cross provability relationships) even though it might be completely consistent.

Since multiple abductions are very common in in real life, our model should take them into account. Given our limited resources and limited time, a large chunk of our knowledge is abduced. Our knowledge bases are continuously updated with new hypotheses which are treated as action-enabling data. In fact, the area of iterated abduction is more central to modelling human behaviour than just one step abduction. This is not to minimise the importantce of one step abduction. As in real life, so too in the logic of abduction, small steps precede large steps.

Abduction admits of complexities beyond the fact of iterability. In a more comprehensive model than we are presenting here, such additional complexities would, of course, be taken into account. Such forms include:

a. *Multiple target abduction*, in which the closure of an agenda by way of a conditional in the form $K(H) \mathrel{\vbox{\hbox{$\scriptscriptstyle\hookrightarrow$}}} V$ is itself a state of affairs that closes a further target.

b. *Compound abduction*, in which for subsets K_1 and K_2 of K, and different payoff propositions V_1 and V_2, we have it that $K_1(H) \mathrel{\vbox{\hbox{$\scriptscriptstyle\hookrightarrow$}}} V_1$ and $K_2(H) \mathrel{\vbox{\hbox{$\scriptscriptstyle\hookrightarrow$}}} V_2$.

c. *Transitive closure abduction*, in which for certain (but not all) interpretations of $\mathrel{\vbox{\hbox{$\scriptscriptstyle\hookrightarrow$}}}$, we have it that $K(H) \mathrel{\vbox{\hbox{$\scriptscriptstyle\hookrightarrow$}}} V_1, V_1 \mathrel{\vbox{\hbox{$\scriptscriptstyle\hookrightarrow$}}} V_2$, hence that $K(H) \mathrel{\vbox{\hbox{$\scriptscriptstyle\hookrightarrow$}}} V_2$. For example, this would work when $\mathrel{\vbox{\hbox{$\scriptscriptstyle\hookrightarrow$}}}$ is construed as causal implication, but not for all interprtations in which $\mathrel{\vbox{\hbox{$\scriptscriptstyle\hookrightarrow$}}}$ is explanatory consequence.

Q1 Our present question is: Do we take into consideration when adding H_2 into $K(H_1)$ that H_1 is an abduced item of data, given that it does not have the same epistemic status as the rest of K? In other words, do we assume, as part of the structures of our knowledge bases, a variety of data with a variety of status degrees and do we let the insertion process take that into consideration?

The answer to (Q1) should be yes, because intuitively a common sense reasoner is sensitive to different kinds of abduced data. If we accept that, then we have to ask question 2.

Q2 Shall we have different types of insertions reflecting different policies of handling the already abduced data in K? In other words, if π denote an insertion policy, should we look at $K_\pi(H)$?

It looks more and more as though the framework we need is that of a Labelled Deductive System (LDS) where data is structured and labelled and different insertion policies can easily be formulated [Gabbay, 1996]. If this is the case, then we can ask our next question.

Q3 Why should we insist on $K(H)$ being consistent? We know from LDS methodology that we can easily (in fact more conveniently) work with a general labelled database, in which there are several notions of inconsistency and where the notion of consistency, although definable, is not central at all.

A more likely notion is that *acceptability*, that we want the database to have a certain kind of structure. It may be inconsistent, but structured in such a way that we can handle it, or it may be consistent but structured in a way that is unacceptable [Gabbay and Hunter, 1991; Gabbay and Hunter, 1993; Woods, 2005; Gabbay and Woods, 2005].

We now ask a very simple question. How do we effectively find this H? Without a proof theoretic algorithm for $\succ\!\!\!\sim$ (i.e. for checking whether $K \succ\!\!\!\sim V$ holds for arbitrary K and V) we cannot find such candidates H. The proposed criteria (AKM 4)–(AKM 6) (namely $K(H)$ is consistent and minimal in satisfying $K(H) \succ\!\!\!\sim V$) is too general and implicit.

We therefore need to assume that some algorithm $\mathcal{A}$ is available for checking whether $K \succ\!\!\!\sim V$ holds and that using this algorithm we can determine that (AKM 2) $\sim (K \succ\!\!\!\sim V)$ and (AKM 3) $\sim (H \succ\!\!\!\sim V)$ hold.

Our strategy in modelling abduction is to assume some general properties of this proof algorithm and define an abductive algorithm as a metalevel abductive mechanism $\mathcal{M}(\mathcal{A})$ which works on $\mathcal{A}$ trying to find a candidate H.

We now need to postulate some basic assumptions on $\mathcal{A}$.

$\mathcal{A}$ operates on data K and goals V. So it manipulates the pair $K \succ\!\!\!\sim ?V$. We can also reasonably assume that whatever we do next in this algorithm at a given point depends also on what we have done up to that point. This means that we also need to take into consideration the history of the algorithm. Call it $\mathbb{H}$. Thus our algorithm manipulates triples like

$$[K \succ\!\!\!\sim ?V; \mathbb{H}] \qquad (*)$$

This means our current data structure is K, our current goal to prove is V and the history of the computation up to this point is $\mathbb{H}$.

The algorithm must tell us at this point how to continue. The following are the options:

Option Fail
$\mathcal{A}$ says we cannot continue. No rule applies.

Option Succeed
$\mathcal{A}$ says we succeed (in this branch of the computation) (e.g. if $K = V$, $\mathcal{A}$ might say that we succeed).

Option Continue
$\mathcal{A}$ may have a stock of rules which might apply. A rule has the form below:

$$\text{General form of a computation rule } \mathbb{R}:$$
$$\frac{[K \succ\!\!\!\sim ?V, \mathbb{H}]}{\text{if}} \qquad (**)$$
$$\bigwedge [K_i \succ\!\!\!\sim ?V_i; \mathbb{H}_i]$$

In other words: to compute $[K \succ\!\!\!\sim ?V; \mathbb{H}]$ successfully we must succeed with all of $[K_i \succ\!\!\!\sim ?V_i; \mathbb{H}_i]$.

We must assume that according to some complexity measure $[K_i \leftarrowtail ?V_i, \mathbb{H}_i]$ are simpler tasks and that $\mathbb{H}_i$ is obtained from $\mathbb{H}$ by further recording of the rule we have just applied, and possibly abandoning some recorded history which is no longer needed (for example we might wish to remember only the last step!)

It is possible that several rules may apply. Success is assured if one of them can lead us to success.

What does it mean then that $\sim (K \leftarrowtail V)$ holds? This means that if we start our algorithm with the task $[K \leftarrowtail ?V; \varnothing]$, then no matter how we continue we always encounter a subtask of the form $[K' \leftarrowtail ?V'; \mathbb{H}']$ which should succeed (in order for the original task to succeed) but which is actually fails (option failure holds for this task).

How do we do abduction $\mathcal{M}(\mathcal{A})$ on top of an algorithm $\mathcal{A}$?

Abduction works as follows:

Given a task $[K_i \leftarrowtail ?V_i, \mathbb{H}_i]$ the abduction machine needs to tell us the following

(D1) Whether it is allowed to abduce at this point

(D2) What to abduce.

Assume that we are told to abduce H_i. It must be a simple and obvious choice (e.g. H_i can be V_i). At this point the intention is not to start a new process of abduction[2] but to choose something simple and immediate.

Once this simple and immediate H_i is chosen, then it is immediately clear what $K_i(H_i)$ is supposed to be and it is also immediately clear that the algorithm $\mathcal{A}$ applied to $[K_i(H_i) \leftarrowtail V_i, \mathbb{H}_i]$ succeeds.

An obvious choice of rules is to allow abduction (D1) only if the sole option available is the failure option (why abduce now if the computation can continue?[3]) and then D2 can choose something which can be calculated in a deterministic way out of K_i, V_i and $\mathbb{H}_i$. It must be a calculation which is sure to terminate and yield something.

(D1)–(D2) are not enough for our purpose. Remember that the original problem was $[K \leftarrowtail ?V; \varnothing]$ and that the current piece of computation is only a cog in a big logical machine. We need to be able to reconstruct H out of all the H_i. How do we do that? The simplest way is to attach with every

---

[2] In more complex systems there may be several algorithms for abduction and a hierarchy of when one process hands over to another. The hierarchy must not be circular.

[3] If the system talks about component failure in some machine or a system, then some components may be known from experience to be weak and likely to fail. In this case we abduce immediately even though we can continue and explore further.

rule $\mathbb{R}$ an abduction rule $\mathcal{M}(\mathbb{R})$ going upwards. So if $\mathbb{R}$ has the form

$$\mathbb{R} : [K \looparrowright ?V; \mathbb{H}] \text{ If } \bigwedge_i [K_i \looparrowright ?V_i, \mathbb{H}_i]$$

then we need a rule $\mathcal{M}(\mathbb{R})$ allowing us to know what to abduce for $[K \looparrowright ?V; \mathbb{H}]$ provided we know what to abduce for each $[K_i \looparrowright ?V_i; \mathbb{H}_i]$.

It may be useful to say that we always abduce something, even when $K \looparrowright V$ is successful, in which case we abduce $\top$ (truth) or something harmless (a *unit* such that $K(unit) = K$).

The reader may think that $\mathcal{M}(\mathbb{R})$ is just a technical rule, but actually there is a much deeper phenomenon here. The algorithm $\mathcal{A}$ is in practice a meaningful algorithm in the application area it addresses. Put differently, $\mathcal{A}$ follows natural lines of reasoning in the application area. This means that the backward propagation of abduction (the family of rules $\mathcal{M}(\mathbb{R})$) can also have a meaning. But let us stop and reflect on what we are saying here! We are saying that the abductive process is concerned not only with finding a hypothesis H when needed, but also with providing lines of reasoning of how to propagate this hypothesis backwards against the flow of deduction. This point is important and, so far as we know, new.

There is an interesting point to observe here. Our own algorithm $\mathcal{A}$ goes backwards, reducing one provability question to another. The abduction algorithm, therefore, goes forwards, in an opposite direction to the provability.

The tableaux method, to take a well known example, works like that. We try to falsify $K \looparrowright V$ and if the tableaux is closed then $K \looparrowright V$ is impossible to falsify. So the abduction process will close the endpoint tableaux and propagate backwards the additional assumptions.

Note that we can as easily (in the tableaux case) abduce on the goal rather than on the data. In other words, if $K \looparrowright V$ fails, we use the same process to abduce an V' such that $K \looparrowright V'$ will succeed! I.e. we change the goal posts. People do that a lot in real life.

The real life abduction changes both K and V in order to succeed. So, for example, if someone undertakes a project and cannot exactly meet the goal he may slightly misinterpret K and justify a slightly different V and hope he can get away with it!

Another point to discuss is the connection of abduction with inconsistency. If $K \looparrowright V$ does not hold, then $K(\neg V)$ is consistent namely it does not prove $\bot$. Find an H such that $K(\neg V)(H)$ is inconsistent (i.e. does prove the 'goal' $\bot$). Then this means $K(\neg V) \looparrowright \neg H$ holds and in many logics this implies that $K(H) \looparrowright V$ holds[4].

---

[4] In practice one can use a theorem prover to generate $Y_1, Y_2, Y_3 \ldots$ such that $K(\neg V) \vdash$

So now conditions on acceptability of H become parallel to conditions on what kind of wffs we want to use to render a theory inconsistent.

Let us take stock of what we have so far. We start with $K \looparrowright ?V$ which does not succeed, i.e. our algorithm $\mathcal{A}$ definitely fails. We have a meta algorithm $\mathcal{M}(\mathcal{A})$ which follows the computation $\mathcal{A}$ and yields an H (or several of them) such that $K(H) \looparrowright V$ succeeds.

This is how we abduce. The reader may ask whether the abduction depends on the choice of $\mathcal{A}$? The answer is yes. Different $\mathcal{A}$'s with the same $\mathcal{M}$ will (or may) give different H's. We are not bothered by that. We think that part of the logic is its proof theory and so it makes sense that the abduction depends on the proof theory. In practice all (or let us say, almost all, to be safe) abduction algorithms for a logic are tagged to some proof theory for that logic.

Also note that according to Gabbay [1996], a system like $\mathcal{M}(\mathcal{A})$ (i.e. a logic with abduction) is also considered a logic. In other words, part of the notion of a logic is to have resident abduction algorithms. This opens the way for several abduction algorithms, a primary one $\mathcal{M}$ and a secondary μ one. We can apply the secondary abduction when the primary one does not yield an answer. Here is an example: Start with $K \looparrowright ?V$ failing. We apply $\mathcal{M}(\mathcal{A})$ to the problem and obtain H. We look at $K(H)$ and it is not acceptable (therefore the abduction fails). We now have a secondary target. Modify the abduction $\mathcal{M}$ to $\mathcal{M}' = \mu(\mathcal{M})$ so that an acceptable H' is obtained[5].

Let us now rewrite (AKM 1)–(AKM 7) in view of the above discussion. The prefix 'N' stands for 'new'.

NAKM 1 V (should succeed).

NAKM 2 $\sim (K \looparrowright V)$, The algorithm $\mathcal{A}$ fails to succeed from K with V as a goal.

NAKM 2.5   We apply $\mathcal{M}(\mathcal{A})$ and get an H.

NAKM 3 $\sim (H \looparrowright V)$. To ensure this we need to look at how $\mathcal{M}$ works and prove it as a theorem.

NAKM 4 $K(H)$ is acceptable.

NAKM 5 $K(H)$ is minimal. Again we need to say what this means and prove it as a theorem.

$Y_i, i = 1, 2, \ldots$ Then any $H_i = \neg Y_i$ can serve as a hypothesis to prove V. Conditions on what kind of H we want become conditions on what kind of Ys we generate.

[5]Going back to a previous footnote, each abduction process $\mathcal{M}$ and $\mathcal{M}'$ may be itself a multiple algorithm.

NAKM 6 $K(H) \looparrowright V$. We prove this as a theorem on $\mathcal{M}$.

NAKM 7   Therefore, H. This statement makes sense in traditional logics where $K(H)$ is $K \cup \{H\}$. In our context we must say "Therefore we insert H into K". If we insert H into K as a structure, in what sense do we say that H holds?

Do we mean that $K(H) \looparrowright H$ holds? We can require it if we want it as part of NAKM 7 but in general it does not need to hold.

NAKM 8   Provide machinery for the backward reasoning of the abduced hypotheses (against the forward deductive flow of H).

General comments

The backward propagation mechanism that we saw we need for the case of abduction is a special case of a general way of propagating metapredicates over proof algorithms. We can imagine that $\mathcal{M}$ gives any value (action, cost, time, etc) to the tasks involved in a rule $\mathbb{R}$ and that $\mathcal{M}(\mathbb{R})$ propagates these values backwards. In fact, such metapredicates can apply to any kind of finitary algorithms (not necessarily having to do with logic) and the backward propagation can be an inductive definition of what the values are.

So in particular abduction can be applied to general (not proof theoretical) algorithms to abduce (find) wasy to make them get unstuck or extract some values from them even though they get stuck. For more on abduction, see [Gabbay and Woods, 2005].

BIBLIOGRAPHY

[Aliseda-LLera, 1997] Atocha Aliseda-LLera. *Seeking Explanations: Abduction in Logic, Philosophy of Science and Artificial Intelligence*. Amsterdam: Institute for Logic, Language and Computation, 1997. PhD dissertation (ILLC Dissertation Series 1997-4).

[Gabbay and Hunter, 1991] Dov M. Gabbay and A. Hunter. Making inconsistency respectable. In Ph. Jorrand and J. Kelemen, editors, *Proceedings of Fundamental of Artifiial Intelligene Research (FAIR '91)*, volume 535 of *LNAI*, pages 19–32. Berlin: Springer-Verlag, 1991.

[Gabbay and Hunter, 1993] Dov M. Gabbay and A. Hunter. Making inconsistency respectable part 2: Meta-level handling of inconsistency. In *LNCS 747*. Berlin: Springer-Verlag, 1993.

[Gabbay and Woods, 2003] Dov M. Gabbay and John Woods. *Agenda Relevance: A Study in Formal Pragmatics*, volume 1 of *A Practical Logic of Cognitive Systems*. Amsterdam: North-Holland, 2003.

[Gabbay and Woods, 2005] Dov M. Gabbay and John Woods. *The Reach of Abduction: Insight and Trial*, volume 2 of *A Practical Logic of Cognitive Systems*. Amsterdam: North-Holland, 2005.

[Gabbay and Woods, to appear] Dov M. Gabbay and John Woods. *Advice on abductive logic*. to appear.

[Gabbay, 1996] Dov M. Gabbay. *Labelled Deductive Systems*. Oxford: Oxford University Press, 1996.

[Kuipers, 1999] Theo A.F. Kuipers. Abduction aiming at empirical progress of even truth approximation leading to a challenge for computational modelling. *Foundations of Science*, 4:307–323, 1999.

[Magnani, 2001] Lorenzo Magnani. *Abduction, Reason and Science: Processes of Discovery and Explanation*. New York: Kluwer, Plenum, 2001.

[Meheus *et al.*, forthcoming] Joke Meheus, Liza Verhoeven, Maarten Van Dyck, and Dagmar Provijn. Ampliative adaptive logics and the foundation of logic-based approaches to abduction. In Lorenzo Magnani, Nancy J. Nersessian, and Claudio Pizzi, editors, *Logical and Computational Aspects of Model-Based Reasoning*. Dordrecht and Boston: Kluwer, forthcoming.

[Woods, 2005] John Woods. Dialectical considerations on the logic of contradiction. *Logic Journal of the IGPL*, 2005.

Dov Gabbay
Department of Computer Science
King's College London
Email: dg@dcs.kcl.ac.uk

John Woods
Department of Computer Science
King's College London
Email: jhwoods@interchange.ubc.ca

Integrating MMASS with a Hybrid Commonsense Spatial Logic

STEFANIA BANDINI, ALESSANDRO MOSCA,
MATTEO PALMONARI, AND GIUSEPPE VIZZARI

ABSTRACT. The ubiquitous computing scenario calls for new models, and in general new ways to conceive computer systems. Research on Multi Agent Systems has extensively investigated this field and provided instruments supporting the modeling, design and development of complex ubiquitous systems. Multilayered Multi-Agent Situated Systems (MMASS) is a multi-agent model providing an explicit spatial structure representation and a set of interaction mechanisms that are strongly related to agents' context. A Commonsense Spatial Model (CSM) is then introduced as a suitable structure on which represent and perform reasoning on facts related to spatial information and knowledge. Since MMASS agents have almost a reactive flavor, the integration of a hybrid logic based on CSM and MMASS provides some of the MMASS agents with a more rational behavior, giving them the opportunity to exploit an explicit model of the environment in their action selection process.

1 Introduction

Ubiquitous computing can be viewed as a paradigm concerned with a new way of conceiving the interaction among humans (users) and computational devices. Mobile devices, sensors and integrated environments depict a scenario in which users will interact with embedded devices, dynamically connected with each other and almost disappearing in the environment. Thanks to the improvement and growing availability of information acquisition and delivery technologies (sensors, personal devices, wi-fi, and so on) computational power can be embedded almost in every object populating the environment. Nevertheless, technological evolution is not combined with an equally rapid evolution of the conceptualization necessary to understand and govern the new situation [Zambonelli and Parunak, 2002].

The fact that computational elements enabling users' interactions and their access to certain services are diffused in the environment requires considering spatial aspects of these systems that were previously not so relevant.

On one hand, said infrastructures should be able to shield users from technical details of the operations required to carry out tasks they may require. From this point of view, the infrastructure should consider, encapsulate and manage mechanisms related to data transmission in complex and growing unstructured and heterogeneous networks. On the other hand, there is a growing need of services which consider elements of users' contexts, in order to produce smarter behaviors which effectively exploit the potential of new technologies. In particular, a user's location is a relevant element of his/her context, and the growing availability and affordability of devices supporting global positioning (e.g. GPS) makes feasible the modeling, design and implementation of location-aware systems. However, both in order to shield users and the application level from technical details of communication, and also to supply context aware services, space should be considered as a first class abstraction to be included in models supporting context aware forms of interaction or reasoning.

The ubiquitous computing paradigm calls thus for new models, and in general new ways to conceive computer systems. The main goal in this new scenario is to supply services through the interaction of computational nodes that are diffused in the environment, and that however should be able to perform specific services in an autonomous way. Research on Multi Agent Systems (MAS) has extensively investigated these topics, and it can be a source of abstractions and instruments (both theoretical and computational) supporting the modelling, design and development of complex ubiquitous systems. In particular, in this paper the Multilayered Multi-Agent Situated Systems (MMASS) [Bandini et al., 2002b] will be introduced as a multi-agent model providing an explicit spatial structure representation and a set of interaction mechanisms that are strongly related to agents' context. Exploiting this structure and these mechanisms makes it possible to define systems of agents which can obtain complex form of coordination by means of a simple behavioural specification.

However, although it is quite simple to model a situation in which an agent is able to move in a spatial structure according to signals that are spread in it by elements of the environment modeled as static agents, it is not so simple to encapsulate into MMASS action specification the deliberative process that governs the autonomous choice of an agent, i.e. to be sensitive to a specific type of signal, and thus its choice to move towards a certain destination. Basically agents follow their perceptions and they have a rather reactive flavor, whereas actual modeling experiences, i.e. in the simulations domain [Bandini et al., 2004b], showed that their decision making process could take advantage from a deeper knowledge of the environment. This leads to investigate the possible integration between a perception-based multi-agent model and commonsense spatial reasoning from a knowledge representation perspective.

The aim of this paper is therefore to present the integration of MMASS with a model-based logical approach to spatial reasoning in order to provide some of the MMASS agents with a more rational behavior, giving them the opportunity to exploit an explicit model of the environment in their action selection process. The MMASS model is introduced in the next section, with respect to its basic elements and concepts and to formal definitions. Section 3 presents a formal model supporting commonsense spatial reasoning; since this Commonsense Spatial Model (CSM) is defined as a relational structure, it is shown how it can be viewed as the semantics specification for a hybrid modal language. The integration of the MMASS model with the CSM is discussed in Section 4, investigating the motivations, the mapping between the two models and some reasoning issues. Concluding remarks end the paper.

2 Multilayered multi-agent situated system model

The MMASS is a formal and computational framework for the definition of systems made up of a set of autonomous entities acting and interacting in a structured environment. This section does not represent a formal description of the model, but will briefly introduce its main concepts, specifically focusing on the environmental structure. In fact the latter deeply influences agents behavior, as the environment is the source of their perceptions, a constraint limiting their actions (e.g. their movement), but it also provides them with a medium to interact with other entities. First of all related works and their relationships with agent environment modeling will be described, then the MMASS and its main concepts will be introduced.

The MMASS model provides an explicit representation of agent environment, that is made up of a set of interconnected layers whose structure is an undirected graph of sites. These layers may represent abstractions of an actual physical environment but can also be related to "logical" aspects as well (e.g. the organizational structure of a company). Between these layers specific connections (*interfaces*) can be specified. The latter are used to specify that a given field type, generated in one of these layers, may also propagate into a different one. This mechanism allows to generate interactions among different aspects and levels of the system. *Field based interaction* is the first mechanism for agent interaction, allowing a multicast form of interaction among agents occupying distant points in their environment. Adjacent agents may also perform a coordinated change of their state through a *reaction*, which is the second mechanism for agent interaction. The presence of an entity emitting a field represents a way to augment the basic environment spatial representation, for instance in order to represent dynamical properties of portions of the environment itself.

The model has been successfully applied to several simulation contexts in which the concepts of space and environment are key factors for the problem solving activity and cannot be neglected (e.g. crowd modeling [Bandini et al., 2004b], localization problems [Bandini et al., 2004a]). In the ubiquitous computing context, instead, active entities of a pervasive system (e.g. users having a computational device, sensors and other sources of information) can be modeled as agents which interact by means of an infrastructure which is somehow mapped to a spatial structure (see, e.g., [Bandini et al., 2005b]). This structure should reflect the dynamics of the actual environment, and thus it should dynamically determine the possible interactions among agents according to spatial relationships (e.g. distance). The MMASS model allows the representation of these elements (i.e. active entities and their environment) in a unified framework. In the following the model will be briefly introduced and formally described.

2.1 An overview of MMASS model

According to the MMASS model, agents are situated in sites, that is, nodes of the graphs related to a layer of the environment. Every site may host at most one agent (according to a non-interpenetration principle: "two agents cannot occupy the same site at the same time"), and every agent is situated in a single site at a given time (non–ubiquity: "at a given time an agent occupies a single site"). Agents inherit the spatial relationships defined for the site it is occupying; in other words an agent positioned in site p is considered adjacent to agents placed in sites adjacent to p.

The adjacency relation among agents is a necessary condition for the applicability of *reaction*, the first kind of interaction mechanism defined by the MMASS model. In fact this operation involves two or more agents that are placed in adjacent sites and allows them to synchronously change their state, after they have performed an agreement. This mechanism resembles the one defined by transition rules in Cellular Automata (CA) [Wolfram, 1986], that also provide an explicit representation of a spatial structure.

CA is the model that has mainly inspired MMASS specification, and one of the main differences between the two models is the possibility to represent *action-at-a-distance*. In fact, the second interaction mechanism defined by the MMASS model provides the possibility for agents to emit *fields*, that are signals able to diffuse through the environment that can be perceived by other agents according to specific rules. This mechanism resembles pheromone approaches to agent communication (see, e.g., [Hadeli et al., 2004]), but fields are not just related to an intensity value and may

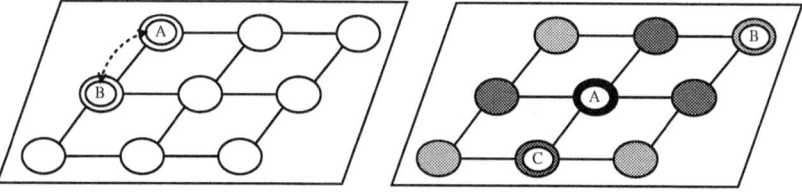

Figure 1. A sketch of MMASS interaction mechanisms: on the left, reaction among adjacent agents A and B, on the right field emission by agent A.

convey more complex kinds of information. Moreover for every field type a *diffusion function* can be specified in order to define how related signals decay (or are amplified) during their diffusion in the environment, from the source of emission to destination sites. Other functions specify how fields of the same kind can be *composed* (for instance in order to obtain the intensity of a given field type at a given site) or *compared*. From a semantic point of view fields themselves are neutral even if they can have related information in addition to their intensity; they are only signals, with an indication on how they diffuse in the environment, how they can be compared and composed. Different agent types may be able to perceive them or not and, in the first case, they may have completely different reactions, according to their behavioral specification. With reference to perception, an agent may perceive a field with a non–null intensity active in the site it is situated on according to two parameters characterizing its type and related to the specific field type. The first one is the *sensitivity threshold*, indicating the minimum field intensity that an agent of that type is able to perceive. The second is the *receptiveness coefficient* and it represents an amplification factor modulating (amplifying or attenuating) field value before the comparison with the sensitivity threshold. Thanks to these parameters it is possible to model dynamism in the perceptive capabilities of agents of a given type, since these parameters are related to agent state. In this way, for instance, the same agent that was unable to perceive a specific field value could become more sensitive (increase its own receptiveness coefficient) as a consequence of a change in its state. This allows to model physical aspects of perception, but also conceptual ones such as agent interests.

Reaction and field emission are two of the possible actions available for the specification of agent behavior, related to the specification of how agents may interact. Other actions are related to the possibility to move (*transport* operation) and change the state upon the perception of a specific field (*trigger* operation). These primitives are part of a language for the specification of MMASS agents behavior [Bandini et al., 2002b]. An important part of the language also provides the possibility to dynamically modify the structure of agent environment, in order to generate new sites and edges (or destroy existing ones) and create (or destroy) agents of a specific type, with a given initial state. *Agent type* is in fact a specification of agent state, perceptive capabilities and behavior.

2.2 MMASS: formal definitions

A *Multilayered Multi–Agent Situated System (MMASS)* is defined as a constellation of interacting *Multi-Agent Situated System* (MASS) that represent different layers of the global system: $\langle MASS_1 \ldots MASS_n \rangle$. A single MASS is defined by the triple $\langle Space, F, A \rangle$ where *Space* models the environment where the set A of agents is situated, acts autonomously and interacts through the propagation of the set F of fields and through reaction operations.

The structure of a layer is defined as a not oriented graph of sites. Every *site* $p \in P$ (where P is the set of sites of the layer) can contain at most one agent and is defined by the 3–tuple $\langle a_p, F_p, P_p \rangle$ where:

- $a_p \in A \cup \{\bot\}$ is the agent situated in p ($a_p = \bot$ when no agent is situated in p that is, p is empty);

- $F_p \subset F$ is the set of fields active in p ($F_p = \emptyset$ when no field is active in p);

- $P_p \subset P$ is the set of sites adjacent to p.

In order to allow the interaction between different MMASS layers (i.e. intra-MASS interaction) the model introduces the notion of *interface*. The latter specifies that a gateway among two layers is present with reference to a specific field type. An interface is defined as a 3–tuple $\langle p_i, p_j, F_\tau \rangle$ where $p_i \in P_i, p_j \in P_j$, with P_i and P_j sets of sites related to different layers (i.e. $i \neq j$). With reference to the diffusion of field of type F_τ the indicated sites are considered adjacent and placed on the same spatial layer. In other words fields of type F_τ reaching p_i will be diffused in its adjacent sites (P_p) and also in p_j.

A MMASS agent is defined by the 3–tuple $< s, p, \tau >$ where τ is the *agent type*, $s \in \Sigma_\tau$ denotes the *agent state* and can assume one of the values specified by its type (see below for Σ_τ definition), and $p \in P$ is the site of the *Space* where the agent is situated. As previously stated, agent *type* is a specification of agent state, perceptive capabilities and behavior. In fact an agent type τ is defined by the 3–tuple $\langle \Sigma_\tau, Perception_\tau, Action_\tau \rangle$. Σ_τ defines the set of states that agents of type τ can assume. $Perception_\tau$:

$\Sigma_\tau \to [\mathbf{N} \times W_{f_1}] \dots [\mathbf{N} \times W_{f_{|F|}}]$ is a function associating to each agent state a vector of pairs representing the *receptiveness coefficient* and *sensitivity thresholds* for that kind of field. $Action_\tau$ represents instead the behavioral specification for agents of type τ. Agent behavior can be specified using a language that defines the following primitives:

- $emit(s, f, p)$: the *emit* primitive allows an agent to *start the diffusion of field* f on p, that is the site it is placed on;

- $react(s, a_{p_1}, a_{p_2}, \dots, a_{p_n}, s')$: this kind of primitive allows the specification a *coordinated change of state* among adjacent agents. In order to preserve agents' autonomy, a compatible primitive must be included in the behavioral specification of all the involved agents; moreover when this coordination process takes place, every agent involved may dynamically decide to effectively agree to perform this operation;

- $transport(p, q)$: the *transport* primitive allows to *define agent movement* from site p to site q (that must be adjacent and vacant);

- $trigger(s, s')$: this primitive specifies that an agent must *change its state* when it senses a particular condition in its local context (i.e. its own site and the adjacent ones); this operation has the same effect of a reaction, but does not require a coordination with other agents.

For every primitive included in the behavioral specification of an agent type specific preconditions must be specified; moreover specific parameters must also be given (e.g. the specific field to be emitted in an emit primitive, or the conditions to identify the destination site in a transport) to precisely define the effect of the action, which was previously briefly described in general terms.

Each MMASS agent is thus provided with a set of sensors that allows its interaction with the environment and other agents. At the same time, agents can constitute the source of given fields acting within a MMASS space (e.g. noise emitted by a talking agent). Formally, a field type t is defined by
$$\langle W_t, \mathit{Diffusion}_t, \mathit{Compare}_t, \mathit{Compose}_t \rangle$$
where W_t denotes the set of values that fields of type t can assume; $\mathit{Diffusion}_t : P \times W_f \times P \to (W_t)^+$ is the diffusion function of the field computing the value of a field on a given space site taking into account in which site (P is the set of sites that constitutes the MMASS space) and with which value it has been generated. $\mathit{Compose}_t : (W_t)^+ \to W_t$ expresses how fields of the same type have to be combined (for instance, in order to obtain the unique value of field type t at a site), and $\mathit{Compare}_t : W_t \times W_t \to \{True, False\}$

is the function that compares values of the same field type. This function is used in order to verify whether an agent can perceive a field value by comparing it with the sensitivity threshold after it has been modulated by the receptiveness coefficient.

3 Commonsense spatial model

3.1 Basic concepts: places and conceptual spatial relations

The literature about space modeling, supporting computational frameworks to be adopted in order to develop reasoning capabilities, is wide and distributed in several areas of Artificial Intelligence such as Automated Vision, Robotics, Knowledge Representation, and so on. Within a rough classification two main classes of approaches can be distinguished: a first one tends to justify commonsense spatial inference with mathematical models such as Euclidean geometry, trigonometry, differential equations systems and so on [Davis, 1990]; in the second one different topological approaches can be considered, ranging from point set and algebraic topology, with the choice of different kinds of primitive entities and relationships (e.g. RCC calculus [Randell et al., 1992], modal logics [Aiello and Benthem, 2002]), to topological route maps (see [Kuipers, 1978; Leisler and Zilbershatz, 1989], and [Gopal et al., 1989]).

Within the second conceptual framework, correlation as commonsense spatial reasoning can be supported by defining a formal model of space that exploits the basic notions of place and conceptual spatial relation. Spatial disposition of information sources distributed in the environment (e.g. close circuit cameras, smart home or complex industrial plant sensor networks) can be mapped into a set of relations among interesting places (i.e. a topology) and high-level reasoning beyond low-level sensors' capabilities can be carried out by reasoning about properties holding at different places.

Imagine having a sensor platform installed in a building in order to monitor a significant portion of it (and, eventually, to take suitable control actions). Sensors distributed in the environment return values that can be interpreted in order to provide *local* descriptions, possibly generating alerts or alarms, of what is happening in the range of each sensor. Architectural issues are out of the scope of this paper, but in [Bandini et al., 2004c] the advantages of distinguishing the detection, local interpretation and correlation levels have been widely discussed, and a four-leveled architecture, which had been fruitfully exploited in the traffic monitoring domain [Bandini et al., 2005a], has been presented.

An example of such an environment, e.g. an apartment, is given in Figure 2. Here, different types of sensors are located in separated rooms: in the corridor, for example, there can be a camera, a smoke/fire detector and

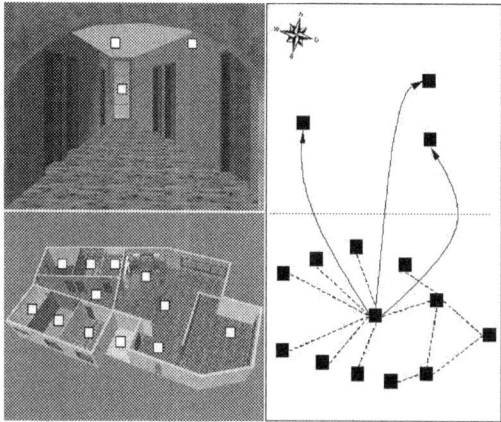

Figure 2. The emergence of a commonsense spatial model in the context of a monitored apartment. On the left a 3D model of the apartment and a cross-section of its corridor are presented. In the right side, the generation of the corresponding spatial model is represented: the nodes are the interesting places (rooms and sensors), while proximity and containment relations are represented by dashed and unbroken lines respectively. Orientation relations can be guessed but have been omitted for sake of clarity.

a broken-glass sensor. Sensors and rooms are related together by means of orientation relations, such as *"to be at north of"*; rooms are linked together by means of proximity relations; and, finally, rooms and sensors are linked together by means of containment relations. In the example proximity between rooms has been defined taking into account "direct access", but the proximity relation can be interpreted differently as well (e.g. as the relation between adjoining rooms). Here sensors and rooms and their reciprocal relations define a commonsense model of space of the monitored area.

A commonsense model of space supporting reasoning about the environment emerges therefore as a topology whose nodes are identified by interesting *places* and whose relations are *conceptual spatial relations* (CSR) arising from an abstraction of the spatial disposition of these places. A place is a conceptual entity completely identified by an aggregation of attributes/properties of different kinds; examples are the type of place (e.g. a place can be a sensor or a room), its internal status properties (e.g. "is_faulty"), its functional role (e.g. a kitchen or a living room), and so on.

Observe that a CSR is grounded on physical space but not "founded" on it: no necessary relationship among CSRs and any objective physical representation of space needs to be assumed as primitive. Nevertheless, theoretical considerations about the epistemological relevance of this notion of "emergent topology" based on these two basic concepts concerns controversial philosophical issues, which would deserve a deeper analysis that goes beyond the aims of this paper.

Once a topological model has been defined, properties holding at different places can be correlated together to provide a more comprehensive understanding of the environment (e.g. neither a broken glass nor a person detected by the camera are per se a proof of intrusion, but those two facts considered together may lead to infer that a stranger has entered into the house passing through the window and walking in the corridor). Observe that a fundamental characteristic of a commonsense model of space is finiteness, that is, the number of places is always limited; this issue is significant for computability and tractability but is also sound with the fact that, when considering a specific situation, any reasoner necessarily selects a limited portion of the context. As it will be stressed out in the conclusions, this work does not deal with dynamical aspects of the environment yet: the interesting places may change in time; nonetheless, this problem is related to the places selection process and concerns how the model forms and changes, but it does not hinder the model finiteness.

3.2 *CSM*, a model for commonsense spatial reasoning

From a representational perspective the conceptual framework introduced naturally recalls the definition of a relational structure, whose nodes are places and relations are CSRs. A relational structure is a non-empty set on which a number of relations have been defined; they are widespread in mathematics, computer science and linguistics. In particular, according to the epistemological framework specified in the previous section, only finite structures are considered and this is a fundamental characteristic with respect to the computational tractability problem as mentioned in the previous section. A general commonsense spatial model is thus defined as follows:

DEFINITION 1 *A commonsense spatial model* $CSM = \langle P, R_\sigma \rangle$ *is a relational structure, where* $P = \{p_1, ..., p_i\}$ *is a finite set of places, and* $R_\sigma = \{R_1, ..., R_n\}$ *is a finite non-empty set of binary conceptual spatial relations, labeled by means of a set of labels* N.

Finiteness and cardinality of P (the domain must contain at least two places) are minimal requirements to have a well-founded commonsense model of space according to the observation reported at the end of the previous section. An *edge labeled multigraph*(a graph with admitted multiple edges between nodes as in [Harary, 1972]), whose nodes and labeled edges are respectively places and CSRs, is a powerful instance of a *CSM*.

A place can be anything that satisfies the informal definition of the previous section. As for R_σ, although R_σ can be any arbitrary set of binary CSRs, some classes of relations significant for a wide reasoning domain will be characterized in the following paragraphs. As far as a commonsense model of space is concerned, it is not possible (nor useful) to identify a minimal set of primitive relations (as for RCC). In fact, this approach is not aimed at providing a mathematical model of space, but rather to define the basic elements for the specification of axioms defining relevant properties of specific environments.

Nevertheless, there are some significant classes of relations that provide a powerful enough model but still general. In particular, a place can be "oriented by" the presence of an other (distinct) place, a place can be "contained in" or can be "proximal to" an other place. Although many different relations can fit here, according to different application domains, it seems natural to identify in *Orientation*, *Containment*, and *Proximity*, the archetypes of any form of commonsense spatial arrangement among entities.

Orientation. First of all, we need some relations to ensure *orientation* in space: assuming reference points is a rudimentary but fundamental way to start orienting into space. Assuming points of reference consists in ordering entities with respect to these particular points. Since many different sources of orientation can be found (stars, magnetic fields, a subjective set of mnemonic sites, and so on), a further step is to choose *good* reference entities and this can be achieved by means of the traditional four *cardinal points*: North, East, South, West. The latter suggests the definition of a set of orientation relations R_N, R_E, R_S, and R_W among places (observe that, from a formal perspective, only two of these relation symbols need to be taken as primitive).

Thus, two relations $R_N \subseteq P \times P$ and $R_E \subseteq P \times P$ are introduced and interpreted in the following way. Let p and q be two places, the relation $R_N(p,q)$ holds *iff* p is *at north of* q ($R_E(p,q)$ is defined analogously). Orientation relations are both *strict partial orders* on the set of places that is, they are *irreflexive*, *asymmetric* and *transitive* relations; the order is "partial" because two places might be incomparable. Moreover, both relations have a *superior* and an *inferior* that coincide respectively with North and South, and with East and West. The relations R_S and R_W are defined as the inverse respectively of R_N and R_E. Other non-primitive relations such as *at north-east of* (R_{NE}), *at north-west of* (R_{NO}), and so on, can be defined by means of usual set theoretic operators from the previous ones, e.g. $R_{NE} = R_N \cap R_E$.

It is important to observe that, for what concerns orientation, the notion of order among entities is more fundamental than the contingent choice of particular reference points in order to enable that ordering. The choice of cardinal points seems quite intuitive, but, if different perspectives are needed, reference points can be easily changed or added preserving the basic structure (a lattice with superior and inferior) and the relations' properties (irreflexivity, asymmetry, transitivity). For instance, higher/lower relations can be represented by orientation relations with suitable entities as superior and inferior of the lattice.

Containment. Since places are arbitrary entities, possibly with different shapes, dimensions and nature (e.g. a room and a printer can both be places), a physical inclusion relation $R_{IN} \subseteq P \times P$ is needed in order to relate different types of places: an object may be in a room that may be in a building (where the object, the room and the building are interesting places of the same topology). The relation $R_{IN}(p,q)$ is interpreted as stating that the place q is *contained* in the place p; R_{IN} is a typical mereological relation: it is a partial order, and, more precisely, a *reflexive, antisymmetric* and *transitive* relation. Here, the stronger antisymmetry (i.e. $\forall p,q (R_{IN}(p,q) \land R_{IN}(q,p) \rightarrow p = q)$) holds because this can be exploited to infer identity between two places for which is said that one is in another and vice versa.

Proximity. Another basic relation useful to characterize space concerns the possibility of accessing one place from another (in both physical and metaphorical sense). Two places are said to be proximal if it is possible to go from one to the other without passing through another place: a *proximity* relation $R_P \subseteq P \times P$ is then introduced, whose meaning is that the place q is directly reachable from place p. This relation can be modeled as an adjacency relation since is *irreflexive* and *symmetric*. However, different criteria of reachability can be adopted to define an adjacency proximity relation. In a network of radio transmitter/receiver devices proximity is a very different notion from the one adopted in crowding dynamic analysis or in molecular morphogenesis.

Therefore, according to the above observations about orientation, containment, and proximity relations, it is possible to define an *elementary* Conceptual Spatial Model CSM_e as a CSM where, at least $\{North, South, East, West\} \in P$ (the upper and lower bounds of the orientation relations), and $R_\sigma = \{R_N, R_E, R_{IN}, R_P\}$.

3.3 Reasoning into space: a hybrid logic approach

Since the commonsense spatial model just introduced is a relational structure, it can be naturally viewed as the semantic specification for a modal logical language. According to a well known modal logic tradition, which relates to Kripke's "possible worlds" semantics, classes of relational structures (such as CSMs) can be considered as "frame", structures whose relations define the meaning of specific sets of modal operators.

Therefore, modal languages turn out to be very useful as far as reasoning about relational structures is concerned, and have been exploited for temporal and spatial logics, for logic of necessity and possibility and many others (see [Blackburn et al., 2000]). Nevertheless, recent studies in Modal Logic lead to further improve its expressiveness and power according to issues coming mainly from research in the Knowledge Representation area. One of the most notable results has been the development of Hybrid Logic. Hybrid languages are modal languages that allow to express (in the language itself) sentences about satisfiability of formulas, that is to assert that a certain formula is satisfiable at a certain world (i.e. at a certain place in our framework). In other words, its syntactic side is a formidable tool to reason about what is going on at a particular place and to reason about place equality (i.e. reasoning tasks that are not provided by basic modal logic).

The definition of a hybrid logic for commonsense spatial reasoning according to the presented CSM requires the assumption of a specific *sort* of atomic formulas (i.e. "nominals") to refer to the interesting selected places. As usual, each place-nominal is true at exactly one place of the CSM and the introduction of the so-called "satisfaction operators" $@_i$ provides the capabilities of reasoning globally on the universe of places. Given a model $\mathcal{W} = \langle CSM, V \rangle$, where CSM is the frame and V is an hybrid valuation, the true condition for a formula $@_i \phi$ (where ϕ can be any arbitrary formula), is given as follows:

$$\mathcal{W}, w \models @_i \phi \text{ if and only if } \mathcal{W}, w' \models \phi,$$

where the place w' is the denotation of i, i.e. $V(i) = w'$. A complete set of symbols for modal operators is then given according to the classification of the basic conceptual spatial relations introduced above. Thus, with respect to the CSM_e, the operators $\Diamond_N$, $\Diamond_E$, $\Diamond_S$, $\Diamond_W$, $\Diamond_{IN}$, and $\Diamond_P$ are introduced; their groundedness in the CSM is guaranteed by the fact that their accessibility relations are defined, respectively, by the CSM's relations R_N, R_E, R_{IN}, and R_P (the semantics of $\Diamond_S$, $\Diamond_W$ is defined over the inverse of the R_N and R_E relations).

According to the aims of the modeled correlation task, a domain dependent set of properties can be chosen and represented in the formal language by means of a suitable set of symbols for propositional letters (e.g. the information "there is a man", coming from a local data processing, can be represented with a proposition "is_man", true or false at some place of the model).

The combination of the multimodal and hybrid expressiveness provides a powerful logical reasoning tool to shift perspective on a specific place by means of a $@_i$ operator, which allows checking properties holding over there; for instance, with respect to Figure 2, when a system devoted to intrusion detection needs to query if "a glass is broken" at the place corresponding to the broken-glass sensor, the satisfiability of the formula

$$@_{window\_sensor} broken\_glass$$

must be checked. Moreover, exploiting this operator, it is possible to define local and internal access methods to explore the spatial model according to the other defined operators – e.g. checking the satisfiability of the formula

$$@_{kitchen} \Diamond_W \Diamond_{IN} smoke$$

formally represents the verification, for the system, that "in" some room "at west of" the kitchen some "smoke" has been detected.

Hybrid Modal logic is particularly useful to model reasoning about correlation in pervasive computing environments and, especially, when correlation is exploited for context-awareness, thanks to the double perspective over reasoning that this logic introduces, that is, both local and global.

In modal logic, in fact, reasoning and deduction start always from a given point of the model, i.e. from what is taken as the "current world". In terms of the interpretation of worlds as places, this means that reasoning is performed by a local perspective, and precisely, from the place in the environment taken as the current one. Since, according to the presented model, devices are places, each device can reason about context from its local perspective but exploiting a shared model of the environment. Taken a device, checking the satisfiability of the formula $\Diamond_P (sensor \wedge broken\_glass)$ from this current place means to query if a broken glass has been detected by a sensor adjacent to the current one (an adjacent place on which *sensor* and *broken_glass* are true). On the other hand, hybrid modal logic, still preserving the same local attitude to reasoning of classic modal logic, allows global queries such as $@_{window\_sensor} broken\_glass$. This, in fact means, that whatever is the device on which reasoning is performed, the query regards a specific place/device, that is, the *window_sensor*.

This twofold approach to knowledge representation and reasoning typical of hybrid logic (which has been well described in [Blackburn, 2000]) allows correlation to be modeled as performed both by a central processing unit that reason globally and by single devices locally: this is consistent with different technological approaches to context-awareness, from more centered-based approaches such as blackboard approaches, to approaches stressing more the autonomy of devices, such as multi-agent based approaches.

A similar modal approach has been already applied to correlation as commonsense spatial reasoning, and in particular, in the design and implementation of the Alarm Correlation Module of SAMOT, a monitoring and control system mainly devoted to traffic anomalies detection (as shown in [Bandini et al., 2002a]).

4 Integration of MMASS and CSM

4.1 Motivations

The previous Sections introduced two different models, one aimed at allowing the definition of multi-agent systems in which the spatial aspects of the environment are a crucial factor and another supplying concepts, abstractions and mechanisms supporting the representation and reasoning on facts related to spatial information and knowledge. The motivations to consider a possible integration of the introduced models derives from experiences of applications of the MMASS to the development of simulations in different domains. The main consideration is that MMASS agents have a rather reactive flavor. The modeling activities generally provide the definition of a spatial structure, the modeling of types of signals which are exploited to define the interaction mechanisms among situated entities and then the specification of the behaviors related to agents which inhabit the system. Some of these entities can be immobile agents which are used to generate (possibly dynamical) properties of portions of the environment through the emission of fields. The latter is generally made up of a set of rules specifying that an action may take place according to certain conditions. The preconditions of these rules may be composed of elements of the model and thus do not allow a simple specification of complex expressions on properties of the environment. Moreover, the behavior of agents depends on their state, which also determines the signals that they may perceive. State changes can thus represent the deliberation of a change in the goals of an agent. Once again, the preconditions of this kind of action cannot easily encapsulate a form of reasoning on beliefs and complex expressions on agent's context.

CSM has been introduced as a suitable structure on which represent and perform reasoning on facts related to spatial information and knowledge. A possible way to integrate CSM and MMASS provides thus the possibility to exploit the former as a language for the definition of parts of agents' state that are related to their beliefs on spatial elements of their context. In this way it would be possible to specify preconditions to agents' actions that include complex expressions supported by this formalism. In this way agents could be endowed with additional abstractions and concepts supporting a more complex form of reasoning supporting choices on their behaviors. This can be achieved by means of model checking techniques for formulas of the hybrid language introduced above, basically because of the homogeneity of the model of space embedded in MMASS and CSMs, the relational structures that provides the semantics to the hybrid language. Both MMASS and CSMs represents the environment through a graph-like model and this allows to draw a first mapping between between basic concepts of MMASS spatial representation and the CSMs relational structure.

4.2 Mapping

Firstly, the fundamental elements of the MMASS model, that is "agent" and "site", are translated in the CSM by means of the unique notion of "place": in CSM both sites and agents of MMASS become places. But, since places are (completely) identified by means of the set of properties holding on it, two properties such "site" and "agent" can be exploited also to qualify two first different types of places: places that are sites in MMASS and places that in MMASS are agents . For simplicity we will refer to those types of places respectively as *site-places* and *agent-places*.

Secondly, however the adjacency relations among sites has been defined in MMASS (adjacency among sites can be defined following different criteria, e.g. arbitrarily, or based on Moore or Von-Neumann neighborhood on the sites sets), this relation is represented by means of a proximity relation R_{P_a} among site-places in the CSM; in fact, proximity relations are characterized exactly by the basic formal properties of adjacency relations. Finally, situatedness of agents, i.e. every agent is situated on a site, is represented by means of a containment relation R_{IN}, that is, agent-places are contained in site-places. In this way, the first mapping from a MMASS model leads to a CSM whose nodes are qualified by means of the distinction among agents and sites, and which includes a proximity relation holding among site-places according to adjacency of sites in MMASS and a containment relation among site-places and agent-places according to the disposition of MMASS agents into the sites' graph.

Now, this can be considered a first mapping between the two models, but how does it reflect on representation and reasoning as far as the Hybrid Language for CSM is concerned? Well, Hybrid Logic turns out to be extremely versatile in such contexts. What we need is to include in the language a set of nominals $i_1, i_2, ..., i_n$ to name both sites and agents we are interested in (for example, in the apartment of fig. 2, nominals can be *smoke_sensor*, *camera_1*, and so on). Then formulas such as $@_i site$ and $@_j agent$ allows to define the type of place, that is, they say that the place i is a site-place and the place j is an agent-place. An operator $\lozenge_{P_a}$ is interpreted over R_{P_a}, such that with a formula $@_i \lozenge_{P_a} j$ it is possible to express in the language that the place j is adjacent to the place i (remember that nominals are, syntactically, formulas and can be used like other formulas, with the difference that they are true always at a single state of the model). Finally, if an agent-place i is situated in a site-place j, this can be asserted by means of a formula $@j\,(site \wedge \lozenge_{IN}(i \wedge agent))$, where $\lozenge_{IN}$ is the containment operator interpreted over the commonsense spatial relation R_{IN}.

Naturally, it is necessary to take into account important structural principles of MMASS which define the interactions among agents and sites such as non-interpenetration and non-ubiquity (see section 2.1). Let i, i' be nominals for site-places and j, j' nominals for agent-places, these two principles can be modeled in the CS Hybrid Language with the following two formulas:

$$@_i\,(\lozenge_{IN}(j \wedge j')) \to @_j j' \qquad (1)$$
$$@_j\,(\lozenge_{\overline{IN}}(i \wedge i')) \to @_i i' \qquad (2)$$

The first formula says that if a site-place contains two different agent-places, these places are the same (i.e. the two nominals name the same place), and therefore guarantees the non-interpenetration principle. Conversely, non-ubiquity is expressed by the second formula, which asserts that an agent-place cannot be contained (i.e. situated) in different site-places, where the meaning of $\lozenge_{\overline{IN}}$ is defined as the inverse of $\lozenge_{IN}$.

Moreover, since a CSM has been defined as an open model, increasing the complexity of the hybrid language associated, it is possible to introduce new places, qualified by properties, and new orientation, containment and proximity relations whose meanings go beyond the original representational objectives in the MMASS model. In this way new, higher level information coming from abstractions of the agents' environment can be included in the commonsense spatial model exploited by deliberative agents.

As for the new places, it is interesting for the deliberative agents to be able to reason about the spatial disposition of places grouping together a large set of sites coming from the MMASS infrastructure, once that these sites are identified as semantic units (e.g. rooms, passageways, buildings, squares, etc.). For example, places as "kitchen room", "corridor" and so on (see Figure 2), can be used to abstract a higher level representation with respect to the MMASS spatial model and named in the language with specific nominals; the site-places in CSM are linked to those new places by a containment relation, so that, e.g. a place "kitchen" contain site-places that on their turn may contain an agent-place. As for the extension of the set of CSRs, orientation relations turn out to be particularly important to distinguish among adjacent places (e.g. the formula $@_{kitchen} \Diamond_W corridor$ say that there exists a place "corridor" at west of the room "kitchen"). Orientation relations are inherited through the containment relation as stated by the following formula:

(3) $\quad @_i (\Diamond_* i' \wedge \Diamond_{IN} j) \rightarrow @_j \Diamond_* i',$

where $* \in \{N, E, S, W\}$ and i, i', j are nominals for places. Moreover, observe that new places and relations introduced in CSM admits the possibility, proper of MMASS, of representing and connecting different layers according to different levels of abstraction.

4.3 Reasoning

Behavioral characteristics of deliberative agents strongly depend on their inferential capabilities. Action selection strategies and sensitivity thresholds to fields are influenced by the present status of the agent and, during his life, the status of the agent is in fact influenced by its spatial model-based reasoning activity. In this framework, MMASS agents that reason upon the spatial disposition of the entities populating the environment, perform their reasoning by means of model checkers for hybrid formulas based on the spatial knowledge corpus of each agent. The model checker verifies the satisfiability of the truth conditions of multi-modal propositional formulas according to a given model (a Kripke structure with a CSM frame *and* a valuation function).

As an example, consider the Smart Home case depicted in Figure 2. In the MMASS computational framework, a graph of sites represents the environment in such a way that the several devices spread over it are modeled as situated agents (agents can be: a sensor, an access point, a PDA, and so on). Given to a specific agent, e.g. the monitoring and control component of the system, an explicit representation of the environment through a CSM, it is possible to add logic-based preconditions to the agent's action selection strategy. Suppose that an alarm must be diffused by this agent (through a field emission) when he infers that some smoke has been detected by a "sensor contained in a site at west of the kitchen", the precondition of this "emit" action can be represented exploiting the higher level representation of the environment by the hybrid formula:

(4) $\quad @_{kitchen} \Diamond_W \Diamond_{IN} smoke$

Given a model $\mathcal{W}'$, the model checker verifies the truth conditions of the above formula exploring the model through the following steps:

$\models @_{kitchen} \Diamond_W \Diamond_{IN} smoke$	iff	there exist a place $w = kitchen$, such that
$\models_w \Diamond_W \Diamond_{IN} smoke$	iff	there exist a place w', with $R_W(w, w')$, and
$\models_{w'} \Diamond_{IN} smoke$	iff	there exist a place w'', with $R_{IN}(w', w'')$, and
$\models_{w''} smoke$		★

It is relevant to observe that the finiteness that always characterizes the CSM model (namely, the set of places and relations is finite), guarantees that the verification algorithm of the formulas satisfiability, implemented in the model checker, terminates.

5 Concluding remarks

This paper presented two different models, and more precisely the MMASS which supports the definition of complex MASs in which the environment, and especially its spatial aspects, can be explicitly represented and exploited by interaction mechanisms, and CSM which represents a structure which represents and performs reasoning on facts related to spatial information and knowledge. The main aim of the paper was to propose an integration of the two models aimed at supplying more expressive abstractions supporting the behavioral specification of MMASS agents. In particular, CSM represents a suitable formalism to express complex conditions involving information related to facts representing properties of elements of a spatial abstraction. These conditions can be exploited for the definition of preconditions to MMASS actions. The next steps of this research are aimed at the design and implementation of applications exploiting the proposed integration in the context of ubiquitous systems and supporting the simulation of crowd behaviors in realistic environments (e.g. shopping centers).

Actually, there are many domains in which time dimension is crucial and a very interesting problem for further formal and theoretical work is how to consider time and dynamism integrated with CSM. On one hand, in fact, considering the dynamical evolution of a MMASS system, the explicit spatial reasoning may need to relate facts true at different places at different time (properties holding over a place change in time). On the other hand, in domains characterized by the presence of wireless technologies, interesting places, properties holding over them and the relations' extension may change, since new interesting places can be discovered (e.g. a mobile agent is identified as a place) and known places can move.

BIBLIOGRAPHY

[Aiello and Benthem, 2002] M. Aiello and J. Van Benthem. A modal walk through space. *Journal of Applied Non-Classical Logics*, 12(3–4):319–363, 2002.

[Bandini et al., 2002a] S. Bandini, D. Bogni, and S. Manzoni. Alarm correlation in traffic monitoring and control systems: a knowledge-based approach. In Frank van Harmelen, editor, *Proceedings of the 15th European Conference on Artificial Intelligence, July 21-26 2002, Lyon (F)*, pages 638–642, Amsterdam, 2002. IOS Press.

[Bandini et al., 2002b] S. Bandini, S. Manzoni, and C. Simone. Heterogeneous agents situated in heterogeneous spaces. *Applied Artificial Intelligence*, 16(9–10):831–852, 2002.

[Bandini et al., 2004a] S. Bandini, S. Manzoni, and G. Vizzari. Multiagent approach to localization problems: the case of multilayered multi agent situated system. *Web Intelligence and Agent Systems*, 2(3):155–166, 2004.

[Bandini et al., 2004b] S. Bandini, S. Manzoni, and G. Vizzari. Situated cellular agents: a model to simulate crowding dynamics. *IEICE Transactions on Information and Systems: Special Issues on Cellular Automata*, E87-D(3):669–676, 2004.

[Bandini et al., 2004c] S. Bandini, A. Mosca, M. Palmonari, and F. Sartori. A conceptual framework for monitoring and control system development. In *Ubiquitous Mobile Information and Collaboration Systems (UMICS'04)*, volume 3272 of *Lecture Notes in Computer Science*. Springer-Verlag, 2004. in press.

[Bandini et al., 2005a] S. Bandini, D. Bogni, S. Manzoni, and A. Mosca. A ST-modal logic approach to alarm correlation in monitoring and control of italian highways traffic. In *Proceedings of The 18th International Conference on Industrial & Engineering Applications of Artificial Intelligence & Expert Systems*. Bari, June 22-25. LNCS, 2005. in press.

[Bandini et al., 2005b] S. Bandini, S. Manzoni, and G. Vizzari. A spatially dependant communication model for ubiquitous systems. In *The First International Workshop on Environments for Multiagent Systems*, volume 3374 of *LNCS*, pages 74–90. Springer-Verlag, 2005.

[Blackburn et al., 2000] P. Blackburn, M. de Rijke, and Y. Venema. *Modal Logic*. Cambridge University Press, 2000.

[Blackburn, 2000] P. Blackburn. Representation, reasoning and realtional structures: a hybrid logic manifesto. *Logic Journal of the IGPL*, 8(3):339–365, 2000.

[Davis, 1990] E. Davis. *Representations of commonsense knowledge*. Morgan Kaufmann Publishers, 1990.

[Gopal et al., 1989] S. Gopal, R.L. Klatzky, and T.R. Smith. Navigator: A psychologically based model of environmental learning through navigation. *Journal of Environmental Psychology*, 9:309–331, 1989.

[Hadeli et al., 2004] K. Hadeli, P. Valckenaers, C. Zamfirescu, H. Van Brussel, B. Saint Germain, T. Hoelvoet, and E. Steegmans. Self-organising in multi-agent coordination and control using stigmergy. In *Engineering Self-Organising Systems: Nature-Inspired Approaches to Software Engineering*, volume 2977 of *Lecture Notes in Computer Science*, pages 105–123. Springer–Verlag, 2004.

[Harary, 1972] F. Harary. *Graph Theory*. Addison-Wesley, Reading, MA, 1972.

[Kuipers, 1978] B. Kuipers. Modelling spatial knowledge. *Cognitive Science*, 2:129–154, 1978.

[Leisler and Zilbershatz, 1989] D. Leisler and A. Zilbershatz. The traveller: A computational model of spatial network learning. *Environment and Behaviour*, 21(4):435–463, 1989.

[Randell et al., 1992] D.A. Randell, Z. Cui, and A.G. Cohn. A spatial logic based on regions and connection. In *Proc. 3rd Int. Conf. on Knowledge Representation and Reasoning*, pages 165–176, San Mateo, CA, 1992. Morgan Kaufmann.

[Wolfram, 1986] S. Wolfram. *Theory and Applications of Cellular Automata*. World Press, 1986.

[Zambonelli and Parunak, 2002] F. Zambonelli and H.V.D. Parunak. Signs of a revolution in computer science and software engineering. In *Proceedings of Engineering Societies in the Agents World III (ESAW2002)*, volume 2577 of *Lecture Notes in Computer Science*, pages 13–28. Springer-Verlag, 2002.

Stefania Bandini, Alessandro Mosca,
Matteo Palmonari, and Giuseppe Vizzari
Dipartimento di Scienze della Comunicazione (DISCo)
Università di Milano, Bicocca, Italy
Email: {bandini, alessandro.mosca,
matteo.palmonari}@disco.unimib.it

On Abductive Equivalence
KATSUMI INOUE AND CHIAKI SAKAMA

ABSTRACT. We consider the problem of identifying equivalence of two knowledge bases which are capable of abductive reasoning. Here, a knowledge base is written in either first-order logic or nonmonotonic logic programming. In this work, we will give two definitions of abductive equivalence. The first one, *explainable equivalence*, requires that two abductive programs have the same explainability for any observation. Another one, *explanatory equivalence*, guarantees that any observation has exactly the same explanations in each abductive framework. Explanatory equivalence is a stronger notion than explainable equivalence, and in fact, the former implies the latter. In first-order abduction, explainable equivalence can be verified by the notion of extensional equivalence in default theories. In nonmonotonic logic programs, explanatory equivalence can be checked by means of the notion of relative strong equivalence. We also discuss how the two notions of abductive equivalence can be applied to extended abduction, where abducibles can not only be added to a program but also be removed from the program to explain an observation.

1 Introduction

Nowadays, abduction is used in many AI applications, including diagnosis, design, updates, and discovery. Abduction is an important paradigm for problem solving, and is incorporated in programming technologies, i.e., *abductive logic programming* (ALP) [Kakas et al., 1998; Denecker and Kakas, 2002]. Automated abduction is also studied in the literature as an extension of deductive methods or a part of inductive systems [Flach and Kakas, 2000; Inoue, 2002], and its computational properties have also been studied [Selman and Levesque, 1996; Eiter and Gottlob, 1995; Eiter et al., 1998].

In this work, we are concerned with such computational issues on abductive reasoning. Despite being a problem-solving paradigm, ALP has a lot of issues which have not been fully understood yet. In particular, there are no concrete methods for (a) evaluation of abductive power in ALP, (b) measurement of efficiency in abductive reasoning, (c) semantically correct simplification and optimization, (d) debugging and verification in ALP, and (e) standardization in ALP. Since all these topics are important for any programming paradigm, the lack of them is a serious drawback of ALP. Then, it can be recognized that all the above issues are related to different notions of identification or *equivalence* in ALP. In particular, the item (c) is related to understanding the semantics of ALP with respect to modularity and *contexts*.

The notion of equivalence between two knowledge bases is also one of the most important problems in knowledge representation based on logic. For example, one axiom set A_1 represents the specification of a device or a program and the other formula set A_2 is a result of the design of a hardware/software system. Then, we should check whether A_2 is equivalent to A_1, intending the verification of the design. Similarly, the notion of equivalence in logic programming has recently become important. Because a logic program is used to represent knowledge of a problem domain [Baral and Gelfond, 1994], we often have to consider whether two logic programs P_1 and P_2 represent the same knowledge. For example, one logic program P_1 may be viewed as a specification of knowledge in some domains, and another representation P_2 may be expected to be a compact form of P_1 which can easily be computed.

Abduction can be formalized in various logics [Levesque, 1989; Flach and Kakas, 2000]. Then, we can consider several notions of equivalence in several logics for abduction. In this paper, we will give two definitions of abductive equivalence in two logical frameworks for abduction. Two logics we consider here are *first-order logic* (FOL) and *abductive logic programming* (ALP). The first abductive equivalence, called *explainable equivalence*, requires that two abductive programs have the same explainability for any observation. Another one, *explanatory equivalence*, guarantees that any observation has exactly the same explanations in each abductive framework. Explanatory equivalence is stronger than explainable equivalence, and in fact, the former implies the latter.

In this paper, we characterize these two notions of abductive equivalence in terms of other well-known concepts in AI and logic programming. In abduction in first-order logic, we will see that explainable equivalence can be verified by the notion of *equivalence* in *default logic* [Reiter, 1980], which is defined for the families of *extensions* of two default theories. On the other hand, abductive equivalence in ALP is more complicated than in the case of FOL due to the nonmonotonicity in logic programs. In fact, equivalence between two abductive logic programs has little been discussed in the literature except that effects of some program transformation techniques in ALP are analyzed in [Sakama and Inoue, 1995]. In this work, by means of the notion of *strong equivalence* [Maher, 1988; Lifschitz et al., 2001] and its relativized extension [Lin, 2002; Inoue and Sakama, 2004; Woltran, 2004], we will show that explanatory equivalence can be checked in ALP.

Finally, we also discuss how the two notions of abductive equivalence can be applied to *extended abduction* [Inoue and Sakama, 1995; Inoue and Sakama, 1999], where hypotheses can not only be added to a program but also be removed from the program to explain an observation. In extended abduction, abductive equivalence can be characterized by the notion of *update equivalence* [Inoue and Sakama, 2004].

The rest of this paper is organized as follows. Section 2 presents two definitions for abductive equivalence. Section 3 considers first-order logic as the representation language, while section 4 considers nonmonotonic logic programming for ALP. Section 5 extends the equivalence results in ALP to extended abduction. Section 6 gives concluding remarks.

2 Abductive equivalence

We start with a question as to when two abductive frameworks are equivalent. As far as the authors know, there is no answer for such a question in the literature of ALP. Moreover, no such a concept can be found in philosophy, either. It is conceivable that there must be several aspects to this question. When can we consider that an *explanation* E is equivalent to another explanation F for an observation? When can we say that an *observation* G is equivalent to another observation H in an abductive framework? In what circumstances, can we say that *abduction by* person A is equivalent to abduction by person B? When can we regard that *abduction with* knowledge P is equivalent to abduction with knowledge Q?

There are also many parameters which should be considered important in defining equivalence notions in abductive frameworks. In the *world*, both background knowledge and observations are surely essential. In an *agent* who performs abduction, on the other hand, its abductive power must depend on its *logic* (language, syntax, semantics) of background knowledge, observations and hypotheses. Moreover, the quality of abduction is relevant to other parameters such as axioms, inference procedures, logics of explanations, and criteria of best explanations. If we would take all such parameters into account, the task of defining the equivalence notion might become combinatorial and too complex.

In the following, we thus consider a rather simple framework for our problem while we try to hold the essence of equivalence notions as much as possible. First, logic, background knowledge and hypotheses are put as input parameters in each abductive framework. Second, a logic of explanations is taken into account in a definition, but its diversity is reflected in different notions of abductive equivalence.

The following definition of abductive frameworks is a standard one [Levesque, 1989; Selman and Levesque, 1996; Eiter and Gottlob, 1995; Eiter et al., 1998]. As a notation, $\Sigma \models_L F$ means that a formula F is derived from a set Σ of formulas in a logic L.

Definition 2.1 Let B and H be sets of formulas in some underlying logic L. An *abductive framework* is defined as a triple (L, B, H), where B is called *background knowledge* and each element of H is called a *candidate hypothesis*.

Definition 2.2 Let (L, B, H) be an abductive framework, and O a formula in L, and E a formula belonging to H. We define that E is an *explanation* of an *observation* O in (L, B, H) if $B \cup E \models_L O$ and $B \cup E$ is consistent in L. We say that O is *explainable* in (L, B, H) if it has an explanation in (L, B, H).

Remark. Definition 2.2 requires that each explanation E must be consistent with the background knowledge B in the logic L. This condition is sometimes too strong in realistic cases, and can be weakened if the logic L is *paraconsistent*.

In the next two sections, we consider two logics for abduction both of which are popular formalisms in AI: first-order logic (FOL) (section 3), and logic programming with negation as failure (ALP) (section 4).

We now give two definitions for abductive equivalence. We assume that the underlying logic L is common when two abductive theories are compared.

Definition 2.3 Two abductive frameworks (L, B_1, H_1) and (L, B_2, H_2) are *explainably equivalent* if, for any observation O, there is an explanation of O in (L, B_1, H_1) iff there is an explanation of O in (L, B_2, H_2).

Explainable equivalence requires that two abductive frameworks have the same *explainability* for any observation. Explainable equivalence may reflect a situation that two theories have different knowledge to derive the same goals.

Definition 2.4 Two abductive frameworks (L, B_1, H_1) and (L, B_2, H_2) are *explanatorily equivalent* if, for any observation O, E is an explanation of O in (L, B_1, H_1) iff E is an explanation of O in (L, B_2, H_2).

Explanatory equivalence assures that two abductive frameworks have the same *explanation power* for any observation. Explanatory equivalence is stronger than explainable equivalence. In fact, the former implies the latter. The two notions coincide if $H_1 = H_2 = \emptyset$.

Proposition 2.1 *If abductive frameworks (L, B_1, H_1) and (L, B_2, H_2) are explanatorily equivalent, then they are explainably equivalent.*

Proposition 2.2 *Two abductive frameworks $(L, B_1, \emptyset)$ and $(L, B_2, \emptyset)$ are explainably equivalent iff they are explanatorily equivalent.*

For explanatory equivalence, we can assume that the hypotheses H are common in two abductive frameworks in Definition 2.4, as the following property holds.

Proposition 2.3 *Suppose that $A_1 = (L, B_1, H_1)$ and $A_2 = (L, B_2, H_2)$ are abductive frameworks. If A_1 and A_2 are explanatorily equivalent, then $H_1' = H_2'$, where $H_i' = \{h \in H_i \mid B_i \cup \{h\} \text{ is consistent in } L\}$ for $i = 1, 2$.*

Proof. Assume that $H_1' \setminus H_2' \neq \emptyset$. Then, for a formula $\varphi \in H_1' \setminus H_2'$, $\{\varphi\}$ is an explanation of φ in A_1 because $B_1 \cup \{\varphi\}$ is consistent in L. However, $\{\varphi\}$ is not an explanation of φ in A_2. Hence, A_1 and A_2 are not explanatorily equivalent. ∎

Note in Proposition 2.3 that any hypothesis h in $H_i \setminus H_i'$ cannot be added without violating the consistency of $B_i \cup \{h\}$ in L. Thus, H_i' is the set of hypotheses that can be actually used in explanations of some formulas.

Example 2.1 Suppose two abductive frameworks, $A_1 = (\text{FOL}, \{a \to p\}, \{a, b\})$ and $A_2 = (\text{FOL}, \{b \to p\}, \{a, b\})$. Then, A_1 and A_2 are explainably equivalent, but are not explanatorily equivalent. On the other hand, $A_3 = (\text{FOL}, \{a \to p\}, \{b\})$ and $A_4 = (\text{FOL}, \{b \to p\}, \{b\})$ are neither explainably equivalent nor explanatorily equivalent.

3 Abduction in first-order logic

Abduction is used in many AI applications, and classical first-order logic (FOL) is most often used as the underlying logic for abduction [Poole, 1988; Levesque, 1989; Eiter and Gottlob, 1995; Selman and Levesque, 1996; Inoue, 2002]. When the underlying logic L is FOL, the relation $\models_L$ becomes the usual entailment relation $\models$. A first-order formula f is *closed* if f contains no free variables. A *ground instance* of a first-order formula f

is a formula obtained by replacing every variable in f with a term containing no variables. In first-order abduction, explanations are usually defined as a set of ground instances from hypotheses as follows [Poole, 1988; Inoue, 2002].

Definition 3.1 Suppose an abductive framework (FOL, B, H), where both the background knowledge B and the hypotheses H are sets of first-order formulas. Given a closed formula O as an observation, a set E of ground instances of elements of H is an *explanation* of O in (FOL, B, H) if $\Sigma \cup E \models G$ and $\Sigma \cup E$ is consistent.

In the following, $Th(\Sigma)$ denotes the set of logical consequences of a set Σ of first-order formulas. That is, $Th(\Sigma) = \{F \mid \Sigma \models F\}$. The next definition is originally given for *default logic* by Reiter [1980].

Definition 3.2 [Reiter, 1987; Poole, 1988] Let B and H be sets of first-order formulas. An *extension of B with respect to H* is $Th(B \cup S)$ where S is a maximal subset of ground instances of elements from H such that $B \cup S$ is consistent.

When an abductive framework (FOL, B, H) is given, we can associate a Reiter's *default theory* $\Delta = (D, B)$ where D is the set of *prerequisite-free normal defaults* $\{\frac{:d}{d} \mid d \in H\}$ such that there is a one-to-one correspondence between the *extensions* of Δ (which are defined in [Reiter, 1980]) and the extensions of B with respect to H [Poole, 1988]. Using the notion of extensions in Definition 3.2, explainable equivalence can be characterized in first-order abduction.

Theorem 3.1 *Two abductive frameworks* (FOL, B_1, H_1) *and* (FOL, B_2, H_2) *are explainably equivalent iff the extensions of B_1 with respect to H_1 coincide with the extensions of B_2 with respect to H_2.*

Proof. First, we claim that the union of the extensions of B with respect to H are exactly the set of formulas explainable in (FOL, B, H). To see this, we can use a well-known theorem [Poole, 1988; Selman and Levesque, 1996] that a formula O can be explained in (FOL, B, H) iff there is a consistent extension X of B with respect to H such that X contains O. Thus, the set of all explainable formulas are precisely those formulas contained in at least one extension of B with respect to H.

Now, let $A_1 = $ (FOL, B_1, H_1) and $A_2 = $ (FOL, B_2, H_2) be two abductive frameworks. Suppose that the extensions of B_1 with respect to H_1 coincide with those of B_2 with respect to H_2. By the above claim, the set of formulas explainable in A_1 is equal to the set of formulas explainable in A_2. This means that A_1 and A_2 are explainably equivalent.

Conversely, assume that there is an extension X_2 of B_2 with respect to H_2 which is not an extension of B_1 with respect to H_1. Let F_{X_2} be a first-order formula which is logically equivalent to X_2. Such a formula actually exists because $X_2 = Th(B_2 \cup S)$ holds for some maximally consistent subset S of H_2, and hence X_2 is logically equivalent to $\bigwedge_{f \in B_2} f \wedge \bigwedge_{g \in S} g$. Since X_2 is consistent, F_{X_2} is consistent too. Then, F_{X_2} is explainable in A_2 because S is an explanation of F_{X_2}.

Now, if F_{X_2} is not explainable in A_1, then obviously A_1 and A_2 are not explainably equivalent. Hence, there is an explanation of F_{X_2} in A_1. Then, there is an extension X_1 of B_1 with respect to H_1 which contains F_{X_2}. Since X_2 is not an extension of B_1 with respect to H_1, $X_1 \neq X_2$ holds. Then, $X_2 \subset X_1$. Let F_{X_1} be a formula which is logically equivalent to X_1. By the same argument as above, F_{X_1} is explainable in A_1. However, this F_{X_1} cannot be explained in A_2. This is because, if F_{X_1} were explained in A_2, there must be an extension X_2' of B_2 with respect to H_2 such that $X_2 \subset X_2'$, which is impossible because any extension is *orthogonal* to another extension in a default theory [Reiter, 1980]. In any case, A_1 and A_2 are not explainably equivalent. ∎

In [Marek et al., 1997], Reiter's default theories $\Delta_1 = (D_1, B_1)$ and $\Delta_2 = (D_2, B_2)$ are said to be *equivalent* if the extensions of Δ_1 are the same as the extensions of Δ_2. Using this notation, explainable equivalence in first-order abduction can also be represented as follows.

Corollary 3.2 *Two abductive frameworks* (FOL, B_1, H_1) *and* (FOL, B_2, H_2) *are explainably equivalent iff the default theories* (D_1, B_1) *and* (D_2, B_2) *are equivalent where* $D_i = \{\frac{:d}{d} \mid d \in H_i\}$ *for* $i = 1, 2$.

Example 3.1 Suppose two abductive frameworks, $A_1 = $ (FOL, B_1, H_1) and $A_2 = $ (FOL, B_2, H_2), where

$$B_1 = \{a \to p, \ b \to \neg p\},$$
$$H_1 = \{a, \ b, \ a \equiv c, \ b \equiv d, \ p \equiv q\},$$
$$B_2 = \{c \to q, \ d \to \neg q\}, \text{ and}$$
$$H_2 = \{c, \ d, \ a \equiv c, \ b \equiv d, \ p \equiv q\}.$$

Then, A_1 and A_2 are explainably equivalent. In fact, the two extensions of B_1 with respect to H_1 are $Th(B_1 \cup (H_1 \setminus \{b\})) = Th(\{a, \neg b, c, \neg d, p, q\})$ and $Th(B_1 \cup (H_1 \setminus \{a\})) = Th(\{\neg a, b, \neg c, d, \neg p, \neg q\})$, which are respectively equivalent to the two extensions of B_2 with respect to H_2, $Th(B_2 \cup (H_2 \setminus \{d\}))$ and $Th(B_2 \cup (H_2 \setminus \{c\}))$.

Logical equivalence of background theories implies explainable equivalence when the hypotheses are common.

Corollary 3.3 *If $B_1 \equiv B_2$, then abductive frameworks (FOL, B_1, H) and (FOL, B_2, H) are explainably equivalent. However, the converse does not hold.*

Proof. If $B_1 \equiv B_2$, then any extension of B_1 with respect to H is an extension of B_2 with respect to H and vice versa. The converse does not hold as Example 2.1 shows. ∎

It is interesting to see that we can transform any abductive framework to an explainably equivalent abductive framework whose background theory is empty. The next property is also derived by the representation theory for default logic [Marek *et al.*, 1997].

Corollary 3.4 *For any abductive framework (FOL, B, H), there is an abductive framework (FOL, $\emptyset, H'$) which is explainably equivalent to (FOL, B, H).*

Proof. Put $H' = \{h \wedge \varphi \mid h \in H\} \cup \{\varphi\}$, where $\varphi = \bigwedge_{f \in B} f$. Then, it holds for any O that, $B \cup E \models O$ iff $E' \models O$ where $E \subseteq H$ and $E' = \{h \wedge \varphi \mid h \in E\} \cup \{\varphi\} \subseteq H'$. ∎

An abductive framework (L, B, H) is called (B, H)-*compatible* if $B \cup H$ is consistent. Explainable equivalence can be easily verified for (B, H)-compatible frameworks.

Corollary 3.5 *Let (FOL, B_1, H_1) and (FOL, B_2, H_2) be (B_i, H_i)-compatible abductive frameworks for $i = 1, 2$. Then, (FOL, B_1, H_1) and (FOL, B_2, H_2) are explainably equivalent iff $B_1 \cup H_1 \equiv B_2 \cup H_2$.*

Proof. For any (B, H)-compatible abductive framework (FOL, B, H), we have that $B \cup H$ is consistent. Then, $Th(B \cup H)$ is the unique extension of B with respect to H. By Theorem 3.1, (FOL, B_1, H_1) and (FOL, B_2, H_2) are explainably equivalent iff $Th(B_1 \cup H_1) = Th(B_2 \cup H_2)$. Hence, the corollary holds. ∎

An abductive framework (FOL, $B, \mathcal{L}$) is called *assumption-free* where $\mathcal{L}$ is the set of all literals in the underlying language. It is known that the complexity of finding explanations in assumption-free abductive frameworks is not harder than that in assumption-based frameworks [Selman and Levesque, 1996]. Explainable equivalence in the assumption-free case can also be simply characterized as follows.

Corollary 3.6 *Abductive frameworks* $(\text{FOL}, B_1, \mathcal{L})$ *and* $(\text{FOL}, B_2, \mathcal{L})$ *are explainably equivalent iff* $B_1 \equiv B_2$.

Proof. For an assumption-free abductive framework $(\text{FOL}, B, \mathcal{L})$, each extension of B with respect to $\mathcal{L}$ is logically equivalent to a model of B. Hence, explainable equivalence implies that the models of B_1 coincide with the models of B_2, and vice versa. ∎

For explanatory equivalence in first-order abduction, logical equivalence of background theories is necessary and sufficient.

Theorem 3.7 *Two abductive frameworks* (FOL, B_1, H) *and* (FOL, B_2, H) *are explanatorily equivalent iff* $B_1 \equiv B_2$.

Proof. If $B_1 \equiv B_2$, then for any E and any O, it holds that, $B_1 \cup E \models O$ iff $B_2 \cup E \models O$, and that, $B_1 \cup E$ is consistent iff $B_2 \cup E$ is consistent. Hence, (FOL, B_1, H) and (FOL, B_2, H) are explanatorily equivalent.

Conversely, suppose that (FOL, B_1, H) and (FOL, B_2, H) are explanatorily equivalent. Then, for any formula O and any E from H, it holds that $B_1 \cup E \models O$ iff $B_2 \cup E \models O$. Then, for any E, we have $Th(B_1 \cup E) = Th(B_2 \cup E)$. That is, $B_1 \cup E \equiv B_2 \cup E$ holds for any E. This implies $B_1 \equiv B_2$ when $E = \emptyset$. ∎

4 Abductive logic programming

Abductive logic programming (ALP) is another popular formalization of abduction in AI [Kakas et al., 1998; Denecker and Kakas, 2002; Eiter et al., 1998]. Background knowledge in ALP is called a *logic program*, and the candidate hypotheses are given as literals called *abducibles*. The most significant difference between abduction in FOL and ALP is that ALP allows the nonmonotonic *negation-as-failure* operator *not* in background knowledge. In abduction, addition of hypotheses may invalidate explanations of some observations if the background theory is nonmonotonic.

Recall that a *(logic) program* is a set of rules of the form

$$L_1; \cdots; L_k; \text{not } L_{k+1}; \cdots; \text{not } L_l \leftarrow L_{l+1}, \ldots, L_m, \text{not } L_{m+1}, \ldots, \text{not } L_n$$

where each L_i is a literal ($n \geq m \geq l \geq k \geq 0$), and *not* is *negation as failure* (NAF). The symbol ; represents a disjunction. The left-hand side of the rule is called the *head*, and the right-hand side is called the *body*. A rule with variables stands for the set of its ground instances. Intuitively, the rule in the above form can be read as follows: if all $L_{l+1}, \ldots, L_m$ are believed and all $L_{m+1}, \ldots, L_n$ are disbelieved then either some L_i ($1 \leq i \leq k$) should be believed or some L_j ($k+1 \leq j \leq l$) should be disbelieved.

In this paper, the semantics of a logic program is given by its *answer sets* [Gelfond and Lifschitz, 1991; Baral and Gelfond, 1994; Inoue and Sakama, 1998], while another semantics can be considered as well in ALP [Kakas et al., 1998; Eiter et al., 1998]. Intuitively speaking, each answer set represents a set of literals corresponding to beliefs which can be built by a rational reasoner on the basis of a program [Baral and Gelfond, 1994]. The answer sets for a program are defined in the following two steps [Gelfond and Lifschitz, 1991; Inoue and Sakama, 1998]. First, let P be a program without NAF (i.e., $k = l$ and $m = n$) and $S \subseteq \mathcal{L}$, where $\mathcal{L}$ is the set of all ground literals in the language of P. Then, S is an *answer set* of P if S is a minimal set satisfying the conditions:

1. S satisfies every rule in P, that is, for any ground rule of the form $L_1; \cdots ; L_l \leftarrow L_{l+1}, \ldots, L_m$ from P, if $\{L_{l+1}, \ldots, L_m\} \subseteq S$ then $\{L_1, \ldots, L_l\} \cap S \neq \emptyset$;

2. If S contains a pair of complementary literals L and $\neg L$, then $S = \mathcal{L}$.

Second, given *any* program P (with NAF) and $S \subseteq \mathcal{L}$, consider the program (without NAF) P^S obtained as follows: a rule $L_1; \cdots ; L_k \leftarrow L_{l+1}, \ldots, L_m$ is in P^S if there is a ground rule of the form

$$L_1; \cdots ; L_k; \text{not } L_{k+1}; \cdots ; \text{not } L_l \leftarrow L_{l+1}, \ldots, L_m, \text{not } L_{m+1}, \ldots, \text{not } L_n$$

from P such that $\{L_{k+1}, \ldots, L_l\} \subseteq S$ and $\{L_{m+1}, \ldots, L_n\} \cap S = \emptyset$. Then, S is an *answer set* of P if S is an answer set of P^S. An answer set is *consistent* if it is not $\mathcal{L}$. A program is *consistent* if it has a consistent answer set. A program has none, one, or multiple answer sets in general. A typical program which has no answer set is $\{p \leftarrow \text{not } p\}$. Problem solving by representing knowledge as a logic program and then computing its answer sets is called *answer set programming* (ASP). In ASP, alternative belief sets of a reasoner are represented by multiple answer sets of a program.

Definition 4.1 An *abductive (logic) program* is defined as a pair $\langle P, \mathcal{A} \rangle$, where P is a logic program and $\mathcal{A}$ is a set of literals called *abducibles*. Instead of using the notation (ALP, $P, \mathcal{A}$), we also use $\langle P, \mathcal{A} \rangle$ to represent an abductive framework whose underlying logic is ALP.

Definition 4.2 Let $\langle P, \mathcal{A} \rangle$ be an abductive program, and G a conjunction of ground literals called *observations*. A set $E \subseteq \mathcal{A}$ is a *(credulous) explanation* of G in $\langle P, \mathcal{A} \rangle$ if every ground literal in G is true in a consistent answer set of $P \cup E$.

Note that both abducibles and observations are restricted to ground literals in ALP. However, it is known for this framework that rules can be allowed in abducibles and that observations can contain NAF formulas as well as literals [Inoue and Sakama, 1996]. We assume that the set of observations includes the special atom $\top$, which represents the empty conjunction of observations. Note that $\top$ is always true in any set of ground literals. Definition 4.2 can also be represented in a different way as follows [Inoue and Sakama, 1996]. A *belief set (with respect to E)* of an abductive program $\langle P, \mathcal{A} \rangle$ is a consistent answer set of a logic program $P \cup E$ where $E \subseteq \mathcal{A}$. Then, $E \subseteq \mathcal{A}$ is an explanation of G if G is true in a belief set of $\langle P, \mathcal{A} \rangle$ with respect to E.

Remark. In Definition 4.2, explanations are defined in a *credulous* way. Another, *skeptical* notion for explanations is defined as $E \subseteq \mathcal{A}$ such that G is true in all consistent answer sets of $P \cup E$. Abductive equivalence relative to skeptical explanations can also be defined in a similar way, but characterization of such notions needs different formalizations. For instance, instead of taking the union of belief sets in the equation of Theorem 4.1, skeptical consequences are computed by taking the intersection of them.

According to section 2, we consider two types of abductive equivalence for ALP.

Definition 4.3 Abductive programs $\langle P_1, \mathcal{A}_1 \rangle$ and $\langle P_2, \mathcal{A}_2 \rangle$ are *explainably equivalent* if, for any ground literal G, G is explainable in $\langle P_1, \mathcal{A}_1 \rangle$ iff G is explainable in $\langle P_1, \mathcal{A}_2 \rangle$.

Definition 4.4 Abductive programs $\langle P_1, \mathcal{A} \rangle$ and $\langle P_2, \mathcal{A} \rangle$ are *explanatorily equivalent* if, for any conjunction of ground literals G, E is an explanation of G in $\langle P_1, \mathcal{A} \rangle$ iff E is an explanation of G in $\langle P_2, \mathcal{A} \rangle$.

Explainable equivalence in ALP guarantees the same explainability for any ground literal as a *single observation*, but it does not matter how each observation is explained. Hence, we do not have to care about whether multiple observations can be *jointly* explained by a common explanation. On the other hand, explanatory equivalence in ALP guarantees that, any *conjunction* (or *set*) *of observations* has exactly the same credulous explanations. Hence, explanatory equivalence implies that any set of abducibles $E \subseteq \mathcal{A}$ should explain the same set of observations in each abductive program. Again, explanatory equivalence implies explainable equivalence.

We now show that explainable equivalence in ALP can be checked by comparing the belief sets of two abductive programs. Because there exist several methods to compute belief sets using ASP [Inoue, 1991; Satoh and

Iwayama, 1991; Inoue and Sakama, 1996; Inoue and Sakama, 1998], checking explainable equivalence is also possible using such methods. In the following, we denote the set of all belief sets of $\langle P, \mathcal{A} \rangle$ as $BS(P, \mathcal{A})$.

Theorem 4.1 *Abductive programs $\langle P_1, \mathcal{A}_1 \rangle$ and $\langle P_2, \mathcal{A}_2 \rangle$ are explainably equivalent iff*

$$\bigcup_{S \in BS(P_1, \mathcal{A}_1)} S = \bigcup_{S \in BS(P_2, \mathcal{A}_2)} S.$$

Proof. Recall that $E \subseteq \mathcal{A}$ is an explanation of a ground literal G iff G is true in a belief set of $\langle P, \mathcal{A} \rangle$ with respect to E. Then, the set of all explainable literals are precisely those literals contained in some belief sets of $\langle P, \mathcal{A} \rangle$ with respect to some E. Hence, the union of the belief sets of $\langle P, \mathcal{A} \rangle$ are exactly the set of literals explainable in $\langle P, \mathcal{A} \rangle$. Therefore, two abductive programs are explainably equivalent iff the unions of the belief sets of two abductive programs coincide. ∎

The next corollary gives a sufficient condition.

Corollary 4.2 *Abductive programs $\langle P_1, \mathcal{A}_1 \rangle$ and $\langle P_2, \mathcal{A}_2 \rangle$ are explainably equivalent if $BS(P_1, \mathcal{A}_1) = BS(P_2, \mathcal{A}_2)$.*

In some case of (B, H)-compatible problems, explanatory equivalence can be easily verified. Here, a logic program is *definite* if every its rule is NAF-free and has exactly one atom in the head and only atoms in the body. A definite program has a unique answer set that is equivalent to its *least model*. An abductive program $\langle P, \mathcal{A} \rangle$ is called *definite* if P is a definite logic program and $\mathcal{A}$ is a set of atoms.

Corollary 4.3 *Suppose that $\langle P_1, \mathcal{A}_1 \rangle$ and $\langle P_2, \mathcal{A}_2 \rangle$ are definite abductive programs. Then, $\langle P_1, \mathcal{A}_1 \rangle$ and $\langle P_2, \mathcal{A}_2 \rangle$ are explainably equivalent if the least model of $P_1 \cup \mathcal{A}_1$ coincides with that of $P_2 \cup \mathcal{A}_2$.*

Example 4.1 Given the common set of abducibles $\mathcal{A} = \{a, b\}$ and three logic programs:

$$\begin{aligned} P_1 &= \{p \leftarrow a, \ q \leftarrow b\}, \\ P_2 &= \{p \leftarrow b, \ q \leftarrow a\}, \\ P_3 &= \{p \leftarrow, \ q \leftarrow a, \ \leftarrow a, b\}, \end{aligned}$$

the three abductive programs $\langle P_i, \mathcal{A} \rangle$ (for $i = 1, 2, 3$) are all explainably equivalent, but none of them are explanatorily equivalent. In particular, the least model of $P_1 \cup \mathcal{A}$ is $\{p, q, a, b\}$, which is identical to that of $P_2 \cup \mathcal{A}$. P_3 is not definite because of the third rule, but $\langle P_3, \mathcal{A} \rangle$ has three belief sets: $\{p\}, \{p, q, a\}, \{p, b\}$, the union of which is equal to that of $\langle P_i, \mathcal{A} \rangle$ for $i = 1, 2$.

Explanatory equivalence in ALP, on the other hand, requires a more semantical notion of logic programming. Note that explanatory equivalence of $\langle P_1, \mathcal{A} \rangle$ and $\langle P_2, \mathcal{A} \rangle$ implies $BS(P_1, \mathcal{A}) = BS(P_2, \mathcal{A})$, but the converse does not hold.

Example 4.2 Suppose $P_1 = \{a \leftarrow , p \leftarrow a\}$, $P_2 = \{a \leftarrow not\, a, p \leftarrow a\}$ and $\mathcal{A} = \{a\}$. Then, $BS(P_1, \mathcal{A}) = BS(P_2, \mathcal{A}) = \{\{a, p\}\}$. However, $\emptyset$ is an explanation of p, a and $\top$ in $\langle P_1, \mathcal{A} \rangle$, but is not an explanation of them in $\langle P_2, \mathcal{A} \rangle$. In fact, P_2 alone has no answer set although $P_2 \cup \{a\}$ has the answer set $\{a, p\}$.

To characterize explanatory equivalence precisely, we need to utilize the concept of equivalence in logic programming and ASP. There are several notions for equivalence in logic programming, and *weak equivalence* and *strong equivalence* are most well known. We say that two programs are *weakly equivalent* if they simply agree with their answer sets. The notion of weak equivalence is similar to that of *logical equivalence* in FOL and other classical logics. Given two abductive programs $\langle P_1, \mathcal{A} \rangle$ and $\langle P_2, \mathcal{A} \rangle$, weak equivalence of P_1 and P_2 is not a sufficient condition for explanatory equivalence of them, and is not even a sufficient condition for explainable equivalence. However, weak equivalence is meaningful when the abducibles are empty.

Proposition 4.4 *Abductive programs $\langle P_1, \emptyset \rangle$ and $\langle P_2, \emptyset \rangle$ are explanatorily equivalent iff P_1 and P_2 are weakly equivalent.*

On the other hand, *strong equivalence* [Maher, 1988; Lifschitz et al., 2001] is a more context-sensitive notion for equivalence of logic programs. Two logic programs P_1 and P_2 are said to be *strongly equivalent* if for any additional logic program R, $P_1 \cup R$ and $P_2 \cup R$ have the same answer sets. Obviously, strong equivalence implies weak equivalence (when $R = \emptyset$). When we allow NAF in logic programs, weak equivalence is too fragile as a criterion. For example, $\{p \leftarrow not\, a\}$ and $\{p \leftarrow \}$ are weakly equivalent with the same unique answer set $\{p\}$, but are not strongly equivalent because the addition of a to both results in the withdrawal of p in the former only. In [Lifschitz et al., 2001], it is argued that strong equivalence can be used to simplify a part of a logic program without looking at the other part. For example, $\{p \leftarrow p\}$ and $\emptyset$ are strongly equivalent, so that the rule in the former can always be eliminated from any program.

For many applications, however, strong equivalence is too strong, and often we can restrict the language for additional programs R to some subset $\mathcal{R}$ of the whole language of programs. Then, two programs P_1 and P_2 are said to be *strongly equivalent with respect to* $\mathcal{R}$ if $P_1 \cup R$ and $P_2 \cup R$ have the same answer sets for any program $R \subseteq \mathcal{R}$ [Inoue and Sakama, 2004]. Such restriction of $\mathcal{R}$ is practicably interesting because knowledge bases are usually divided into invariable and variable parts such that only variable parts are changed in updates. The equivalence notion with such restriction is called *relative strong equivalence* [Lin, 2002; Inoue and Sakama, 2004; Woltran, 2004]. Using this notion, explanatory equivalence can be characterized as follows.

Theorem 4.5 *Two abductive programs $\langle P_1, \mathcal{A}\rangle$ and $\langle P_2, \mathcal{A}\rangle$ are explanatorily equivalent iff P_1 and P_2 are strongly equivalent with respect to $\mathcal{A}$.*

Proof. Suppose that $\langle P_1, \mathcal{A}\rangle$ and $\langle P_2, \mathcal{A}\rangle$ are explanatorily equivalent. Then, for any conjunction G of literals and any $E \subseteq \mathcal{A}$, it holds that, E is an explanation of G in $\langle P_1, \mathcal{A}\rangle$ iff E is an explanation of G in $\langle P_2, \mathcal{A}\rangle$. The latter equivalence then implies that, for any G and any E, we have that, G is true in a belief set of $\langle P_1, \mathcal{A}\rangle$ with respect to E iff G is true in a belief set of $\langle P_2, \mathcal{A}\rangle$ with respect to E. Then, for any G and any E, G is true in an answer set of $P_1 \cup E$ iff G is true in an answer set of $P_2 \cup E$. That is, for any E and any set S of literals, S is an answer set of $P_1 \cup E$ iff S is an answer set of $P_2 \cup E$. Hence, P_1 and P_2 are strongly equivalent with respect to $\mathcal{A}$. The converse direction can also be proved by tracing the above proof backward. ■

Example 4.3 Given the common set of abducibles $\mathcal{A} = \{a, b\}$, consider three programs

$$\begin{aligned} P_1 &= \{p \leftarrow a, \ a \leftarrow b\}, \\ P_2 &= \{p \leftarrow a, \ p \leftarrow b, \ a \leftarrow b\}, \\ P_3 &= \{p \leftarrow b, \ a \leftarrow b\}. \end{aligned}$$

Then, the three abductive programs $\langle P_i, \mathcal{A}\rangle$ (for $i = 1, 2, 3$) are explainably equivalent. Although $\langle P_1, \mathcal{A}\rangle$ is explanatorily equivalent to $\langle P_2, \mathcal{A}\rangle$, it is not to $\langle P_3, \mathcal{A}\rangle$ [Sakama and Inoue, 1995]. In fact, P_1 and P_2 are strongly equivalent with respect to $\mathcal{A}$, while P_1 and P_3 are not because the addition of a derives p in P_1 but this is not the case in P_3. This example shows that unfold/fold transformation [Tamaki and Sato, 1984] does not preserve explanatory equivalence in ALP [Sakama and Inoue, 1995] even when P_1 and P_2 are definite.

5 Abduction with removal of hypotheses

The two notions of abductive equivalence in section 4 can be applied to *extended abduction* [Inoue and Sakama, 1995; Inoue and Sakama, 1999] in ALP, in which abducibles can not only be added to a program but also be removed from the program to explain an observation. Extended abduction is defined by Inoue and Sakama [Inoue and Sakama, 1995] in autoepistemic logic for formalizing dynamics of abductive theories, and is then incorporated in ALP [Inoue and Sakama, 1999]. The intuition behind extended abduction is that, when the underlying logic is nonmonotonic, removal of some formulas makes other formulas become true. Hence, explanations are caused not only by addition of new hypotheses but also by deletion of old hypotheses.

To characterize abductive equivalence in extended abduction, we need to extend both the definition of belief sets and the notion of relative strong equivalence by taking removals of literals and rules into account.

Definition 5.1 Let $\langle P, \mathcal{A} \rangle$ be an abductive program, and G a conjunction of ground literals. A pair (E, F) where $E, F \subseteq \mathcal{A}$ is a *(credulous) explanation* of G (in $\langle P, \mathcal{A} \rangle$) if G is true in some consistent answer set of $(P \setminus F) \cup E$.

The notion of *normal* abduction, which has been discussed in section 4, can be defined as the task of finding explanations with $F = \emptyset$ in extended abduction.

Remark. In extended abduction, Inoue and Sakama also define the notion of *anti-explanations* as follows [Inoue and Sakama, 1995; Inoue and Sakama, 1999]. A pair $(E, F) \in 2^{\mathcal{A}} \times 2^{\mathcal{A}}$ is an *anti-explanation* of G if G is not explainable in $(P \setminus F) \cup E$. The notion of anti-explanations is useful when there are negative observations which should not exist in the world. Because explanations are defined in a credulous way in section 4, anti-explanations are defined in a skeptical way: (E, F) is an anti-explanation of G if G is not true in any consistent answer set of $(P \setminus F) \cup E$. Abductive equivalence relative to anti-explanations can also be defined in a way similar to that for explanations in this section, but characterization of such notions needs different formalizations.

We now define the notion of abductive equivalence for extended abduction. This can be done as straightforward extensions of Definitions 4.3 and 4.4.

Definition 5.2 Abductive programs $\langle P_1, \mathcal{A}_1 \rangle$ and $\langle P_2, \mathcal{A}_2 \rangle$ are *explainably equivalent in extended abduction* if, for any ground literal G, G is explainable in $\langle P_1, \mathcal{A}_1 \rangle$ iff G is explainable in $\langle P_1, \mathcal{A}_2 \rangle$.

Definition 5.3 Abductive programs $\langle P_1, \mathcal{A} \rangle$ and $\langle P_2, \mathcal{A} \rangle$ are *explanatorily equivalent in extended abduction* if, for any conjunction of ground literals G, (E, F) is an explanation of G in $\langle P_1, \mathcal{A} \rangle$ iff (E, F) is an explanation of G in $\langle P_2, \mathcal{A} \rangle$.

The notion of belief sets also needs to be extended by taking removal of abducibles into account. That is, a *belief set (with respect to (E, F))* of $\langle P, \mathcal{A} \rangle$ is a consistent answer set of $(P \setminus F) \cup E$ where $E, F \subseteq \mathcal{A}$. Then, a pair (E, F) is an explanation of G iff G is true in a belief set of $\langle P, \mathcal{A} \rangle$ with respect to (E, F).

Like Theorem 4.1, explainable equivalence in extended abduction can be checked by comparing the unions of the belief sets of two abductive programs.

Example 5.1 Suppose the set of abducibles $\mathcal{A} = \{a\}$ and two logic programs:
$$P_1 = \{p \leftarrow not\, a,\ a \leftarrow \},$$
$$P_2 = \{p \leftarrow a\},$$
the abductive programs $A_1 = \langle P_1, \mathcal{A} \rangle$ and $A_2 = \langle P_2, \mathcal{A} \rangle$ are explainably equivalent in extended abduction. In fact, p has the unique explanation $(\emptyset, \{a\})$ in A_1 and the explanations $(\{a\}, \emptyset)$ and $(\{a\}, \{a\})$ in A_2, and a has the explanations $(\emptyset, \emptyset)$, $(\{a\}, \emptyset)$ and $(\{a\}, \{a\})$ in A_1 and the explanations $(\{a\}, \emptyset)$ and $(\{a\}, \{a\})$ in A_2. Obviously, A_1 and A_2 are not explanatorily equivalent in extended abduction.

To characterize explanatory equivalence in extended abduction, we use the equivalence criterion called *update equivalence* [Inoue and Sakama, 2004]. Given two sets of rules $\mathcal{Q}$ and $\mathcal{R}$, two logic programs P_1 and P_2 are said to be *update equivalent with respect to* $(\mathcal{Q}, \mathcal{R})$ if $(P_1 \setminus Q) \cup R$ and $(P_2 \setminus Q) \cup R$ have the same answer sets for any two logic programs $Q \subseteq \mathcal{Q}$ and $R \subseteq \mathcal{R}$. Here, two parameters $\mathcal{Q}$ and $\mathcal{R}$ correspond to the languages for deletion and addition, respectively. Update equivalence is suitable for taking program updates into account when two logic programs are compared. Clearly, the notion of strong equivalence is a special case of update equivalence where $\mathcal{Q}$ is empty and $\mathcal{R}$ is the set of all rules in the language. The notion of update equivalence is strong enough to capture explanatory equivalence as the next theorem shows.

Theorem 5.1 *Two abductive programs $\langle P_1, \mathcal{A} \rangle$ and $\langle P_2, \mathcal{A} \rangle$ are explanatorily equivalent in extended abduction iff P_1 and P_2 are update equivalent with respect to $(\mathcal{A}, \mathcal{A})$.*

Proof. Suppose that $\langle P_1, \mathcal{A} \rangle$ and $\langle P_2, \mathcal{A} \rangle$ are explanatorily equivalent in extended abduction. Then, for any conjunction G of literals and any $E, F \subseteq \mathcal{A}$, it holds that, (E, F) is an explanation of G in $\langle P_1, \mathcal{A} \rangle$ iff (E, F) is an explanation of G in $\langle P_2, \mathcal{A} \rangle$. The latter equivalence then implies that, for any G and any (E, F), we have that, G is true in a belief set of $\langle P_1, \mathcal{A} \rangle$ with respect to (E, F) iff G is true in a belief set of $\langle P_2, \mathcal{A} \rangle$ with respect to (E, F). Then, for any G and any (E, F), G is true in an answer set of $(P_1 \setminus F) \cup E$ iff G is true in an answer set of $(P_2 \setminus F) \cup E$. That is, for any (E, F) and any set S of literals, S is an answer set of $(P_1 \setminus F) \cup E$ iff S is an answer set of $(P_2 \setminus F) \cup E$. Hence, P_1 and P_2 are update equivalent with respect to $(\mathcal{A}, \mathcal{A})$. The converse direction can also be proved by tracing the above proof backward. ∎

Example 5.2 Suppose that two programs P_1 and P_2 are given as

$$P_1 = \{p \leftarrow a, q, \quad q \leftarrow \mathit{not}\, b, \quad b \leftarrow \},$$
$$P_2 = \{p \leftarrow a, \mathit{not}\, b, \quad q \leftarrow \mathit{not}\, b, \quad b \leftarrow \}.$$

Let $\mathcal{A}_1 = \{a, b\}$ and $\mathcal{A}_2 = \{a, b, p, q\}$. Then, $\langle P_1, \mathcal{A}_1 \rangle$ and $\langle P_2, \mathcal{A}_1 \rangle$ are explanatorily equivalent, while $\langle P_1, \mathcal{A}_2 \rangle$ and $\langle P_2, \mathcal{A}_2 \rangle$ are not explanatorily equivalent. In fact, P_1 and P_2 are update equivalent with respect to $(\mathcal{A}_1, \mathcal{A}_1)$, but are not with respect to $(\mathcal{A}_2, \mathcal{A}_2)$. For the latter claim, we see that the answer sets of $P_1 \cup \{a, q\}$ are $\{\{a, b, p, q\}\}$, which are not the same as the answer sets of $P_2 \cup \{a, q\}$, i.e., $\{\{a, b, q\}\}$.

6 Discussion

We have introduced the notion of abductive equivalence in this paper. We have considered two definitions of abductive equivalence in two logics. Two important differences between FOL and ALP as the underlying logics are that (1) explainability is considered for all formulas in FOL while only literals are considered as observations in ALP, and that (2) nonmonotonicity by NAF appears in ALP while this is not the case in FOL. Intuitively, the restriction of observations to literals in ALP gives more chances for two abductive programs to be equivalent, but the existence of nonmonotonicity in ALP makes comparison of abductive programs more complicated.

We have observed that logical equivalence of background theories in FOL or weak equivalence of logic programs does not simply imply abductive equivalence except for some very simple cases. That is why we need to characterize abductive equivalence in terms of other known concepts in classical or nonmonotonic logics. Having such characterizations in this paper, the next target will be to develop transformation techniques which preserve abductive equivalence.

We have considered a rather simple framework for abductive equivalence. In future work, further parameters should be considered in defining abductive equivalence. For example, we can consider another underlying logic for background theories, hypotheses and observations. The criteria of *best explanations* are also important. It is easy to show that explanatory equivalence implies coincidence of the *minimal* explanations. However, the converse does not hold.

Another future topic is to define the concept of *generality/specificity* or *strength/weakness* for abductive frameworks. These concepts are useful for comparing two abductive frameworks, and generality is related to induction too. It might be natural for such relations to be *anti-symmetric*, that is, two abductive frameworks are explainably/explanatorily *equivalent* iff one is both stronger and weaker than another at the same time. Once such a notion is formalized, suppose we know that an abductive program $\langle P_1, \mathcal{A}_1 \rangle$ is weaker than another abductive program $\langle P_2, \mathcal{A}_2 \rangle$. This means, for example, that there is a literal G which cannot be explained in the former but can be in the latter. Then, we expect that P_1 may have less knowledge than P_2 or $\mathcal{A}_1$ may have less hypotheses than $\mathcal{A}_2$. However, the situation is more complicated in nonmonotonic programs. Hence, relationships between amounts of background theories and hypotheses should be important in these concepts for abductive frameworks.

Abduction has been used in the process of *scientific discovery*. We should update our theory in accordance with situation change and discovery of surprising facts. The notions of equivalence and generality are always important in evaluating such scientific processes. We hope that our work can serve a basis for the theory of abductive change.

BIBLIOGRAPHY

[Baral and Gelfond, 1994] C. Baral and M. Gelfond. Logic programming and knowledge representation. *J. Logic Programming*, 19/20:73–148, 1994.

[Denecker and Kakas, 2002] M. Denecker and A. Kakas. Abductive logic programming. In A.C. Kakas and F. Sadr, editors, *Computational Logic: Logic Programming and Beyond – Essays in Honour of Robert A. Kowalski, Part I*, pages 402–436. LNAI 2407, Springer, 2002.

[Eiter and Gottlob, 1995] T. Eiter and G. Gottlob. The complexity of logic-based abduction. *J. ACM*, 42:3–42, 1995.

[Eiter et al., 1998] T. Eiter, G. Gottlob, and L. Leone. Abduction from logic programs: semantics and complexity. *Theoretical Computer Science*, 189:129–177, 1998.

[Flach and Kakas, 2000] P.A. Flach and A.C. Kakas, editors. *Abduction and Induction – Essays on their Relation and Integration*. Kluwer Academic, 2000.

[Gelfond and Lifschitz, 1991] M. Gelfond and L. Lifschitz. Classical negation in logic programs and disjunctive databases. *New Generation Computing*, 9:365–385, 1991.

[Inoue and Sakama, 1995] K. Inoue and C. Sakama. Abductive framework for nonmonotonic theory change. In *Proc. IJCAI-95*, pages 204–210. Morgan Kaufmann, 1995.

[Inoue and Sakama, 1996] K. Inoue and C. Sakama. A fixpoint characterization of abductive logic programs. *J. Logic Programming*, 27:107–136, 1996.

[Inoue and Sakama, 1998] K. Inoue and C. Sakama. Negation as failure in the head. *J. Logic Programming*, 35:39–78, 1998.
[Inoue and Sakama, 1999] K. Inoue and C. Sakama. Computing extended abduction through transaction programs. *Annals of Mathematics and Artificial Intelligence*, 25:339–367, 1999.
[Inoue and Sakama, 2004] K. Inoue and C. Sakama. Equivalence of logic programs under updates. In *Proc. 9th European Conference on Logics in Artificial Intelligence*, pages 174–186. LNAI 3229, Springer, 2004.
[Inoue, 1991] K. Inoue. Extended logic programs with default assumptions. In *Proc. 8th International Conference on Logic Programming*, pages 490–504. MIT Press, 1991.
[Inoue, 2002] K. Inoue. Automated abduction. In A.C. Kakas and F. Sadri, editors, *Computational Logic: Logic Programming and Beyond – Essays in Honour of Robert A. Kowalski, Part II*, pages 311–341. LNAI 2408, 2002.
[Kakas et al., 1998] A.C. Kakas, R.A. Kowalski, and F. Toni. The role of abduction in logic programming. In D.M. Gabbay, C.J. Hogger, and J.A. Robinson, editors, *Handbook of Logic in Artificial Intelligence and Logic Programming*, volume 5, pages 235–324. Oxford University Press, 1998.
[Levesque, 1989] H.J. Levesque. A knowledge-level account of abduction (preliminary version). In *Proc. IJCAI-89*, pages 1061–1067. Morgan Kaufmann, 1989.
[Lifschitz et al., 2001] V. Lifschitz, D. Pearce, and A. Valverde. Strongly equivalent logic programs. *ACM Transactions on Computational Logic*, 2:526–541, 2001.
[Lin, 2002] F. Lin. Reducing strong equivalence of logic programs to entailment in classical propositional logic. In *Proc. 8th International Conference on Principles of Knowledge Representation and Reasoning*, pages 170–176. Morgan Kaufmann, 2002.
[Maher, 1988] M.J. Maher. Equivalence of logic programs. In J. Minker, editor, *Foundations of Deductive Databases and Logic Programming*, pages 627–658. Morgan Kaufmann, 1988.
[Marek et al., 1997] V.W. Marek, J. Truer, and M. Truszczyński. Representation theory for default logic. *Annals of Mathematics and Artificial Intelligence*, 21:343–358, 1997.
[Poole, 1988] D. Poole. A logical framework for default reasoning. *Artificial Intelligence*, 36:27–47, 1988.
[Reiter, 1980] R. Reiter. A logic for default reasoning. *Artificial Intelligence*, 13:81–132, 1980.
[Reiter, 1987] R. Reiter. A theory of diagnosis from first principles. *Artificial Intelligence*, 32:571–95, 1987.
[Sakama and Inoue, 1995] C. Sakama and K. Inoue. The effect of partial deduction in abductive reasoning. In *Proc. 12th International Conference on Logic Programming*, pages 383–397. MIT Press, 1995.
[Satoh and Iwayama, 1991] K. Satoh and N. Iwayama. Computing abduction by using the tms. In *Proc. 8th International Conference on Logic Programming*, pages 505–518. MIT Press, 1991.
[Selman and Levesque, 1996] B. Selman and H.J. Levesque. Support set selection for abductive and default reasoning. *Artificial Intelligence*, 82:259–272, 1996.
[Tamaki and Sato, 1984] H. Tamaki and T. Sato. Unfold/fold transformation of logic programs. In *Proc. 2nd International Conference on Logic Programming*, pages 127–138. MIT Press, 1984.
[Woltran, 2004] S. Woltran. Characterizations for relativized notions of equivalence in answer set programming. In *Proc. 9th European Conference on Logics in Artificial Intelligence*, pages 161–173. LNAI 3229, Springer, 2004.

Katsumi Inoue
National Institute of Informatics
Tokyo, Japan
Email: ki@nii.ac.jp

Chiaki Sakama
Department of Computer
and Communication Sciences
Wakayama University
Wakayama, Japan
Email: sakama@sys.wakayama-u.ac.jp

A Transconsistent Logic for Model-Based Reasoning

JOSEPH E. BRENNER

ABSTRACT. Computational and dynamic approaches to cognitive science use fundamentally different models. The former applies algorithms of various kinds; the framework of the latter is that of dynamic systems theory and emergence. Advances in logic are usually associated with the former. Of the published contributions to the previous Conference on *Model-Based Reasoning*, held in Pavia in 2001, the volume of papers centered on philosophical, epistemological and cognitive questions is sub-titled *"Science, Technology, Values"*. The other major volume is sub-titled *"Logical and Computational Aspects of Model-Based Reasoning"*, and its papers are correspondingly oriented.

In the paper, we describe the "transconsistent" logical system developed by Stéphane Lupasco (1900–1988) and extended by Basarab Nicolescu (1940–). It suggests an alternative, dynamic framework for model-based reasoning in science, especially where creativity, values and non-sentential accounts of mental structure are involved. It may offer a new approach to some outstanding questions, such as the interface of models and reality.

Its application requires a shift from focus on the axioms and formalism of classical and non-classical propositional or mathematical logic as the criteria of a valid logical system. The Lupasco approach, as a supplement to existing methods of inquiry, could provide insights for new notions of reasoning based on differences in the respective domains of application of the two forms of logic.

1 Introduction

Reasoning is a real dynamic process, *par excellence*. Discussion of reasoning thus requires recognition of the metaphysical assumptions being made about reality and the philosophy of process, as well as the dynamics of human thought, with which reasoning is contiguous. Model-based reasoning (MBR) is a sub-category of reasoning, and the interpretations, logics and metaphysics that apply to reasoning can also be analyzed as models of it.

This paper inquires into the relation between aspects of model-based reasoning (MBR) and the logic or logics that may be applicable to them. My starting point is the rough division of the published contributions to the previous Conference on *Model-Based Reasoning*, held in Pavia in 2001: one volume of papers, centered on philosophical, epistemological and cognitive questions, is sub-titled *"Science, Technology, Values"* [Magnani and Nersessian, 2002]; the other major volume is sub-titled *"Logical and Computational Aspects of Model-Based Reasoning"* [Magnani et al., 2002], and its papers are correspondingly oriented.

As in computational and dynamic approaches to cognitive science, the two target domains involve fundamentally different models. Broadly speaking, the latter applies algorithms of various kinds; the framework of the former is that of dynamic systems theory and emergence. Advances in logic are generally associated with the latter.

Lorenzo Magnani stated in his prefaces to both volumes that many ways of performing reasoning cannot be described using notions derived from classical logic. This point applies differently to the two target domains of MBR under consideration. In particular, we will need to decide if logic, as generally conceived, 1) applies to one target domain only; 2) if different aspects of one logic applies to both; or 3) different logics exist for the two domains.

The objection can be made that there is no formal method of establishing whether one is within one regime or the other, computational or non-computational, in other words, whether one is reasoning about simple phenomena – facts or opinions - or about how to survive another four years of uncertainty. One can show that there can be a dynamic relation between these domains, in line with Peirce's intuition that abduction is always followed by deduction and that simple or complex creative reasoning occurs in both domains, albeit to a different degree.

The key word in the citation is "performing". The distinction implied necessary to define these domains is between MBR as a process and the structure of classes of proper objects (which may also be processes) towards which this process tends.

Where one is concerned primarily with the computational aspects of models, binary classical logics, suitably modified to handle the necessary incompleteness of real-world situations are adequate. I will suggest, however, that a different logic applies to MBR that results in the emergence of new entities, what Magnani has called the "iterative processes of creativity".

2 Non-classical logics and reasoning

Among the large number of logics that have been developed and studied in the last half-century, the ones most relevant for reasoning may be extensions of classical logic, e.g., ampliative adaptive logics and paraconsistent logics.

2.1 Ampliative adaptive logic

Diderik Batens, Joke Meheus and their colleagues at the University of Ghent have analyzed in depth the relationship between logic and reasoning, in regard to both computational and scientific aspects [Batens, 2005]. They focus on abductive reasoning, viewed as a form of inductive forward reasoning from effect to possible cause, and demonstrate the advantages of ampliative adaptive logics as the preferred method of treating it [Meheus et al., 2002]. Adaptive refers to their ability to handle inconsistent sets of premises and ampliative to their provision for inference patterns that go beyond the information contained in the premises, that is, permit the addition of new conclusions or the withdrawal of old ones. Such logics have a proper proof theory, and are considered closer to natural reasoning than existing systems, while retaining the advantages of classical logic.

Following Rescher, Meheus also distinguishes [Meheus, 2005] between reasoning in respect of commonsense vs. scientific knowledge and beliefs. The relative vagueness and imprecision of the former results in the conclusions of the corresponding reasoning process being less secure, that is, highly non-monotonic.

2.2 Paraconsistent logic

Paraconsistent logics were developed to permit inference from inconsistent or contradictory information by another method, namely, by relaxing the condition of the 2nd classical axiom or law of non-contradiction that the conjunction of *e* and *non-e* entails triviality. The paraconsistent relation of logical consequence is supported, for example in the work of Graham Priest, by the demonstration of the existence of real contradictions, dialetheias, in thought and in interpretations of time and quantum phenomena [Priest, 1987].

Priest has described the kinds of contradiction that can occur in scientific reasoning. Meheus accepts that the study of scientific reasoning often requires a paraconsistent logic. However, she has also argued that paraconsistent logics do not provide the proper tools to study commonsense or natural reasoning. In particular, they are deductive only and not ampliative. The treatment of negation in monotonic paraconsistent logics is too weak to handle the underlying information structure, which is considered incomplete, but not inconsistent. For natural reasoning, however, this metaphysical position cannot be supported. In the philosophy of the adaptive logics of Batens and Meheus, inconsistencies and/or contradictions are managed, circumscribed or eliminated through appropriate choices of boundary conditions or other formal logical mechanisms. In my view, such logical approaches are adequate to provide a structure for a discontinuous "state-transition" domain of reasoning processes, commonsense or scientific, which tend toward one or the other static extreme of belief or judgment.

It is not correct, however, to compare ampliative adaptive logics and paraconsistent logics directly. Once one admits that there is no mandatory requirement for consistency, in the empirical sciences, for example, then, Priest claims, the only situations about which it makes sense to reason classically are consistent ones. Paraconsistent logicians, and transconsistent ones, may apply classical logic in simple, consistent situations as intuitionist logicians can when reasoning about finite mathematics. Priest admits, however, [Priest, 2002a] that the set of logical truths in his paraconsistent Logic of Paradox is identical to that of classical logic, and all of the more complex varieties of paraconsistent logics add truth operators or relations whose value as descriptions of the real world are open to question.

In our view, neither paraconsistent logics nor adaptive logics answer the questions of why a choice is made or what are in fact the driving forces for change. They thus fail to provide an adequate physical/metaphysical picture of reality, including natural reasoning. Another logic seems required that can provide a model of, also, the "irrational" and non-sentential aspects of reasoning that reflect those forces.

3 Logic as reasoning and logic in action

Logic has been almost universally conceived of as the study of reasoning and the construction of adequate, formal descriptions of its operations and the processes that characterize it: induction, deduction, implication and, most recently, relevance. The domain of logic has been limited almost entirely to linguistic and mathematical problems. It is clear that all possible linguistic models, for example of common-sense reasoning, require a degree of abstraction. In my view, the discussion should center on ways to retain, in what is left after abstraction, as much as possible of the original dynamic phenomenon. Despite the large number of important practical applications of standard logic, for example in computer science and artificial intelligence, the underlying body of logic has not undergone major modifications in this direction. An improved reflection of phenomenal reality is in a nutshell what the logical system described in this paper may bring to the table.

Due to their complexity, diversity and appearance of the operation of randomness or, at the least, of chaotic behavior, phenomena as such have thus been generally considered outside the purview of logic. Since they, in contrast to the abstractions dealt with classically by logic, depend directly on human observation and human ratiocination, in the disciplines corresponding to them, phenomenology and epistemology, the introduction of logical considerations, for example by Husserl and Quine, has proven extremely complicated. Attempts have been made to capture aspects of action and dynamics, as in the logics of von Wright and van Bentham. Close inspection shows again, however, that these logics do not talk to the essential processes of change or becoming that are present, in terms of the actual energies involved in them.

It is becoming apparent that systems involving binary transition relations are inappropriate for modeling physical dynamics, even classically. A different type of duality, both in quantum mechanics and elsewhere, seems required, and [Coecke et al., 2002] have provided a critique of action logics in this regard.

4 A transconsistent logic for MBR

The "transconsistent" logical system (TCL) developed by Stéphane Lupasco (Bucharest, 1900 – Paris, 1988) and extended by Basarab Nicolescu (Bucharest, 1940 -) suggests an alternative framework for discussion of model-based reasoning. Specifically, it suggests notions of the dynamics of the objects and process of reasoning that apply to the target domains of MBR and may offer a new approach to some outstanding issues and questions [Badescu and Nicolescu, 1996], such as the interface of models and reality and a way of interpreting certain intuitions about different classes of models and types of MBR.

Lupasco grounded his system in 20th Century science - the quantum mechanics of Planck, Pauli and Heisenberg, developmental biology and cosmology. He proposed that the dialectical characteristics of energy - extensive and intensive; continuous and discontinuous; entropic (tendency toward identity or homogeneity, governed by the 2nd Law of Thermodynamics) and negentropic (tendency toward diversity or heterogeneity, following the Pauli Exclusion Principle) - could be formalized as a structural logical principle of dynamic opposition, an antagonistic duality inherent in the nature of energy and accordingly in all phenomena, including information, propositions and judgments, etc. [Lupasco, 1987] The fundamental postulate is that every phenomenon, element or event e is always associated with an anti-phenomenon, anti-element or anti-event $non\text{-}e$, such that the actualization of e entails the potentialization of $non\text{-}e$ and *vice versa*, alternatively, without either ever disappearing completely. The point of equilibrium or semi-actualization and semi-potentialization is a point of maximum antag-

onism or "contradiction" from which, in the case of complex phenomena, a
T-state (*T* for "*tiers inclus*", included third term) emerges, resolving the
contradiction (or "counter-action"), as proposed by Nicolescu, at a higher
level of reality. In the notation developed by Lupasco, and as far as I know
used only by him:

(1) $\quad e_A \supset \bar{e}_P,\ e_T \supset \bar{e}_T,\ \bar{e}_A \supset e_P$

This picture of reality, as I will show, has several, although not all, of the
characteristics of a standard logic, but it is a logic of an *included* middle,
consisting of axioms for describing the dynamic state of the terms involved
in a phenomenon.

(2) $\quad \begin{aligned} (e_A \supset \bar{e}_p) &\supset \bar{C}_A \supset C_p \\ (\bar{e}_A \supset \bar{e}_p) &\supset C_A \supset C_p \\ (\bar{e}_T \supset e_T) &\supset \bar{C}_p \supset C_A \end{aligned}$

In this notation[1], *e* actual implies *non-e* potential implies *non-contradiction* actual; similarly, *non-e* actual implies *e* potential also implies *non-contradiction*; and *e-Tstate* implies *non-e-Tstate* implies *contradiction*. (I use the term "dynamic" in its physical sense, as related to phenomenological change rather than to formal change, e.g., the facility of changing rules or conclusions.) A comparison of the axioms of this and classical logic follows:

Overall, this calculus does not have a set of formal rules other than the "proposal" that phenomena should be analyzed by the dynamic criteria suggested. This is equivalent to saying, in other words, that it is not a standard logic but an extension of logic to reality.

4.1 Truth and falsity

The transconsistent logical system includes a radical difference from the standard concepts of truth and falsity. The key postulate, also formulated originally by Lupasco [Lupasco, 1987], is that what is true is non-contradictory and that the notion of truth is defined logically by non-contradiction. According to this concept, there is always a possibility of relative non-contradiction, since for each of the two elements, *e* and *non-e*, each can be actualized and potentialized, resulting in *four* possible "trues", $Te_a,\ T \sim e_a,\ Te_p,\ T \sim e_p$. The "true" is thus a function of the actualizations and potentializations of phenomena that have a finite probability. As they are, according to the fundamental postulate of TCL, constitutionally relative (that is, never anything without its contradiction), the quadruple

---

[1] I am grateful to Jean-Paul Bertrand, former Editor of Éditions du Rocher, Paris, for permission to reproduce these and other formulas from his editions of Lupasco's works.

	ABSTRACT IDEAL LIMIT	REAL WORLD
1.	Identity: $A = A$	No real element is rigorously identical to itself (imperfect circularity or autopoesis)
2.	Non-contradiction: not (A and not-A) at the same time	Change: not "contradiction" (paraconsistent logic), but dynamic opposition
3.	Excluded middle: no third term = A and not-A at the same time	Resolution of oppositions; emergence of an included middle at another level of reality (Nicolescu; transconsistent logic)

Table 1. The logic of the included middle: expanding the classical axioms.

true is relative in the same sense as the non-contradictions that it signifies or which produce it. An object, a true proposition or a truth in the above sense, as a predominately actualized phenomenon, is never found in isolation. To every actualization, which is true as a consequence of its actualization, that is, to every truth, always corresponds a potentialization, also true as a consequence, of the contradictory element, giving rise to the contradictorial truth of the potentialization. The term truth is to be preferred over validity since the latter implies less of a real existence of the respective states and could be confused with the semantic and proof-theoretic notions of validity in standard propositional logic.

One can thus say that there are two truths, again, one positive or affirmative or "of identity" and the other negative or "of non-identity", attributed to the logical values (elements) themselves, e and non-e respectively. Since these have the possibility of being respectively actual, potential or neither, under the conditions indicated above, the notion of falsity as such disappears, as follows: the two truths produce non-contradiction when one of them is actual and the other potential. This non-contradiction can be considered as a true relative non-contradiction or as a truth of relative non-contradiction. The same two truths produce contradiction when they inhibit themselves reciprocally and prevent themselves from being able to actualize themselves more fully with respect to one another; this contradiction appears (or emerges) accordingly as a relative (non-absolute) contradiction. What is understood as "false" in classical logic is redefined in this way as negative truth or the truth of (the existence of) contradiction.

In any case, a truth cannot be absolute, because it can never be rigorously (totally) actualized; the contradictory truth can be potentialized as much as one wants theoretically without it ever completely disappearing in reality. The non-contradiction that is present will accordingly never be absolute either and will always include an irreducible amount of contradiction. Similarly, contradiction can never be considered as absolute, because it never takes place between two rigorously actual, absolutely contradictory elements. Contradiction never occurs except between antagonistic dynamisms. Since they are dynamisms, no matter how far they are from their potential states, they still inhibit one another's full instantiation and therefore possess an irreducible residual margin of respective actualization. This approach to truth and falsity is compatible with, but complements, a context theory of truth, the idea that no proposition about the real world can be viewed in isolation from its context.

In TCL formalism, the reciprocally determined "reality" values of the degree of actualization A, potentialization P and T-state T replace the truth values in standard truth tables.

(3)
e	$\bar{e}$
True	False
False	True

e	$\bar{e}$
A	P
T	T
P	A

(4)
e	$\bar{e}$	$\bar{C}$	C
A	P	A	P
T	T	P	A
P	A	A	P

The fundamental postulate and its formalism are also applied to the logical operations, answering a potential objection that the operations themselves would imply or lead to rigorous non-contradiction. Thus: e implies e is impossible rigorously because e also implies not-e, or else, because e implies e implies at the same time that e excludes e. Inversely, e excludes e is impossible rigorously since e implies, at the same time, e. We **the** write, in place of $e_A \supset \bar{e}_P$, the actualization of e implies the potentialization of non-e, the symbol $\bar{\supset}$ for exclusion or non-implication, or, preferably, negative implication, such that:

(5)
$(e \supset e)_A \supset (e \bar{\supset} e)_P$ or $(e \supset_A e) \supset (e \bar{\supset}_P e)$. Similarly,
$(e \bar{\supset} e)_A \supset (e \supset e)_P$ or $(e \bar{\supset}_A e) \supset (e \supset_P e)$, and for the T-state
$(e \supset e)_T \supset (e \bar{\supset} e)_T$ or $(e \supset_T e) \supset (e \bar{\supset}_T e)$

A Transconsistent Logic for Model-Based Reasoning

Actualization and potentialization thus applying to the operation of implication, e can be neglected, since it is any element, and the reciprocal contradictional implications can be written as follows:

(6) $(\supset_A) \supset (\bar{\supset}_P); (\bar{\supset}_A) \supset (\supset_P); (\supset_T) \supset (\bar{\supset}_T)$

This should be read: if an implication is actualized, an exclusion or negative implication is potentialized; if a negative implication is actualized, an affirmative or positive implication is potentialized, and if a positive implication can be neither actualized or potentialized, a negative implication cannot be potentialized or actualized.

One can therefore proceed, as previously for elements, to construct a table of values for contradiction and non-contradiction:

(7)
$\supset$	$\bar{\supset}$	$\bar{C}$	C
A	P	A	P
T	T	P	A
P	A	A	P

Written for implication, the contradictional conjunctions and disjunctions are, therefore:

(8) $\supset_A \cdot \bar{\supset}_P; \ \bar{\supset}_A \cdot \supset_P; \ \supset_T \cdot \bar{\supset}_T;$ and
$(\supset_A \ \bar{\supset}_P) \vee (\bar{\supset}_A \supset_P) \vee (\supset_T \ \bar{\supset}_T)$

These develop into a transfinite series of disjunctions of implications:

(9) $(\supset_A \ \bar{\supset}_P \vee \bar{\supset}_A \supset_P) \vee (\supset_A \ \bar{\supset}_P \vee \supset_T \ \bar{\supset}_T) \vee (\bar{\supset}_A \supset_P \vee \supset_T \ \bar{\supset}_T)$
etc..., etc...

However, every implication implies a contradictory negative implication, such that the actualization of one entails the potentialization of the other and that the non-actualization non-potentialization of the one entails the non-potentialization non-actualization of the other. This leads to the tree-like development of chains of implications, of which one example is indicated in the following diagram:

(10) $(\supset_A) \supset (\bar{\supset}_P) \begin{cases} [(\supset_A) \supset_A (\bar{\supset}_P)] \supset [(\supset_A) \bar{\supset}_P (\bar{\supset}_P)] \begin{cases} \text{etc.} \\ \ldots \\ \ldots \end{cases} \\ \ldots \\ \ldots \end{cases}$

This development in chains of chains of implications must be finite but unending, that is, transfinite, since it is easy to show that if the actualization of implication were infinite, one arrives at classical identity (tautology): $(e \supset e)$. Any phenomenon, insofar as it is empirical or diversity or negation, that is, not attached, no matter how little, to an identifying implication of some kind, $(e \supset e)$ suppresses itself. One can use this logical symbolism to show this by reducing the implication $(\bar{\supset} A) \supset (\supset P)$ to: $\bar{e}_A \supset e_P$. One then sees that if A or P are infinite and, accordingly, P disappears, there is nothing left but $\bar{e}_{A\infty}$, that is, an absolute and definitive negation, nothing. Both identity and diversity must be present in existence.

4.2 Contrary, contradictory or "countervailing"

There is an endless discussion in the literature of the difference between contrary and contradictory that harks back to the triangle of Carneade, the tree of Porphyry, and the Aristotelian Square of Oppositions. This semiotic square describes the relation between contrary and contradictory terms. The major relations are between the terms on the diagonals, which are contradictories; those on the same horizontal level, which are contraries; and the verticals, which are implications. These concepts are useful for the analysis of simple terms, but they fail as both a deeper analysis of semantics and above all of phenomena involving dynamics, that is, phenomena involving some internal metabolism or energetic change.

In our opinion, part of the confusion in this area stems from the use of the word contradiction itself, by Lupasco and countless others, given its root in *dicere*, to speak. It would desirable to replace it with a term that conveys the essentially *non-linguistic* character of the conflicting energy states, physical or non-physical, for example "counteraction"[2]. Another possible neologism for contradiction – "counter-being" – should be avoided due to potential confusion with discussions of being in an ontology from which contradiction may be excluded. Perhaps the most appropriate word in English to describe the dynamic opposition of two elements or terms in a phenomenon is *"countervailing"*. This has the same meaning as counteraction, but with the additional idea present in its root – *valere* – of value as well as energy (strength). The use of *countervalence* to describe the dynamic equilibrium of the T-state would be consonant with the use of *prevalence* to describe the dominant process in the pairs: actualization/potentialization and homoge-

---

[2] The term "anti-A" can be considered, but it should not be confused with the same term used by Smarandache in his fuzzy (neutrosophic) logic and by Dov Gabbay in his paper "Anti-formulas, Anti-elements and the logic of Deletion", *Computational Linguistics Colloquium*, Winter, 2001. "The job of anti-x is to delete x." The resulting epistemic logic is another coherent (consistent) logic with applications primarily in AI and automated reasoning.

nizing/heterogenizing. For simplicity, however, the term contradiction will continue to be used with the understanding that it involves this additional dimension.

The structure of the included middles (T-states) is highly complex. Phenomena at any level of reality can be characterized by differing actualization of primary trends toward non-contradiction (identity, homogeneity or diversity, heterogeneity) or toward contradiction (emergence of new entities). To distinguish them, I will call the former contradictional and the latter contradictorial. These trends are themselves actualized or potentialized to a different degree, but never completely. In the resulting emergent elements that enter into further contradictorial relations, either homogeneity and heterogeneity is predominant, but the other is also always present.

4.3 Transconsistent logic

The term "transconsistent" was first introduced by the paraconsistent logician Graham Priest [1987] to describe "the realm beyond the consistent". However, I use it in a different sense from Priest, since in my logic all three axioms of classical logic are modified, and the significance of contradiction is retained but expanded to include the dynamic opposition of real elements, grounded in physics. This logic contains that of the excluded middle as a limiting case, approached asymptotically but never instantiated except in simple situations.

Opposing aspects of phenomena that are generally considered independent can be understood as being in the dynamic relationship suggested, namely, as one is actualized, the other is potentialized. This critical concept does not appear in any theory since and including Aristotle's own view of potentiality and actuality. Problems due to the assumption of an absolute independence or separation between terms (e.g., local/global, part/whole, set/member of set, knower/known, rational/irrational, etc.) can be approached from this standpoint. It provides a realistic explication of intuitions such as "things in nature are neither absolutely the same nor absolutely different", every analogy carries disanalogies with it or Piaget's "reciprocity" between subject and object. Ideas similar to those here were expressed by Sheldon, a follower of Whitehead, who talked about the "creative opposition of polar opposites". Without the grounding in physics and metaphysics indicated here, however, such concepts remain purely speculative.

The Lupasco system is dialectical, but the idea that dialectics is passé is justified only by the incomplete view of dialectics that has prevailed since Hegel. The concept of dialectics in the TCL system is that of a contradiction between two conflicting forces viewed as the determining factor in their continuing interaction. It thus subsumes both the original idea of a representation of perceived conflicts in nature and civilization, and their resolution, which have always been involved in human existence. "Beings and things seem to exist and are able to exist only in function of their successive and contradictory conflicts." [Lupasco, 1979] This notion of dialectics covers both the Hegelian and Marxist concepts of the process of change involving ideal entities or real forces respectively, and it provides a physical basis for talking about the forces involved.

The comparison between Lupasco and Hegel is thus instructive: both *started* from a contradictory picture of reality; developed complex logical systems to handle contradiction beyond formal logic; and applied their ideas to consciousness, art, history, ethics and politics. However, Hegel was an idealist, seeing the basis of existence in Spirit; Idea or Concept as primary ontological terms and one dialectics ascending toward Spirit as Unity (hidden identity) without contradiction.

In Lupasco's realist system, there are two dialectics, ascending and descending toward non-contradictory limits of identity and diversity and a third dialectics converging toward contradiction. Contradiction is established at the basic physical level, with energy as the primary ontological term. It is not, however, "dialectical realism". The critical principle of logical dynamic opposition is absent in current attempts to effect a synthesis between modern physics and a neo-Hegelian or neo-Marxist dialectical materialism.

Reasoning, in our model, does not involve a more-or-less abstract, idealized subject making judgments between alternative and separate choices, but a real cognitive process in which the choices themselves, as conceptual "subject-objects" alternately gain and lose in actuality and potentiality as the underlying systems gain and lose strength. If these underlying systems, however, are themselves composed of essentially inert, static elements, involving little or no exchange of energy (as information or otherwise), the process reduces to the classical model of oscillations between binary limits of which the canonical example is the Liar Paradox.

This picture can be further illustrated by analogy with views of ontology as the decomposition of reality into categories vs. the decomposition of sentences about reality into categories.

Thus:

- classical and neo-classical logics decompose reality into static propositions - premises and conclusions (and equations) using independent rules of deduction and/or inference;

- the dynamic logic of opposition or antagonism decomposes reality into elements of change using rules of induction and deduction or implication that follow the same dynamics, as discussed below. Both types of logic are models of reality, but the second is closer to natural process, behavior and reasoning.

4.4 Formalization of TCL. Logics of formal inconsistency

I claim that the logic of reality I propose is a valid extension of logic, and that only tradition supports the standard definition limiting the content and structure of logic to aspects of language and mathematics, following linguistic and mathematical "turns". In my view, the logic of reality is also a logic *in* reality, that is, it really is a logic. It has some, although not all, of the structural characteristics generally ascribed to a logic of propositions, e.g., the axioms above, a syntax and a semantics.

The further objection can be raised, which goes back to Hegel, that what I have described is not a logic, as it fails to have the required formal structure. Mathematical model demonstration is often portrayed as superior on the basis that it is grounded in "formal logic", but this begs the question. Jean-Yves Béziau has shown [Béziau, 2004] that the essence of logic is not its formality, mathematical or other, and one is best off in speaking about logic *tout court*. At this time, it is not clear whether a mathematization of TCL is possible or whether *any* of the many approaches that have been developed for propositional or multi-valued logics could be applied. The Logics of Formal Inconsistency of Carnielli, which include methods for explication of epistemic paradoxes and conflicting scenarios *via* an operator developed to handle inconsistency, can codify any linguistic classical or paraconsistent reasoning [Carnielli, 2005b]. Carnielli's logics cannot formalize the non-linguistic logic of reality. However, there is no reason why another type of formalization might not be possible [Carnielli, 2005a].

4.5 Semantics

A semantics for TCL can be written, based on the "reality values" of actualization and potentialization, rather than standard truth values.

I consider first that the function of a standard semantics is to insure truth preservation as a basis for the validity of the logical reasoning. However, as Priest suggests, the meaning or sense of truth can change and indeed I have shown how it does so in my system. Let us then try to formulate accordingly the components of a "dynamic" semantics and indicate the conceptual modifications that are required by TCL by comparing the elements of a possible semantics for it with those of a classical logic (CL).

1. *Domain of Interpretation*
   - CL: some set of propositions or other language-like entities.

- TCL: the empirical world of physical and non-physical (mental) phenomena

2. *Symbols of the Object Language of the Calculus*

    - CL: an infinite number of propositional parameters or variables, the connectives, and the punctuation marks.
    - TCL: a transfinite number of reality parameters corresponding to real-world events (phenomena) and their accompanying actualized and potentialized contradictions, $e0_A, e1_A, \ldots, non\text{-}e0_P, non\text{-}e1_P, \ldots$, the connectives and the punctuation marks.

3. *Formulas*

    - CL: the (well-formed) formulas of the language comprise all, and only, strings of symbols that can be generate recursively from the propositional parameters by the following rule:
    - TCL: TCL: the formulas of thelanguage comprise the strings of symbols that can be generated from the parameters using the rules that
        - e_A (*e* actualized) implies $non\text{-}e_P$ (*non-e* potentialized) and *vice versa*; both imply that contradiction is potentialized and non-contradiction actualized;
        - the parameter e_T implies $non\text{-}e_T$ which implies that contradiction is actualized and non-contradiction is potentialized.

4. *Interpretation of the Language*

    - CL: a function ν, which assigns to each propositional parameter either 1 (true) or 0 (false).
    - TCL: a function ν which assigns to each pair of reality parameters A and P a value that is greater than 0 and less than 1, such that each is 1 minus the other. The formulas in the interpretation do not have truth values other than the values of the reality parameters. They *are* the "truth" values.

5. *Premises and Conclusion. Consequence*

    - CL: for any set of formulas (the premises) S, then A, the conclusion is a *semantic* consequence of S iff (if and only if) there is no interpretation that makes all the members of S true and A false, that is every interpretation that makes all the members of S true makes A true.

- TCL: no set of formulas can be considered as a set of independent premises permitting a conclusion as a semantic consequence (or non-consequence), given the relation of opposition between e and non-e. The *dynamic* consequence of e and non-e is a T-state (see Axiom 3 above), the emergent result of an interpretation that has given each of the reality parameters the value of 1/2.

6. *Logical Truth (Tautology)*

    - CL: Many (but not all) classical logics make extensive use of tautologies, based on classical Axiom 1.
    - TCL: The empty set of premises does not exist in this logic, and logical truth as tautology is meaningless (meta-logical).

The semantics is, obviously, not truth-functional, given the change noted in the concept of truth. But the major aspect of this semantics, the *sense* of truth that the semantics gives is the dynamic state of the event, phenomenon, judgment, etc, where the event is "on the way", more or less, as the case may be, between its actualization and the potentialization of its contradiction.

4.6 Many-valued and fuzzy logics

The possible significance of many-valued and fuzzy logics to this discussion is that they provide for multiple truth values. Fuzzy logic [Priest, 2001] is related to the original 3-valued logic of J. Lukasiewicz in which sentences can take on the truth values of 0, 1/2 and 1. This is supposed to enable the description of the state of uncertainty or vagueness existing in the absence of complete knowledge of "something", by avoiding the dichotomy between truth and falsity. Why then should TCL not simply be another fuzzy logic or system for fuzzy reasoning? My answer, which I am sure will not satisfy everyone, is that such logics are still applicable only to the essentially idealized, abstract entities of binary systems from which dynamic interactions are (almost completely) absent[3]. The only similarity between many-valued logics and TCL is that both deal with multiple values of "something".

The existence of truth coming in multiple or continuous degrees, in my view, is not sufficient to resolve Sorites paradoxes (e.g., the point at which a "child" becomes an "adult"). In this scheme, there is still a point in the

---

[3]On the other hand, it is impossible to ignore the vast number of practical applications that fuzzy logics have found. In fact, without them, it would be difficult to imagine the management of real uncertainty in global economic and technical decision-making, "soft computing" and knowledge engineering, as well as other areas in which computer science is essential such as artificial intelligence and neural networks. Books on fuzzy-logic-based programming and fuzzy logic for business, finance and management are bestsellers.

Sorites transition where the truth value changes from *completely* true to *less than completely* true, and the existence of such a point seems to be intuitively problematic. This suggests that there is something fundamentally wrong, or at the very least not sufficiently general, with the concepts and properties of truth and falsity that continue to be used. The partial truth-values of the related propositions in the logics of Lukasiewicz and his followers do not change the basically binary characteristics of the logics derived. In TCL, the partial truth-values are replaced by values corresponding to the degrees of actualization and potentialization of the phenomena themselves, as discussed ("reality values"). The continuity of such values does not pose a problem, since at the "point" of maximum contradiction, the discontinuity involves another level of reality. The term "orthogonal" for this situation, however, is not felicitous as it implies an absence of the relation between the T-state and its precursors.

The above remarks hold, with minor modifications in the sense of their having their own systems of calculi, for modal logics and relevance logics. However, the basic notions of interest, e.g. of truth, are unchanged in these non- or neo-classical systems.

4.7 Consistency, inconsistency and paracompleteness

The TCL system (which I believe is one in the sense of Gabbay, 1994) provides an alternative to the handling of fundamental inconsistencies by inconsistency-adaptive logics *vis à vis* their critique of paraconsistent logics mentioned above, namely, that they are too weak due to their invalidation of the axiom of non-contradiction. In my picture, A and *non-A* are both true at the same time, but only in the sense that if A is actual, *non-A* is potential and *vice versa*. Conflict is foundational, and the TCL systems model does not require consistency in the classical sense. It applies most clearly to systems that *cannot* be described by algorithms in the sense of E.O. Wilson [2000].

The quantum physicist Diederik Aerts [Aerts *et al.*, 1991] suggests paracompleteness as a third possible world-view or attitude towards inconsistency:

i) *The Regularizing Attitude*

- Inconsistency is temporary. One contradictory term will be found to be false and subsequently eliminated by new theoretical or experimental findings, resolving the contradiction.
- Classical logic

ii) *The Paraconsistent Attitude*

- Inconsistency is accepted. Both contradictory elements are retained in the model.
- Paraconsistent logic.

iii) *The Paracomplete Attitude*

- Inconsistency is interpreted as pointing to a fundamental incompleteness of the model. Neither of two contradictory parts is considered absolutely true. Theoretical completeness is sacrificed for consistency, but the former is temporary or virtual; a new concept is introduced that closes the gap, embracing the previously contradictory concepts.
- Transconsistent logic? (included middle)

Table 2. The Aerts model.

The Lupasco system could thus also be considered transcomplete, but this term is, perhaps, even more unfamiliar.

The Aerts' view of paracompleteness has been cited because of its greater similarity to the TCL system than suggested by some more recent reviews [Béziau, 1999]. A logic that admits non-trivial incomplete theories is paracomplete, and paracompleteness is in a sense the reverse of paraconsistency, with the former involving truth "gluts", i.e., paradoxes of self-reference. All these essentially semantic issues are, however, not relevant to the logic of reality that TCL represents.

4.8 Why TCL is a logic

The view that TCL is not a logic is supported by the fact that there are no proofs in TCL in the semantic sense. The demonstrations are closer to those in science, in that they purport to describe in a more or less coherent manner processes and changes that are occurring or have occurred, by reference to a model of the elements involved and their interrelationships. TCL could be considered a science, due to its grounding in physics, plus an ontology, in which energy is the primary ontological category.

In my opinion, the fact that TCL is *grounded* in physics may be novel (I think it is), but as a method of making reasoned inferences about reality, including reasoning and models and mechanisms of reasoning themselves, TCL is worthy of consideration, at least, as *also* a logic.

TCL is a formal construction that reduces to standard logics where actualization and potentialization are total (ideally). There are other many-valued logics, cf. section 3.6, but the values of TCL are of a different type. The difficulties encountered in its acceptance as a logic are due to its unfamiliarity as a new type of extension of logic, rather than for other reasons.

5 Non-computational aspects of MBR

The situation with regard to philosophical, epistemological and cognitive aspects of MBR is complicated, but I was struck by the pattern of dualities to which attention was called in the Magnani-Nersessian compendium.

- Natural, internalized mental vs. externalized formal or constructed models
- Internal and external dynamics in experimental models
- Models located between statements about the world and the world itself, mediating between theory and reality
- Theory, together with models and simulations on one end of a spectrum; experiment and the world on the other
- Synchronic constitution and diachronic change in conceptual hierarchies in the interplay between taxonomy and model
- "Intertwined categories" of production and science, resulting in the emergence of new forms of reasoning
- Distribution of the cognitive process between person and external representation.

Table 3. Dualities in MBR.

Distinctions were made, for example, between natural, especially internalized mental vs. externalized formal or constructed models; formal vs. visual mental models; internal and external dynamics in experimental models [Morgan, 2002] and others as indicated. I suggest this pattern is not only not accidental, but reflects something basic about reasoning as a process involving real change.

It is not possible to discuss, in this paper, all of these dualities individually. However, it is easy to show that they instantiate aspects of a dynamic opposition between elements to which my transconsistent logic applies. In the TCL conception of time and space, simultaneity and succession are contradictorily related, and synchronic and diachronic processes can coexist on an alternating, reciprocal basis.

5.1 Process aspects of MBR

Following Nersessian and Magnani, I use the term model-based reasoning to indicate the process aspects of the construction and manipulation of various kinds of mental representations, primarily not sentential, mathematical or formal, related functionally to aspects of the environment. In my view, the "memory system" or cognitive process that crosses the boundary between person and environment is correctly described as being distributed between them, not metaphorically but physically. As put by Ward [Ward, 2002], neither maps nor models have simple binary, true or false relationships to the world.

TCL supports Ward's view that many of the representational structures constituting the psychological architecture of human beings need not be understood as having the form of truth-functional sentences. The ternary dynamic structures proposed here can be seen as examples of non-discursive representations. Pragmatic, real-world values of actuality or potentiality replace, as noted above, static concepts of truth embedded in the assumptions of classical and neo-classical logics. These include, as noted by Dummett, the concept underlying quantification itself.

5.2 Chains of implication

The critical formal notion linking transconsistent logic and the domains of reasoning is that of chains of implication, for which the formal notation developed by Lupasco was given above. The point is that implication in classical logic is not an operation but a definitive and static determination, ultimately resulting either in tautology or in its own suppression *via* exclusion. Only if implication is itself considered as a dynamics, a real action because it cannot be absolute and implies a contradictory implication, is there a development possible from one implication to another, in fact into a transfinite series of implications, which Lupasco terms ortho-deduction. This gives an exact meaning to the expression that p entails q; the antecedent p *really* entails the consequent q. In this approach, logical implication signifies a real physical causality and physical causality becomes in this approach a real logical causality.

In the notation used, the three chains of implication of primary interest are as follows:

1. positive or identifying ortho-deduction, which approaches, asymptotically, absolute actualization or classical (tautological) deduction, equivalent to the logic of causality of classical science;

2. negative or diversifying ortho-deduction. This represents a new concept, a logic and causality of irrationality and diversity, but that can be related to induction and the Axiom of Choice in the set theory of Zermelo.

3. contradictorial ortho-deduction, involving the amplification of the T-states of the previous chains, tending toward maximum contradiction and emergence, e.g., of new ideas, concepts, art, etc..

6 Deduction, induction and abduction in the Lupasco formalism

The relationship of deduction, induction and/or abduction is critical to the understanding of the fit between logics and reasoning. The ampliative adaptive logic of Meheus and her associates, referred to above, permits an integration of deduction and induction, making more reasonable the concept that deduction and induction are opposite processes, occurring sequentially. However, it is difficult to relate the formal concepts involved to what happens in real reasoning. Ampliative logics add (or subtract) conclusions stepwise, one at a time. They, like other non-monotonic logics, recognize the existence of change, but they do not explain its real source or mechanism.

In TCL, every operation is a logical process, tending either toward a non-contradiction of identity or diversity, or toward an emergent contradiction [Lupasco, 1947]. Induction and deduction are thus not only opposing operations, they correspond to opposing real processes in the mind. They are linked because all real elements or processes instantiate or actualize identity or diversity, the general and the particular, continuity and discontinuity, but never completely and only to the extent that the opposite element is potentialized, as in Table 3:

Consequently, in induction, as in classical science, facts are contingent and what seems most real is some underlying identity, a "law", until, by the actualization of new facts by experiment, a new identity, a new hypothesis is constituted by the dynamic contradiction that links them. Deduction corresponds to the processes of applied science, in which, after physics, for example, has operated inductively, it becomes deductive to verify the effective actualization *via* machines, instruments, etc., of the potential identities of theory. Somewhat counterintuitively, what is potentialized is objective and made present to the conscious mind, and what is actualized is subjective and is repressed or less evident.

A third state of affairs, corresponding to the reciprocal repression of inductive and deductive processes, results in the establishment of the T-states referred to earlier. Rather than a simple equilibrium, this state should be looked at as the basis for the emergence of complex phenomena. This picture supports the differentiation made by Magnani between scientific reasoning and complex forms of creative reasoning. The key word in Magnani's development of action-based abduction in science [Magnani, 2004] is "extra-theoretical". In my view, this term effectively captures the real, continuous

METHOD = PROCESS	ACTUALIZATION	POTENTIALIZATION (Conscious Cognition)
INDUCTION		
Classical Science Verifies existence of deductive potentialities = life sciences	Subjectivity of Diversity	Objectivity of Identity
	Facts as "Appearance"	"Law" as Reality (Hypothesis)
	Particular	General
DEDUCTION		
(Tautology) Applied Science Verifies existence of inductive theoretical identities = physical sciences	Subjectivity of Identity	Objectivity of Diversity
		Facts as "Irrationalities"
	General	Particular

Table 4.

and processual aspects of reasoning and complex mental activity in general. The transconsistent system I outline does nothing more (nor less) than provide a physical/metaphysical framework for including his approach within a definition of logic, essentially, extended to reality. This, in turn, captures the underlying dynamics better than ampliative logics that are hampered by ingrained classical presuppositions derived from sentential constructs. My logic can be seen, thus, as an improvement on a *classical* model in the sense of describing how the satisfaction of a set of positive and negative constraints can take place without necessary total elimination of incoherence or contradiction.

From another logical standpoint, Burger and Heidema [Burger and Heidema, 2002] looked at the relations of inference, as entailment or consequence and conjecture, as abduction or antecedence, as opposing dynamic processes. My transconsistent approach founds their proposed "merger" into a single new relation and provides a rigorous notion of the "varying strengths of the two components in the blend". Further, it is not limited to a propositional language that excludes cases of physical dynamic interest.

The dualities indicated above thus receive natural explanations. Non-elimination of subjectivity does not imply elimination of objective truth.

The computational viewpoint about human processes, as in the ϕ-calculus of Finkeissen [Finkeissen, 2002], does not provide a philosophically solid foundation for reasoning or other forms of human theory formation. My approach resolves the problem of the relation of an objective model to reality in the sense of Finkeissen. In conceptual space, model and reality overlap contradictorially, that is, share one another's characteristics in the alternating way described. The term contradictorial is correctly applied here, since I am talking about cognitive concepts as real energetic elements of the process of reasoning.

7 Emergence and creativity. The cyclic model of the universe

The transconsistent approach to emergence can be related further to emergent creative reasoning. In my metaphysics, I postulate 1) that all higher-level scientific principles reflect, and can be derived from, the same basic antagonistic properties of energy that constitute the fundamental principles of existence, including those of quantum physics; and 2) the emergence of complex phenomena can be described by the former and are explicable in terms of the latter, in the sense that matter at each level 'inherits' not only the actualities but the *potentialities* of its substratum. Phenomena at each level can be characterized by differing major trends toward identity or diversity. Their equality results in emergence, specifically at the microphysical level, at the transition from non-life to life, consciousness and higher-level human mental activity, such as creative reasoning.

[Meheus *et al.*, 2002] has analyzed the reasoning leading to the discovery of the outer planets, and there seems to be much interest in other models from the history of astronomical science. A relevant example from current cosmology is the cyclic model of the universe of Steinhardt [Steinhardt and Turok, 2002].

- Big Bang
    - 1 singularity → infinite temperature and density; finite, closed
- Big Bang + Big Crunch
    - 2 singularities → infinite temperature and density; finite, closed (curved); driver for reversal absent
- Cylic Model
    - no singularities → finite temperature and density; infinite, flat alternate expansion and contraction between asymptotic approaches to Bang and Crunch states; theory for "bounce"

- Current State: Expansion
  - driven by increase in (actualization of) negative matter/energy (70%) at the expense of (potentialization of) dark (20%) and ordinary (5%) matter/energy

Table 5. Models of the Universe.

In this model, the universe is infinite and flat, in states of alternate expansion and contraction between only asymptotic approaches to a Big Bang or Big Crunch singularities. The universe is currently expanding, driven by an increase in negative matter/energy (70%) at the expense of dark (20%) and ordinary (5%) matter/energy. It is tempting to see the model as consistent with a rejection of absolute limits (singularities) and the alternating actualization and potentialization of two opposing forms of energy, following the scheme outlined here.

8 Summary and conclusions

I have described a transconsistent logic of reality, TCL, that is relevant to current advances in model-based reasoning, especially where creativity, values and non-sentential accounts of mental structure are concerned. Based on a semi-formal view of reasoning as a process, it suggests a separation of reasoning into domains, in one of which selection between specific alternatives and/or stepwise addition of others (with consequent selection between these and previous ones) predominates. In the other domain, a dynamic process of creative interaction or opposition results in the emergence of new, more complex entities. The logics applicable to the first, ampliative adaptive logics or paraconsistent logics, provide an effective reductive structure for analyzing the domain. The transconsistent logical system represents a further bridge or intermediate explanatory structure between these logics and reality.

The acceptance of transconsistent logic as a valid logical system requires a shift from focus on the axioms and formalism of classical and non-classical propositional or mathematical logics. These logics use tools that were originally developed for handling consistency, and inconsistency is handled with essentially the same tools. The initial existing formalisms, only some of which are indicated in this paper, are in fact an attempt to model reality by reference to the dynamic aspects of the elements undergoing change. They are certainly not the final word, and there is no reason that further mathematization and formalization should not be possible.

The logical system I have proposed deals with models of reasoning and other phenomena by considering them as energetic events taking place in reality. It remains a logic, however, and not a scientific theory, since the domain of reference is at a higher level of interpretation. My TCL approach, as a supplement to existing methods of inquiry, could provide insights leading to new notions of reasoning based on differences in the domains of application of the two forms of logic.

BIBLIOGRAPHY

[Aerts et al., 1991] D. Aerts, J. Broekaert, and S. Smets. Inconsistencies in constituent theories of world views: Quantum mechanical examples. *Foundations of Science*, 3:2, 1991.

[Badescu and Nicolescu, 1996] H. Badescu and B. Nicolescu, editors. *Stéphane Lupasco; L'homme et l'úuvre*. Peter Lang, New York, 1996.

[Batens, 2005] D. Batens. The adaptive logics home page, 2005. http://logica.rug.ac.be/adlog/al.html.

[Béziau, 1999] J.-Y. Béziau. The future of paraconsistent logic. *Logical Studies*, 2:1–17, 1999.

[Béziau, 2004] J.-Y. Béziau. What is formal logic? paper submitted for publication, 2004.

[Burger and Heidema, 2002] I.C. Burger and Heidema. Degrees of abductive boldness. In L. Magnani, N.J. Nersessian, and C. Pizzi, editors, *Logical and Computational Aspects of Model-Based Reasoning*, pages 163–180. Kluwer Academic Publishers, Dordrecht/Boston/London, 2002.

[Carnielli, 2005a] W. Carnielli, 2005. private communication to the author.

[Carnielli, 2005b] W. Carnielli. Logics of formal inconsistency. *CLE e-Prints*, 5(1), 2005.

[Coecke et al., 2002] B. Coecke, D.J. Moore, and S. Smets. Logic of dynamics & dynamics of logic; some paradigm examples. *arXiv:math.LO/0106059 v6*, 6, 2002.

[Finkeissen, 2002] E. Finkeissen. Combining strategy and sub-models for the objectified communication of research programs. In L. Magnani, N.J. Nersessian, and C. Pizzi, editors, *Logical and Computational Aspects of Model-Based Reasoning*, pages 313–330. Kluwer Academic Publishers, Dordrecht/Boston/London, 2002.

[Gabbay, 1994] D. Gabbay. *What is a Logical System?* Clarendon Press, Oxford, 1994.

[Largeault, 1993] J. Largeault. *La logique in the series 'Que sais-je?'*. Presses Universitaires de France, Paris, 1993.

[Lupasco, 1947] S. Lupasco. *Logique et Contradiction*. Presses Universitaires de France, Paris, 1947.

[Lupasco, 1979] S. Lupasco. *L'univers psychique*. Denol/Gonthier, Paris, 1979.

[Lupasco, 1987] S. Lupasco. *Le principe d'antagonisme et la logique de l'Énergie*. Editions du Rocher, Paris, 1987. 2nd edition.

[Magnani and Nersessian, 2002] L. Magnani and N.J. Nersessian, editors. *Model-Based Reasoning: Science, Technology, Values*. Kluwer, Dordrecht, 2002.

[Magnani et al., 2002] Magnani, L., Nersessian, N.J., and C. Pizzi, editors. *Logical and Computational Aspects of Model-Based Reasoning*. Kluwer Academic Publishers, Dordrecht, 2002.

[Magnani, 2002] L. Magnani. Epistemic mediators and model-based discovery in science. In L. Magnani and N.J. Nersessian, editors, *Model-Based Reasoning: Science, Technology, Values*, pages 305–330. Kluwer Academic Publishers, Dordrecht/Boston/London, 2002.

[Magnani, 2004] L. Magnani. Action-based abduction in science. In A. Aliseda and D. Pearce, editors, *Workshop Scientific Reasoning in Artificial Intelligence and Philosophy of Science, ECAI 2000 Workshop Notes*, pages 46–51, Berlin, 2004.

[Meheus et al., 2002] J. Meheus, L. Verhoeven, M. Van Dyck, and D. Provijn. Combining strategy and sub-models for the objectified communication of research programs. In L. Magnani, N.J. Nersessian, and C. Pizzi, editors, *Logical and Computational Aspects of Model-Based Reasoning*, pages 39–72. Kluwer Academic Publishers, Dordrecht/Boston/London, 2002.

[Meheus, 2005] J. Meheus. Do we need paraconsistency in commonsense reasoning?, 2005.

[Morgan, 2002] M.S. Morgan. Model experiments and models in experiments. In L. Magnani and N.J. Nersessian, editors, *Model-Based Reasoning: Science, Technology, Values*, pages 41–58. Kluwer Academic Publishers, Dordrecht/Boston/London, 2002.

[Priest, 1987] G. Priest. *In Contradiction*. Martinus Nijhoff Publishers, Dordrecht/Boston/Lancaster, 1987.

[Priest, 2001] G. Priest. *An Introduction to Non-Classical Logic*. Cambridge University Press, Cambridge, 2001.

[Priest, 2002a] G. Priest. Inconsistency in the empirical sciences. In J. Meheus, editor, *Inconsistency in Science*, pages 1–11. Kluwer Academic Publishers, Dordrecht/Boston/London, 2002.

[Priest, 2002b] G. Priest. Paraconsistent logic. In D. Gabbay and F. Guenthner, editors, *Handbook of Philosophical Logic*, page 259. Kluwer Academic Publishers, Dordrecht, 2002. 2nd edition.

[Steinhardt and Turok, 2002] P.J. Steinhardt and N. Turok. A cyclic model of the universe. *Science*, 296:1436, 2002.

[Ward, 2002] A. Ward. Conceptual models, inquiry and the problem of deriving normative claims from a naturalistic base. In L. Magnani and N.J. Nersessian, editors, *Model-Based Reasoning: Science, Technology, Values*, pages 243–258. Kluwer Academic Publishers, Dordrecht/Boston/London, 2002.

[Wilson, 2000] E.O. Wilson. *Consilience*. Knopf, New York, 2000.

Joseph E. Brenner
International Center for Transdisciplinary
Research and Study, Paris
Email: joe.brenner@bluewin.ch

An Inductionless and Default-Based Analysis of Machine Learning Procedures

EDOARDO DATTERI, HYKEL HOSNI,
AND GUGLIELMO TAMBURRINI

ABSTRACT. It has been suggested that AI investigations of mechanical learning undermine sweeping anti-inductivist views in the theory of knowledge and the philosophy of science. Contrary to this view, we argue that no trace of epistemically justified induction is to be found within a rather representative class of learning agents and we outline an alternative deductive account of these learning procedures. Finally, the opportunity of developing an induction-free logical analysis of reasoning processes involved in learning machines is emphasized by a broad reflection on some families of non-monotonic, albeit deductive, consequence relations.

1 Introduction

The idea that work in artificial intelligence enables one to adjudicate epistemic issues about induction has attracted ever growing attention over the last two decades. In the early 1980s, for instance Michalski made bold claims on the impact of the emerging field of machine learning on the mechanization of inductive inference:

> [...] [T]here was even doubt whether it would ever be possible to formalize inductive inference and perform it on a machine [...]. The above pessimistic prospects are now being revised. [Michalski, 1984, p. 87]

If Howson recently emphasized the relevance of machine learning developments for philosophical and logical reflections on induction [Howson, 2000], the idea that computational learning systems can aptly perform inductive inference is simply taken for granted in a very influential book in the subject:

> The Theory of Machine Inductive Inference (or "Computational Learning Theory", etc.) attempts to clarify the process by which

a child or adult discovers systematic generalizations about her environment. [Jain et al., 1999, p. 28]

Finally, suggestions are being made (see e.g. [Gillies, 1996]) that the upshot of the recent advances in concept and rule learning is that the Popperian radical brand of anti-inductivism (as developed in [Popper, 1972]) must be deeply reconsidered, if not abandoned.

We argue here that expectations and claims of this sort are not supported by the current work on mechanical learning: no matter how significant for understanding or attaining mechanical intelligence, learning machines fail to bolster the inductivist case in epistemology and the philosophy of science.

To state more precisely this claim, and to set the stage for an alternative logical analysis of mechanical learning to be sketched in the final part of this paper, let us start by recalling the salient features of the commonly agreed notion of inductive inference. Woods presents in [Woods, 2002, pp. 106–107] a comprehensive analysis in which the characterization of induction is given in terms of a "non-deductive argument or inference from a sample to a conclusion which *projects* the sample in some way" where projection is further qualified as either "generalization" or "prediction". Notice that, in this characterization, inductive inference is assumed to be distinct from, and indeed opposed to, the deductive one.

Against this background the fundamental epistemic problem of inductive inference amounts to discerning what sorts (if any) of constraints make it reasonable to accept the outcomes of these generalizations and predictions.

With these notions in mind, we can now state more precisely the issue addressed by the present paper in terms of the following, interrelated questions:

1. Do learning machines (as developed in artificial intelligence and robotics) perform induction?

2. Do results in machine learning provide support to the inductivist claims, that is to say, against the anti-inductivist position in epistemology and the philosophy of science?

We submit that the answers are a qualified "yes" and a sound "no", respectively. As to the former, we will claim that learning machines indeed perform generalizations and predictions, thus exhibiting two essential features of inductive inference. However, another fundamental feature of induction – the distinction from deduction – is missing: as we shall put forward, the generalizations and predictions performed by some representative classes of learning systems *can* be justified *without* appealing to any sort of non-deductive inference. As to the latter, we shall argue that current work in machine learning does not afford a positive solution to the epistemic problem of induction.

Moreover the development of our analysis will allow us to make a third, constructive point, namely that a variety of allegedly inductive procedures in learning machines, – i.e., learning procedures giving rise to the above mentioned projective behaviors from observed samples – can be accounted for in terms of default-based, deductive reasoning, without appealing to as yet unjustified principles of mechanical induction.

We substantiate those conclusions by referring to some representative classes of learning machines. We first analyze, in section 2, a representative behavior-based robotic system provided with reinforcement-based learning capabilities. Our analysis reveals that crucial, if implicit, background assumptions about the environment play a central role in the inferential capabilities of these systems; their predictions and generalizations are clearly not supported by perceived data alone in a bottom-up, non-deductive fashion.

The epistemic problem of induction is directly addressed in section 3, where we consider the symbolically richer ID3-style machine learning algorithms, which provide the main basis for Gillies's rebuttal of radical anti-inductivism in the theory of knowledge and the philosophy of science [Gillies, 1996]. We point out that the overfitting of the training data, which reminds one that a good approximation to the target concept (or rule) on the training data is not, in itself, diagnostic of a good approximation over the whole instance space of that concept (or rule), and thus jeopardizes the idea that epistemically justified inductive processes are at work there [Tamburrini, 2005]. Accordingly, we advance a different view of ID3-like projective behaviors which essentially rely on deductive reasoning from a variety of heuristic hypotheses about concept spaces and current target concept.

This re-interpretation of ID3-style learning brings to the fore the central role of deductive trial and error-elimination processes in autonomous learning mechanisms, which interleave the default-based introduction of projective hypotheses about observed samples, retraction of falsified hypotheses, and selection of new default (background) hypotheses for more effective learning. In section 4, we consider the opportunity of developing an induction-free logical analysis of this multifaceted reasoning process by reflecting on some families of non-monotonic, albeit deductive, consequence relations. These provide an abstract, natural framework to embed mechanical projections of learning hypotheses from observed samples into more comprehensive inference processes, which allow agents to retract falsified learning hypotheses and to modify underlying knowledge bases in suitable ways.

2 Empirical assumptions of behavior-based learning

Behavior-based robotics is a relatively recent area of robotics research, that focuses on the development of adaptive robots inhabiting unknown, real (not simulated) environments. Some behavior-based systems, provided with suitable learning mechanisms, exhibit interesting aspects of the projective behavior which is customarily associated to inductive agents. Of particular interest to us is their capability to "acquire new knowledge from experience".

Let us begin from a brief description of these systems, and then move on to analyze their allegedly inductive capabilities. The functional structure of behavior-based agents is appropriately described as a set of n layers, called *behaviors*, each layer computing a function from sensor data to motor actions [Arkin, 1998]. The response r_i of behavior b_i at time t is

$$r_i = \gamma \times b_i(s_t) \tag{1}$$

where s_t is the state perceived by the sensor associated with behavior b_i at time t, γ is a real-valued gain and $0 \leq i \leq n$. Additionally, to each behavior is associated a set of preconditions, that is to say, conditions of activation.

Behaviors run asynchronously and in parallel [Brooks, 1986]. If m behaviors control the same actuator, a coordination function C, taking as input the outcomes of those m behaviors, outputs a single command for the actuator. A typical coordination function (termed *cooperative*) blends together (typically sums) multiple responses, previously weighted by means of a set of m real valued gains. The function C, together with the various gains and mappings constituting the behaviors, determines how the system acts in response to perceived environmental conditions. This overall perception-action mapping is usually referred to as the agent's control function f [Nehmzow and Mitchell, 1995]. Note that distinct sets of gains determine distinct reaction policies to environmental stimuli, i.e. distinct control functions.

In order to act properly and survive, these agents require an appropriate tuning of the perception-action control algorithms. When the appropriate perception-action control algorithm, i.e. the appropriate sets of gains (for given agents, environments, and tasks) is hard to choose *a priori*, then a viable option is to let the robot learn these values by itself [Mahadevan and Connell, 1991]. Learning, in behavior-based agents, is ordinarily achieved by changing gain values over time, either those associated with each behavior or those associated with the coordination function C; these changes result into a new system control function.

Reinforcement learning is particularly suited when the system has to operate in unknown environments [Kaelbling *et al.*, 1996]. Started with default gains, the agent builds up an appropriate control function by stepwise parameter modifications, which depend on positive or negative action rewards, that is, on how the outcome of these actions are evaluated with respect to the agent's ultimate goal.

In a sense, reinforcement learning relieves system designers from the task of specifying the perception-action control function in every detail. At first sight, the system starts with no knowledge of the world, it is set free in its environment, and comes across positive and negative rewards. It learns "directly" from the environment, acquiring knowledge and skills in a completely autonomous and unbiased way, thereby exhibiting the main distinguishing features of a genuine inductive learning mechanism: "reinforcement learning studies the problem of inducing by trial and error a policy from states to actions that maximizes a fixed performance measure (or reward)" [Kaelbling et al., 1996]. This *prima facie* plausible account, however, fails to underscore the crucial role played by *default assumptions* in this reinforcement learning mechanism. An understanding of this role, we submit, paves the way to a more satisfactory account of this learning mechanism in terms of a particular kind of deductive process.

To begin with, let us see how the learning mechanism is biased by *a priori* assumptions on how the result of the learning process should look like. As an example, consider the behavior-based system described in [Clark et al., 1992] (*A* from now on). It consists of three behaviors, each one associated with a set of gains: an attractive behavior, that forces the robot to head toward the target position, a repulsive behavior, that drives the robot away from perceived obstacles, and a noise behavior, that selects a random heading (useful to escape from local minima). A cooperative coordination function computes a weighted sum of the outcomes of each behavior, determining the heading of the robot; the contribution of the attractive, the repulsive and the noise behavior to the resulting heading depends not only on the sensory data, but also on the gains associated with each behavior.

The designers of *A* emphasized the need of endowing this system with learning capabilities:

> [...] the robot may be required to navigate in unfamiliar environments where the appropriate values for the behaviors cannot be known in advance. Our research has concentrated on extending reactive controllers to include the ability to learn these values.

The learning strategy devised for this system is meant to allow *A* to learn the appropriate gain values (consequently, the function f) "assuming no knowledge of the world". Default behavioral parameters are set at the beginning of the navigation. During the navigation, the learning module (L for short) continuously evaluates the amount and speed of the progress that A has made towards the achievement of its goal. On this basis L evaluates whether the system (i) is blocked in a stalemate, or if (ii) it is moving towards the goal, or if (iii) there are obstacles in the vicinity and the robot is not moving towards the goal, or finally if (iiii) there aren't obstacles and the robot is not moving towards the goal. All these cases require different adjustments of the behavioral parameters. In the first situation, for example, an effective strategy is to increase the gains associated with the noise behavior and decrease the other gains; the noise behavior will increase in a stepwise fashion, and will ultimately dominate every other system behavior; consequently, the system will wander randomly, possibly finding a way out of the stalemate.

In the case (iii), two adjusting strategies suggest themselves. In so-called *ballooning strategy*, the gains of the noise and of the repulsive behaviors are increased; this makes the repulsive behavior drive the agent away from obstacles, inhibiting the effect of the attractive behavior. This is effective when the system comes across a large cluster of obstacles including no pathway through them. But consider now a very cluttered environment, in which obstacles are homogeneously distributed rather than grouped together. The distance between obstacles is small but sufficient for the agent to pass without colliding. Increasing the repulsive behavior, as in the *ballooning* strategy, could have the effect of preventing the agent from getting close enough to find a path, as all free space should be inside the obstacles' sphere of influence. The *ballooning strategy* is clearly unsuitable for this environment. A better option is to decrease the repulsive behavior in favor of the attractive one. As a result, A would no longer escape vigorously from obstacles, as determined by the *ballooning strategy*, and would rather look for a cleft among the obstacles while still avoiding them. Let us call *squeezing* strategy this alternative modification of parameters.

The *ballooning strategy* and the *squeezing* strategy have enabled us to illustrate different learning biases, and to reveal the myth of *tabula rasa* learning in quite simple navigation tasks. Let us now assume that the two types of environments, the one full of sparse clusters of obstacles or canyons, the other one full of sparse, uniformly distributed obstacles, can be defined more precisely – say, by taking into account the mean dimensions of the obstacle clusters, the distance between their centers of mass, or other well-defined parameters. Let us call these types of environment E_1 and E_2 respectively. Furthermore, let L_1 and L_2 be the learning algorithms provided with the sets of adjustment values corresponding to the *ballooning* strategy, and the *squeezing* strategy, respectively. If e is the environment in which A is operating, conceptual analyses of L_1 and L_2 and related experimental results described in [Ram et al., 1997] suggest the following thesis

- due to its adjustment values, L_1 will output a control function adequate for A to survive in environments of type E_1: actions generated after learning in E_1 will be positively rewarded;

- due to its adjustment values, L_2 will output a control function adequate for A to survive in environments of type E_2: actions generated after learning in E_2 will be positively rewarded.

These assumptions state a correlation between actions and positive rewards, as long as the right algorithm was chosen, based on the knowledge of e's properties:

If e is of type E_1, then choose L_1, and if e is of type E_1, then choose L_2.

This discussion shows that no induction from *tabula rasa* is involved in the generation of a control function suitable for e. Indeed, the only way to make A survive in e is not just to let it act and learn. The fact that A will produce, at a certain point in time, a *positively rewarded action* (i.e., allowing its survival in e) crucially depends on the adequacy of the assumptions expressed above, and not on sensory data alone.

Addressing the epistemic problem of induction in these cases amounts to finding appropriate justification for the empirical adequacy (the plausibility, reliability, and so on) of newly learnt behavioral rules in the light of the default assumptions underlying the chosen reinforcement policies. In the next section, we argue that this epistemic problem is not solved by representative inductive learning algorithms developed in machine learning.

3 Machine learning and the epistemic problem of induction

In AI agents that learn concepts from examples, hypothesis spaces are often construed as sets of Boolean-valued functions over concept instances. In order to learn target concept h, the algorithm examines a training set, that is a finite subset of the whole instance space X formed by positive or negative instances of h[1].

---

[1] This theoretical framework is adequate to represent also some robotic behavior-based learning systems [Nehmzow and Mitchell, 1995]. In typical AI machine learning systems the "learning" phase is preliminary to the "testing" phase: the former runs through a set of pre-computed perception-action pairs whilst the latter consists in the application

The assumption that the projective behavior of computational systems that learn concepts from examples is epistemically justified (the so-called "inductive claim", IC for short) can be schematically stated as follows:

> (IC) Any hypothesis found to approximate the target function well over a sufficiently large set of training examples will also approximate the target function well over unobserved examples[2].

A thorough examination of this assumption requires an extensive survey of learning systems that goes well beyond the scope of this paper. Hence, for present purposes, we will focus on versions of (IC) concerning the decision tree algorithm ID3. Decision tree learning is a widely used method in concept learning, and Quinlan's ID3 reflects crucial features of this method [Quinlan, 1983; Quinlan, 1993]. Moreover, ID3 has been widely appealed to in order to undercut anti-inductivist claims in the theory of knowledge and the philosophy of science.

Thus, let us focus on the following:

> (IC-ID3) Any hypothesis constructed by ID3 which fits the target function over a sufficiently large set of training examples will approximate the target function well over unobserved examples.

To begin with, let us recall some distinctive features of (the ID3) decision tree learning. Decision trees provide classifications of concept instances in a training set, formed by conjunctions of attribute/value pairs. Each nonterminal node in the tree stands for a test on some attribute, and each branch descending from that node stands for one of the possible values assumed by that attribute. Each path in the tree represents a classified instance. The terminal node of each path in the tree is labeled with the yes/no classification. The learnt concept description can be read off from the paths which terminate into a "yes" leaf. Such description can be expressed as a disjunction of conjunctions of attribute/value pairs[3]. An instance in the training set is classified by starting at the root of the tree, testing the

---

of the learnt function for action. In behavior-based learning systems the learning and testing phases are usually mixed: every positively or negatively rewarded instance is used both for learning (because it causes a positive or negative reward) and for acting (because it leads to an action). Nevertheless, the two phases are conceptually distinct and easily identified in their algorithmic structure. The argument developed at the end of this section is applicable , with possible minor modifications, to many behavior-based learning systems too.

[2]Cp. [Mitchell, 1997, p. 23].

[3]Concept descriptions that make essential use of relational predicates (such as "ancestor") cannot be learnt within this framework. Hence ID3 decision trees amount to nothing but propositional binary decision diagrams.

attribute associated to this node, selecting the descending branch associated to the value assumed by this attribute in the instance under examination, repeating the test on the successor node along this branch, and so on until a leaf is reached. Each concept instance in the training set is associated to a path in a tree, which is labeled "yes" or "no" at the terminal node. ID3 places closer to the tree root attributes which better classify positive and negative examples in the training set. This is done by associating to each attribute P mentioned in the training set a measure of how well P alone separates the training examples according to their being positive or negative instances of the target concept. Let us call this preference in tree construction the ID3 "*informational* bias".

There is another bias characterizing the ID3 construction strategy. ID3 stops expanding a decision tree as soon as an hypothesis accounting for training data is found. In other words, simpler hypotheses (shorter decision trees) are singled out from the set of hypotheses that are consistent with training data, and more complicated ones (longer decision trees) are discarded. On account of this *simplicity* bias, longer decision trees that are compatible with the training set are not even generated, and thus no conflict resolution strategy is needed to choose between competing hypotheses.

We are now in the position to state more precisely inductive claim (IC-ID3), by reference to the main background hypotheses used by ID-3 to reduce its hypothesis space:

> (IC-ID3: second version): Any hypothesis constructed by ID3 on the basis of its informational and simplicity biases which fits the target function over a sufficiently large set of training examples will also approximate the target function well over unobserved examples.

Scepticism about this claim is fostered by the *overfitting problem*. An hypothesis $h \in H$ is said to overfit the training set if another hypothesis $h' \in H$ performs better than h on the instance space X, even though h' does not fit the training set better than h. Overfitting in ID3 trees commonly occurs when the training set contains an attribute P unrelated to the target concept, which happens to separate well the training instances. In view of this "informational gain" P is placed close to the tree root.

> Overfitting is a significant practical difficulty for decision tree learning and many other learning methods. For example, in one experimental study of ID3 involving five different learning tasks with noisy, nondeterministic data, [...] overfitting was found to decrease the accuracy of learnt decision trees by 10-25% on most problems. [Mitchell, 1997, p. 68]

Unprincipled expansions of the original training set may not prevent the generation of overfitting trees, for a larger training set may bring about additional noise and coincidental regularities. Accordingly, claim (IC-ID3) is to be further qualified: the "sufficiently large set of training examples" mentioned there must be "sufficiently representative of the target concept" as well. This means that conjectures about the representativeness of concept instance collections play a central role in successful ID3 learning.

Confronted with this difficulty, that the sceptic consistently interprets as symptoms that inductive claim (IC-ID3) cannot be convincingly argued for, let us try and see the extent to which ID3 fits into a deductive framework.

We have already formed a vague picture of ID3 as a component of a trial and error-elimination cycle: on the basis of assumptions guiding both training set construction and selection of some concept c, ID3 makes predictions about the classification of concept instances that are not included in the training set.

Let us now provide a deductive account of ID3 predictive behavior, drawing on the above distinction between the preferences or biases embedded in ID3 proper (which determine both the language for expressing concepts and the construction of decision trees) on the one hand, and the presuppositions that are used to select training sets on the other hand. If the presuppositions of the first kind (ID3 biases) are suitably stated in declarative form, a concept learning algorithm such as ID3 can be redescribed as a theorem prover. This is brought out by the following definition of the inductive bias of a concept learning algorithm (see [Mitchell, 1997, p. 43]).

> Definition: Consider a concept learning algorithm L for the set of instances X. Let c be an arbitrary concept defined over X, and let $D_c = \{\langle x, c(x) \rangle\}$ be an arbitrary set of training examples of c. Let $L(x_i, D_c)$ denote the classification assigned to the instance x_i by L after training on the data D_c. The inductive bias of L is any minimal set of assertions B such that, for any target concept c and corresponding training examples Dc,
>
> $(\forall x_i \in X)[L(x_i, D_c)$ is logically derivable from $(B \wedge D_c \wedge x_i)]$.

One is provisionally entitled to preserve B and D_c as long as the classifications coming in through L or its equivalent deductive system are satisfactory. Suppose, however, that for given i's $L(x_i, D_c)$ is an incorrect prediction. Then, this consequence of B and D_c is to be retracted, and either B or D_c are to be appropriately *revised* in order to obtain a correct classification in those cases. This trial and error-correction behavior is essentially non-monotonic. ID3-like systems perform only the "trial" part of this process, but a fully autonomous learning machine should be capable of carrying out the "error-correction" part as well. As a consequence, these machines should be capable of performing non-monotonic reasoning, insofar as autonomous learning involves, in addition to advancing learning hypotheses, the possibility of retracting empirically inadequate hypotheses and revising one's knowledge base accordingly.

One may wonder whether this requirement for non-monotonic reasoning takes autonomous learning machines outside the realm of deductive reasoners. If this were the case, the behavior of fully autonomous learning machines, unlike ID3-like systems, could not be accounted for in terms of deductive procedures only. However, as we shall see, the theories of *non-monotonic reasoning* and *belief revision*, which provide comprehensive and logically well-understood models of inference processes under conjectural knowledge, pave the way to a deductive account of autonomous learning machines too.

4 From learning projections to constrained monotonicity

We have just seen that the allegedly inductive behaviors of machine learning algorithms *à la* ID3, involving generalizations and predictions from sample data, are sensibly construed as default-based, deductive inferences from theories and past observations. More specifically, these projections depend on background assumptions capturing both the inductive bias of ID3, which is invariant over the class of ID3 learnable concepts, and more "local" assumptions about the target concept c under examination. These local assumptions are embodied into the system's empirical experience, that is, in the training set selected for c. Thus, the retaining of any ID3 identification of target concept is non-monotonically conditional on the empirical adequacy of both sorts of assumptions. Moreover, in the behavior-based systems examined above, similar issues arise about the empirical adequacy of background assumptions. There, we have seen, one can choose between different learning strategies about obstacle avoidance and navigation according to the local information/hypotheses about the spatial distribution of obstacles.

The aim of this section is to suggest that a family of non-monotonic consequence relations provide an appropriate *deductive* logical framework for investigating reasoning patterns of this sort which, in addition to ID3-like generalization and prediction, allow one to retract learning hypotheses and to revise background theories.

Before going any further, however, the somehow common misconception that non-monotonic inference falls outside the realm of deductive reasoning must be addressed. Indeed, non-monotonic inference is occasionally dubbed as "quasi-deductive" [Flach, 2002], and more customarily as an outright non-deductive mode of inference (see e.g. [Oaksford and Chater, 2002, p. 176], where this assumption is made in the context of cognitive science studies on human rationality). *From a logico-mathematical point of view*, however it cannot simply be maintained that non-monotonic inference, unlike "deduction", fails to capture "sound inference". In fact, all the key non-monotonic logics (one of which is to be shortly introduced) are based on rules of inference which, when coupled with appropriate semantics, are provably sound (and complete) just like the inference rules of classical logic. In other words, within these logics the "truth" of the premises (classically) implies the "truth" of the conclusion. Indeed it is easily shown, that *semantically*, non-monotonic consequence relations arise naturally from considerations of truth-preservation that *generalize* those introduced by A. Tarski in the 1930's, the back-bone of the formal characterization of deduction.

This possible source of confusion being dismissed, we are now ready to address our main point. Let $\mathcal{SL} = \{\theta, \phi, \ldots\}$ be the set of sentences built up from a classical propositional language $\mathcal{L}$ in the usual way. As usual $2^{\mathcal{L}}$ denotes the set of classical valuations on $\mathcal{L}$, that is the set of all maps from $\mathcal{L}$ to $2 = \{0, 1\}$. We shall follow the custom of writing Cn for the *classical consequence operation* (sometimes called Tarskian), namely, an operation which is reflexive, transitive, and monotonic, the corresponding notation for the classical consequence relation being $\vdash$. As we shall be concerned with *finitary* consequence only (i.e. with premises being finite sets of $\mathcal{SL}$) we adopt the usual convention of freely swapping the notions of consequence operation and consequence relations[4]. Hence Cn and $\vdash$ represent classical logical inference which we assume to be deductive. Consequence relations on $2^{\mathcal{SL}} \times \mathcal{SL}$ other than the classical one will be denoted by $\vdash\!\!\sim$, possibly with decorations. Note that all consequence relations we are going to consider below rest on the same underlying language $\mathcal{L}$.

We now briefly review some basic concepts of non-monotonic reasoning. We shall, as usual in the area, be assuming that $\mathcal{SL}$ represents the information possessed by an agent, and interpret the consequence relation on it as the agent's reasoning process from premisses to conclusions. Then the classical rule of monotonicity just says that the introduction of additional information to the premises of some derivation does not force one to reject any of the conclusions previously drawn from the initial set of premises, *no matter what this new information turns out to be*. More compactly (for

---

[4]Indeed, given $\Gamma \subseteq \mathcal{SL}$, $\theta \in \mathcal{SL}$ and the operation Cn, one can define a relation $\vdash \subseteq 2^{\mathcal{SL}} \times \mathcal{SL}$ as the set of ordered pairs $\langle \Gamma, \theta \rangle$ such that $\theta \in Cn(\Gamma)$, and conversely $Cn(\Gamma) = \{\theta \mid \Gamma \vdash \theta\}$.

$\Gamma, \Delta \subseteq \mathcal{SL}$):

$$\text{if} \quad \theta \in Cn(\Gamma) \quad \text{then} \quad \theta \in Cn(\Gamma \cup \Delta). \tag{MON}$$

The emphasized lack of qualification is an immediate consequence of the fact that no formal constraint is imposed on the additional set Δ. We call this the *unconstrained* form of monotonicity.

Since monotonicity does not impose any constraint on the enlargement of the set of premises, Δ can indeed be *any* subset of $\mathcal{SL}$. Hence, any possible addition to Γ will be *a priori* irrelevant. To put it into more graphic terms, once a monotonic agent draws a conclusion from a given set of premises, nothing the agent might possibly come to learn will cause it changing its mind. In realistic (i.e. "world-like") environments, however, agents do change their mind.

Still, as far as the characterization of intelligent reasoning is concerned, monotonic patterns of reasoning have a number of desirable properties, so that throwing away monotonicity as a whole will not do. A purely anti-monotonic agent, indeed, would waste an enormous amount of resources questioning over and over again each of the previously drawn conclusions. Thus, precepts of informational economy suggest a sensible balance between unconstrained monotonicity and anti-monotonic behaviors: *agents should not change their mind unless they have good reasons to do so.*

The approach to nonmonotonic reasoning based on consequence relations focuses precisely on the characterization of suitable constraints which capture intuitively appealing precepts of informational economy. The seminal papers by Gabbay [Gabbay, 1985], Makinson [Makinson, 1989] and Kraus, Lehmann and Magidor [Kraus *et al.*, 1990] have provided an abstract (proof-theoretic and model-theoretic) framework, suited for the logical investigation of nonmonotonic inference in general, and nonmonotonic consequence relations in particular.

A crucial difference between monotonic and nonmonotonic logics, as far as consequence relations are concerned, is that the latter admit no "smallest" consequence relation: various notions of nonmonotonic consequence arise as soon as certain conditions are fixed[5]. However, this distinguishing feature of nonmonotonic logics, and the corresponding variety of formal developments, are not symptomatic of a lack of shared underlying intuitions about core features of nonmonotonic reasoning. On the contrary, there is wide consensus that formal systems like the one based on Preferential consequence relations (so-called system P [Kraus *et al.*, 1990] or, after the initials

---

[5](with the remarkable property, however, that the intersection of those various relations brings one back to classical monotonic logic [Makinson, 1994; Makinson, 2003], the logic used in metatheoretic investigations of nonmonotonic logics.)

of its authors' names, system KLM) do capture key features of nonmonotonic reasoning [Bochman, 2001; Rott, 2001; Friedman and Halpern, 2001; Makinson, 2005].

In the first place, the intuitive justification of KLM-like formal constraints (to be briefly presented below) in terms of "rationality", "minimality" and "informational economy" connects the investigations on constrained monotonicity to commonsense approaches to uncertain reasoning (see, e.g. [Paris, 1994; Paris, 1999]), where rational reasoning under imperfect information is mathematically characterized as obedience to "common-sense" principles, some of which turn out to be remarkably close the KLM rules/conditions[6]. Moreover, the KLM system can be proved to be sound and complete with respect to various semantic interpretations of reasoning, respectively based on usualness, typicality, normality, very high probability, and so on. Moreover Friedman and Halpern [Friedman and Halpern, 2001] recently suggested a formal explanation of this deep connection between the proof-theoretic characterization and a wide class of semantics for which nonmonotonic consequence is complete, based on the general framework of *plausibility measures*. Particularly relevant to us is the semantical interpretation according to which the relation $\theta \mathrel{|\!\sim}_P \phi$ holds if ϕ is (classically) satisfied in all the *most preferred* worlds in which θ is (classically) satisfied. Finally, there is a tight connection, indeed a genuine translation, between non-monotonic rules of inference and the classic postulates for *Belief Revision* introduced in [Alchourrón *et al.*, 1985]. The main idea underlying this translation is the so-called Makinson-Gärdenfors Identity, according to which inferring nonmonotonically a conclusion from a sentence θ amounts to revising a certain set of background assumptions by θ. For precise details, see [Makinson and Gärdenfors, 1991; Gärdenfors and Makinson, 1994; Rott, 2001; Kern-Isberner, 2001; Bochman, 2001; Makinson, 2005].

Let us now briefly recall the formal rules/conditions characterizing preferential consequence relations. There are interpreted intuitively as constraints that consequence relations should satisfy in order to account for "sensible patterns of reasoning". The inference system is presented here, as usual, as a Gentzen-style system with individual sentences as arguments of the consequence relations[7]. Relations satisfying the following conditions are called *preferential consequence relations*.

---

[6]The first formal aspects of this connection have been put forward by relating nonmonotonic logics to maximum entropy reasoning. See e.g. [Kern-Isberner, 2001; Hill and Paris, 2002].

[7]The first generalization to the infinitary case is studied in [Freund and Lehmann, 1993].

Reflexivity is the only axiom scheme:

$$\theta \mathrel{\mid\!\sim} \theta \quad (\mathbf{REF})$$

Three kinds of rules (or conditions) are imposed on relation $\mathrel{\mid\!\sim}$. Rules of the first kind are "pure conditions": Left Logical Equivalence and Right Weakening[8]:

$$\frac{\vdash \theta \leftrightarrow \phi, \quad \theta \mathrel{\mid\!\sim} \psi}{\phi \mathrel{\mid\!\sim} \psi} \quad (\mathbf{LLE})$$

$$\frac{\theta \mathrel{\mid\!\sim} \phi, \quad \phi \vdash \psi}{\theta \mathrel{\mid\!\sim} \psi} \quad (\mathbf{RWE})$$

Rules of the second kind are the usual Conjunction in the conclusions and Disjunction in the premises:

$$\frac{\theta \mathrel{\mid\!\sim} \phi, \quad \theta \mathrel{\mid\!\sim} \psi}{\theta \mathrel{\mid\!\sim} \phi \wedge \psi} \quad (\mathbf{AND})$$

$$\frac{\theta \mathrel{\mid\!\sim} \psi, \quad \phi \mathrel{\mid\!\sim} \psi}{\theta \vee \phi \mathrel{\mid\!\sim} \psi} \quad (\mathbf{OR})$$

Finally Cautious Monotonicity, imposes a formal constraint on monotonicity:

$$\frac{\theta \mathrel{\mid\!\sim} \phi, \quad \theta \mathrel{\mid\!\sim} \psi}{\theta \wedge \phi \mathrel{\mid\!\sim} \psi} \quad (\mathbf{CMO})$$

Cautious monotonicity puts a very natural condition on the application of monotonicity: the addition of a sentence that was previously derived from some given premise should not induce one to abandon any consequence of that premise only. This is an important conservative feature of monotonicity that need not be rejected in characterizations of sensible reasoning[9].

For the purpose of illustrating how the reasoning of autonomous learning agents can be abstractly interpreted in the light of this system of nonmonotonic reasoning, we now draw on Makinson's approach to bridging classical and (various) nonmonotonic consequence relations [Makinson, 2003; Makinson, 2005]. By considering the imposition of suitable constraints on monotonicity Makinson examines three monotonic consequence relations which naturally provide

---

[8]Note that any consequence relation $\mathrel{\mid\!\sim}^*$ such that $\vdash \; \subseteq \; \mathrel{\mid\!\sim}^*$ is called *supraclassical*, and Reflexivity and Right Weakening entail that.

[9]For more extended discussions on the motivations and justifications for these rules see [Kraus *et al.*, 1990; Lehmann and Magidor, 1992] and see [Rott, 2001; Kern-Isberner, 2001; Bochman, 2001] for general surveys nonmonotonic logics.

three different ways of getting out of a set of premises more than is authorized by straightforward application of classical consequence, without amplifying the language in which these premises are stated, which remains that of classical logic. [Makinson, 2003, p. 74].

In particular, we focus on Makinson's method of generating nonmonotonic consequence relations from the monotonic, *pivotal-assumption* consequence relation.

Suppose that a distinguished subset of $\mathcal{SL}$ – K – sums up an agent *background assumptions*. We could intuitively think of K as to the corpus of sentences which is taken for granted by the agent when deriving a certain conclusion from an explicit set of premises. The *pivotal-assumption* consequence relation is constructed by reserving to the distinguished set K a special role. Indeed, it is just a classical consequence relation *modulo K*. Formally:

$$\Delta \vdash_K \phi \quad \text{iff for no} \quad v \in 2^{\mathcal{L}}, \quad v(\Delta \cup K) = 1 \quad \text{and} \quad v(\phi) = 0.$$

In other words, ϕ is a pivotal-assumption consequence of the set Δ relative to the pivotal set K, just if ϕ is a *classical* consequence of $\Delta \cup K$. Thus, a consequence relation is called a pivotal-assumption consequence if it is of the form $\vdash_K$ for some $K \subseteq \mathcal{SL}$.

Now, pivotal-assumption consequence relations behave in a decidedly classical way to the effect that the logico-mathematical properties of $\vdash_K$ cast no doubts about its deductive nature. Indeed, as determined by a characterization theorem due to Rott ([Rott, 2001, p. 117]), any closure operation which is also supraclassical, compact and satisfies the classical "disjunction in the premises" *is* a pivotal-assumption consequence[10]. It is precisely this relation that provides a deductive bridge between monotonic and non-monotonic reasoning allowing one to reach an unprecedented clarity in the understanding of nonmonotonic consequence relations.

A crucial motivation underlying the idea of marking K as a distinguished subset of $\mathcal{SL}$ is that of reserving to K the special role of a relatively entrenched set of background assumptions. The general heuristics or inductive bias for constructing any ID3-learning trees is clearly a case in point in view of their independence of local information about particular target concepts. In this vein, the set on the left of $\vdash_K$ is interpreted as the set of "local" assumptions (or hypotheses), that is, information specific to a certain situation, choice, decision, etc. (in ID3-learning, this role is played by specific

---

[10]Note however that pivotal-assumption consequence differ from the classical one in that the former need not satisfy uniform substitution of arbitrary formulae for the elementary letters in a formula.

training sets). A characteristic feature of $\vdash_K$, then, is that the set of background assumptions is assumed *not* to vary in relation to the local assumptions. (This clearly follows from the fact that each K determines a distinct relation $\vdash_K$.) Up to suitable abstraction, $\vdash_K$ characterizes any agent whose behavior is strictly determined by heuristics that are invariant over sets of local assumptions. In case of ID3, for example, a change of training set for concept c determined by a revision of local assumptions about the more representative training data for c does not affect the inductive bias of ID3 proper.

This much for monotonic, assumption-based reasoning, which is characterized by assigning a special status to some unchanging set of background assumptions K. Far more interesting, however, are learning systems that adapt the set of background assumptions according to their actual "experience". These systems would qualify as more flexible and adaptable learners. And yet, under certain formal conditions, adaptable learners of this kind do not transgress the boundaries of deductive inference. This is indeed the scenario in which constrained-monotonic inference is at its best. The non-monotonic consequence relation $\mathrel{\mid\!\sim}_K$, in fact, can be derived from $\vdash_K$ just by allowing the set of background assumptions to vary with the set of local assumptions. This "small" variation is all that there is between monotonic and non-monotonic consequence relations.

The main rationale for allowing the set of background assumptions to vary according to "experience" is that a certain set of local hypotheses Δ might be logically inconsistent with K. Consider the case of a behavior-based system which is capable of detecting that the current perceptual data clash with some background hypothesis driving its learning procedure, say, with the hypothesis that the environment is cluttered with obstacles. If such inconsistencies arise, $\mathrel{\mid\!\sim}_K$ behaves non-monotonically by adjusting K so as to remove the inconsistency with local premises. The idea of *default assumption consequences*, thus, is that K should be replaced by its maximal subsets consistent with Δ[11]. Formally:

$$\Delta \mathrel{\mid\!\sim}_K \phi \quad \text{iff} \quad K' \cup \Delta \vdash \phi, \quad \forall K' \subseteq K \text{ which is maxiconsistent with } \Delta.$$

Among the various properties of $\mathrel{\mid\!\sim}_K$, of particular interest to us is the fact that it satisfies Cautious Monotonicity, which, we have seen, is usually regarded as the key ingredient of "core" nonmonotonic reasoning[12]. Subsequent expansions of K' (say, for effective learning by the above behavior-

---

[11] A subset $Y \subseteq X$ is maximal if and only if for all $Z \subseteq X$, $Y \subseteq Z$ implies $Y = Z$.

[12] Preferential reasoning is also accounted for semantically by means of other monotonic bridges and in particular the *pivotal-valuation* consequence discussed in section 3 of [Makinson, 2003].

based system in an uncluttered environment) should be taken care of by means of appropriate heuristics.

5 A final remark

The general notion of an autonomous learning system which in addition to advancing learning hypotheses is capable of retracting empirically inadequate hypotheses and revising its background knowledge base accordingly, does not require one to introduce notions of consequence that exceed or otherwise cannot be captured within the framework of deductive reasoning. And indeed, it is precisely the *extra-logical* consideration about the certainty, or the lack thereof, of information that underlies what is traditionally seen as a logical difference between classical deduction and various forms of uncertain reasoning. In the case of the classical consequence relation, what warrants the validity of the inference is its truth-preservation, or more precisely the preservation of the postulated truth of the initial statements. The same can –and must– be required from logical systems which are intended to capture patterns of reasoning in which the premises are not necessarily taken as stable truths, but only as elements of a revisable body of knowledge. This fact changes –and usually considerably complicates– the sort of constraints that are to be formalized, yet it does *not* change the logical form of the problem. As it has been effectively pointed out long ago by Kleene:

> Whether $\theta_1, \theta_2, \ldots, \theta_n$ are true or not may be a matter of empirical fact, or of belief, or may rest on earlier assumptions under which the argument is being pursued and which make $\theta_1, \theta_2, \ldots, \theta_n$ available for the purpose of the argument. Soundness is thus relative to whatever criteria or standards are being presupposed in the claim of the argument that $\theta_1, \theta_2, \ldots, \theta_n$ are available, and a full statement on the matter of soundness would include such reference. Also it seems convenient to recognize graduations by calling an argument simply *plausible* when it is valid but we can only say that $\theta_1, \theta_2, \ldots, \theta_n$ are plausible. [Kleene, 1967, pp. 67–68].

Acknowledgments

We are grateful to Jeff Paris and David Makinson for useful comments on earlier versions of this work. We would also like to thank Alberto Mura and David Miller for fruitful discussions on this topic over the past few years.

BIBLIOGRAPHY

[Alchourrón et al., 1985] C.E. Alchourrón, P. Gärdenfors, and D. Makinson. On the logic of theory change: partial meet functions for contraction and revision. *Journal of Symbolic Logic*, 50:510–530, 1985.

[Arkin, 1998] R.C. Arkin. *Behavior-based robotics*. The MIT Press, May 1998.

[Beierle and Kern-Isberner, 2002] C. Beierle and G. Kern-Isberner. Using institutions for the study of qualitative and quantitative conditional logics. In S. Flesca, S. Greco, and N. Leone, editors, *Lecture Notes in Artificial Intelligence*, volume 2371, pages 161–172. JELIA 02, Springer, 2002.

[Bochman, 2001] A. Bochman. *A Logical Theory of Nonmonotonic Inference and Belief Change*. Springer, 2001.

[Brooks, 1986] R.A. Brooks. A robust layered control system for a mobile robot. *IEEE Journal of Robotics and Automation*, pages 14–23, 1986.

[Clark et al., 1992] R.J. Clark, R.C. Arkin, and A. Ram. Learning momentum: online performance enhancement for reactive systems. In *Proceedings of the 1992 IEEE International Conference on Robotics and Automation*, volume 1, pages 111–116, Nice, France, May 1992.

[Flach, 2002] P.A. Flach. Modern logic and its role in the study of knowledge. In D. Jacquette, editor, *A Companion to Philosophical Logic*, pages 680–693. Blackwell Publishers, 2002.

[Freund and Lehmann, 1993] M. Freund and D. Lehmann. Nonmonotonic inference operations. *Bullettin of the Interest Group in Pure and Applied Logics*, 1(1):23–68, 1993.

[Friedman and Halpern, 2001] N. Friedman and Y. Halpern. Plausibility measures and default reasoning. *Journal of the ACM*, 48(4):648–685, 2001.

[Gabbay, 1985] D.M. Gabbay. Theoretical foundations for non-monotonic reasoning in expert systems. In *Proceedings NATO Advanced Institute on Logics and Models of Cuncurrent Systems*, pages 439–457, Berlin, 1985. Springer.

[Gärdenfors and Makinson, 1994] P. Gärdenfors and D. Makinson. Nonmonotonic inferences based on expectations. *Artificial Intelligence*, 55:1–60, 1994.

[Gillies, 1996] D. Gillies. *Artificial Intelligence and Scientific Method*. Oxford University Press, 1996.

[Hill and Paris, 2002] L.C. Hill and J.B. Paris. When maximising entropy gives the rational closure. *Journal of Logic and Computation*, 12(1):1–19, 2002.

[Howson, 2000] C. Howson. *Hume's Problem: Induction and the Justification of Belief*. Oxford University Press, 2000.

[Jain et al., 1999] S. Jain, D. Osherson, J.S. Royer, and A. Sharma. *Systems that learn*. MIT Press, Cambridge, 1999.

[Kaelbling et al., 1996] L.P. Kaelbling, M.L. Littman, and A.P. Moore. Reinforcement learning: A survey. *Journal of Artificial Intelligence Research*, 4:237–285, 1996.

[Kern-Isberner, 2001] G. Kern-Isberner. Conditionals in nonmonotonic reasoning and belief revision – considering conditionals as agents. In *Lecture Notes in Computer Science*, volume 2087. Springer, 2001.

[Kleene, 1967] S.C. Kleene. *Mathematical Logic*. Wiley and Sons, 1967.

[Kraus et al., 1990] S. Kraus, D Lehmann, and M. Magidor. Nonmonotonic reasoning, preferential models and cumulative inference. *Artificial Intelligence*, 44(1):167–207, 1990.

[Lehmann and Magidor, 1992] D. Lehmann and M. Magidor. What does a conditional knowledge base entail? *Artificial Intelligence*, 55:1–60, 1992.

[Mahadevan and Connell, 1991] S. Mahadevan and J. Connell. Automatic programming of behavior-based robots using reinforcement learning. In *Proceedings of AAAI-91*, pages 8–14, Pittsburgh, PA, 1991.

[Makinson and Gärdenfors, 1991] D. Makinson and P. Gärdenfors. Relations between the logic of theory change and nonmonotonic logic. In A. Fuhrmann and M. Morreau, editors, *Proceedings of the Workshop on The Logic of Theory Change*, volume 465, pages 185–205. LNCS, Springer-Verlag, 1991.

[Makinson, 1989] D. Makinson. General theory of cumulative inference. In *Lecture Notes in Artificial Intelligence*, volume 346, pages 1–18. Springer, 1989.

[Makinson, 1994] D. Makinson. General patterns in nonmonotonic reasoning. In D.M. Gabbay, C. Hogger, and J. Robinson, editors, *Handbook of Logic in Artificial Intelligence and Logic Programming*, pages 35–110, 1994.

[Makinson, 2003] D. Makinson. Bridges between classical and nonmonotonic logic. *Logic Journal of the IGPL*, 11(1):69–96, 2003.

[Makinson, 2005] D. Makinson. *Bridges Between Classical and Nonmonotonic Logic*, volume 5 of *Texts in Computing*. King's College Publications, London, 2005.

[Michalski, 1984] R.S. Michalski. A theory of methodology of inductive learning. In R.S. Michalski, J. Carbonell, and T.M. Mitchell, editors, *Machine Learning: An Artificial Intelligence Approach*, pages 83–134, Berlin, 1984. Springer Verlag.

[Mitchell, 1997] T.M. Mitchell. *Machine Learning*. McGraw Hill, 1997.

[Nehmzow and Mitchell, 1995] U. Nehmzow and T. Mitchell. The prospective student's introduction to the robot learning problem. Technical report, UMCS95 -12-6, the Department of Computer Science, Manchester University, 1995.

[Oaksford and Chater, 2002] M. Oaksford and N. Chater. Commonsense reasoning, logic and human rationality. In R. Elio, editor, *Common Sense, Reasoning, and Rationality*, pages 174–214, New York, 2002. Oxford University Press.

[Paris, 1994] J.B. Paris. *The Uncertain Reasoner's Companion: A Mathematical Perspective*. Cambridge University Press, Cambridge, England, 1994.

[Paris, 1999] J.B. Paris. Common sense and maximum entropy. *Synthese*, 117(1):73–93, 1999.

[Popper, 1972] K.R. Popper. *Objective Knowledge: An Evolutionary Approach*. Clarendon Press, Oxford, 1972.

[Quinlan, 1983] J.R. Quinlan. Induction of decision trees. *Machine Learning*, 1(1):81–106, 1983.

[Quinlan, 1993] J.R. Quinlan. *C.4.5: Programs for Machine Learning*. Morgan Kaufmann, San Mateo, 1993.

[Ram et al., 1997] A. Ram, R.C. Arkin, K. Moorman, and R.J. Clark. Case-based reactive navigation: a method for on-line selection and adaptation of reactive robotic control parameters. *IEEE Transactions on Systems, Man and Cybernetics, Part B*, 27(3):376–394, June 1997.

[Rott, 2001] H. Rott. *Change, Choice and Inference : A Study of Belief Revision and Nonmonotonic Reasoning*. Oxford University Press, 2001.

[Tamburrini, 2005] G. Tamburrini. AI and Popper's solution to the problem of induction. In I. Jarvie, K. Milford, and D. Miller, editors, *Karl Popper, A Centenary Assessment*. Ashgate, 2005. Forthcoming.

[Woods, 2002] J. Woods. Standard logics as theories of argument and inference: Induction. In D.M. Gabbay, R.H. Johnson, H.J. Ohlbach, and J. Woods, editors, *Handbook of the Logic of Argument and Inference*, pages 105–171. North-Holland, 2002.

Edoardo Datteri
Dip. di Filosofia
Università di Pisa
Email: edoardo@arts.sssup.it

Hykel Hosni
School of Mathematics
University of Manchester
Email: `hykel@maths.man.ac.uk`

Guglielmo Tamburrini
Dip. di Scienze Fisiche
Università di Napoli "Federico II"
Email: `tamburrini@na.infn.it`

The Pragmatic Logic of Ordered Representations

HELMUT PAPE

ABSTRACT. The aim of this paper is to bring out the notion of formal order as an systematic, overall notion in Peirce's pragmatism. This is done by showing that two on the first glance quite unrelated theories, the methodology of pragmatism and the system of visual logic, the so-called existential graphs, have been constructed with the same ordertheoretic and semantic notion in mind, that has its roots in an analogy of how visual processes take place: the practical ordering of valid representations and logical formulas. Furthermore, I show that this idea is also a motivating force behind pragmatism's logic of science. In this way, an aspect of the formal unity that motivates Peirce's pragmatic philosophy is made explicit.

1 Formal concepts in Peirce's approach to visual logic and to pragmatism

Peirce was both – a pragmatist and a logician. This paper studies the overlap and interaction between these two fields. It argues for the primacy of some formal concepts coming from the mathematical theory of order, the logic of relations and mathematical logic. In using concepts and thoughts developed in these formal theories they are not only adapted to the phenomena at hand but are reshaped and serve to interpret problems in logic, epistemology, philosophy of mind and metaphysics. This paper argues that both pragmatism and visual logic are the result of adapting and interpreting formal concepts to an individual domain of discourse.

But are there really any formal – mathematical and logical – properties that can be used both to construct pragmatism and visual logic? And if there are, what is the philosophical rationale behind such a method of practicing philosophy? This is a complicated topic and I will start with an overview of the conclusion the argument of this paper tries to establish.

Let us turn to the philosophical rationale of Peirce's pragmatic use of formal concepts. I want to show that his logical views concentrate on *ordered representations*. Why is it philosophically important to use the same formal concept of an order relation to structure pragmatism and visual logic? There may be many formal properties which pragmatism and visual logic share which are trivial or at least unimportant for our philosophical understanding of thought, knowledge and reality. Both pragmatism and visual logic are theories intended to clarify, on the level of methodology and on the level of visually instantiated logical relations. On different levels, both theories try to explain and to improve the way we understand truth and reality in some of our thoughts and experiences. This view of pragmatism and logic is one of Peirce's most general philosophical projects: His idealistic logical realism is the thesis that formal properties describe reality, i.e. they characterize the most fundamental ontological entities, the events that make-up reality. That is to say: if reality can be characterized by logical - i.e. order-theoretic - properties, what we get is a sort of "logical idealism" of events and processes, describing experience and external reality by the same type of structural features. This is what Peirce was driving at when in 1898 he points out:

> What is reality? Perhaps there isn't any such thing at all. [...] But if there is any reality, then, so far as there is any reality, what that reality consists in is this: that there is in the being of things something which corresponds to the process of reasoning, that the world *lives*, and *moves*, and *has its being*, in a logic of events. [Peirce, 1960, NEM 4: 343-5]

If we take this realistic ontology of the logic of events as our backdrop, what is its counterpart in philosophical and scientific methodology? Surely, an ontology in terms of a logic of events can only be machted by an epistemology that relies on those logical features which the events we experience convey to us.

The thesis of this paper is that pragmatism is one option to formulate such a methodology. For pragmatism is an epistemological and methodological theory about the truth-generating ability of some methods and strategies of an open process of research. There is no pragmatic methodology without the stress on ongoing processes of experimentation, experience and interpretation. Pragmatism's methodological message is: If you want to know the truth, you have to interpret your theoretical thoughts in terms of relevant particular cases and other practical consequences which follow from your theories. You have to take this theory-practice connection seriously because it is the only reality oriented option: It gives us the crucial, even decisive evidence for or against the viability and fruitfulness of your theoretical thought. The formal concept able to describe the epistemological lay-out and setting of pragmatism's methodology are all coming from the logic of relations.

Here, I will have to ignore many other important concepts. I will concentrate on the concept of half-order that is at the conceptual bases of the logic of relations and lattice-theory. Such a relation, adapted to epistemology, orders two beliefs p and q only if there is a connection between them which is transitive, anti-symmetric and reflexive. But the reshaping of such a formal concept does not just subject the domain of discourse to these formal requirements: It takes into account characteristic features of particular phenomena and connects them to a more comprehensive vision of how all things remain together. In particular, how human beings have to go about to acquire knowledge. For example, transitivity becomes the most important feature and a dynamic requirement. On the other hand, the property of reflexivity can be understood in terms of a special requirement for ordered beliefs: we have to assure the identity of the objects of the beliefs related to be in this way.

But how is it possible to use the concept of an order relation to connect the methodology of pragmatism with the construction of Peirce's visual logic the existential graphs? Whereas it makes sense to interpret the method of pragmatism as a set of requirements for relations between beliefs, it is by no means obvious that there is an order relation in visual perception.

So this will be the the first issue this paper addresses: Is it possible to show that implicit in seeing there are sometimes formal properties of relations at work? We want to show (1) that there is visual logic in which the same order relation is at work as in other type of knowledge. And (2) we assume that there is uniform realistic ontology of processes and events connecting the acts and objects of visual perception.

2 Formal properties of visual perception and the iconic logic of visual representations

Let us start with the case most important for our issue: Seeing that something happens. Take an example: Yesterday, I saw how a man was run over by a car. So I saw that there was a man and a car in the street and what happened between the two them. This can be described by expressing the relation using the verbal form "run" over was applicable to what I saw. In general, we perceive things, situations, and in this case an event by grasping the changing relations between variable forms and their properties in our visual field. What I was seeing was that two objects were coming into sudden contact with one another on the street. In addition, I saw a sequence of changing geometrical forms and neighborhood-relations in which these objects of sight stood for each other. What we identify as the same (e.g. the car, the man, a scenery) are the objects of a number of changing views and sights that together make up our experience of seeing that the event

of a man was run over by a car. Let us compress these reflections into our
first thesis about the visual perception of events:

> (TVP 1) What is normally involved in understanding what we
> see to be happening before us, is that we see how visual experi-
> ential contents can be interrelated in such a way that we are able
> to identify those objects which are independent of the individual
> visual experience itself.

One of the functions of linguistic representation and of signs of all sorts
in general is this: To interrelate experiential contents both of the author
and of the interpreter of the respective representation or sign. If we want
to represent explicitly how visual contents may be interrelated in communication, we have to represent this in a schematic, graphical form. This
form may be a sort of diagram or picture that represents the relation between some number of visual experiences. (TVP1) captures these features
of referential interrelatedness of visual perception with other sorts of representation. What constitutes the representative quality of a photograph
is the possibility of "seeing" how it fits into some of its characteristic relations between some *sequences* of visual representations of its topic. For this
reason, if you look at the foto of an accident you will sometimes not really
remember the specific visual properties (form, color, brightness etc.) and,
being asked, you will probably say such things as "I had not really saw the
picture but what it represented". So (TVP 1) has a twin in a corresponding
thesis about how we see and interpret pictures:

> (TVP2) When we look at pictures our understanding of them is
> never only determined by the visual experience of an *individual*
> picture but by the way in which a picture may possibly relate
> experiential contents to one another and to the represented topic
> (event, person, thing etc.).

If visual contents occur always embedded in their relation to one another
and to the objects we consider them to represent, what are the properties of
the relation between sequences of pictures and visual objects? One option
would be to represent the formal features of the relation between visual
experiences in visual form. This leads to the project of devising a system of
visual representation relations that allows for a number of different solutions
for the general shape of such a relation. Let us look at a simple option
of visual segregation in line with the logical realism. This is what I call a
system of *semantic pictures*. It is already implicit in the semantical attitude
of seeing described in (TVP 1 & 2):

We lay down rules how to construct pictures in such a way that the properties of the relation between the represented topics (events, persons, physical objects etc.) are represented by the relation between visual properties in the pictures. That is to say, by the forms, lines and points on the picture.

Obviously, many drawings, maps, flow diagrams, etc. are to some extent semantic pictures in this sense. In all these cases it holds that they are semantic pictures because the form of the relation between the objects represented "is the very form of the relation between the [...] corresponding parts of the diagram" [Peirce, 1931-1958, 4.530]. Whereas diagrams and maps are truly *semantic* pictures, they are not yet *logical* pictures. That is to say, they cannot represent the logical relations that may be said to hold between representations and their objects. In particular the order-relation that pragmatism prescribes for beliefs that stand in information-enhancing ordered relations to one another. But if the realistic approach is viable, the thing we have to do now is to stick to the semantic picture option and to represent the logical relation between propositional contents by visual forms:

> In a graphical logic, visual properties and shapes used are controlled by logical relations building on the semantical relation to the universe of discourse represented by the diagram. All inferential- and formation rules, then, are expressed as rules for the transformation of visual (graphical) forms. (E.g. the most basic rules being rules of insertion and deletion of visual forms.)

Indeed, for Peirce this representation of logical relations between beliefs in the visual format of a semantic picture is one of the most valuable achievements of his system of the existential graphs. This is what he calls its "iconical" character. Peirce's semiotic concept of an icon is a logically truth-value neutral concept to describe the way in which a sign may be related to an object without describing, but by encorporating it. An icon is defined as a quality which becomes a sign of something because it is possible that the sign's quality may be instantiated by an independently existing object. Neither the object's actual existence (this is what indicators assume) nor a general rule or habit of interpretation in community of sign-users is the crucial link for the semantic relation between sign and object.

Another way to describe the semantics of the icon is this: the icon is build upon and generalizes the relation of exemplification of a quality. The icon does not represent any possible objects *in persona*: It instantiates the possibility of an interpretative act that will single out an object having the quality which the icon instantiates. In this way "having the same quality" may be understood as a relation between objects – the sign and its object.

The icon makes a crucial contribution to the logical analysis of language and communication: From this perspective the logical semantics for concepts of relations and concepts of qualities can be identified. The concept of an icon assumes that visual qualities may be structured by formal relations implicitly guiding our activity of constructing or reading a diagram. In our everyday epistemic life, if the successful interpreter, looks at an icon, i.e. at a diagram, he does not see an object, but a possible logical form which might involve a relation to some individuals. Peirce, in tracing out the consequences of his logical realism, turns the icon into a representation of the form of the semantic relation. He claims that "a diagram [...] is [...] an icon of the forms of relations in the constitution of its Object, the appropriateness of it for necessary inference is easily seen" [Peirce, 1931-1958, 4.531].

Now the next step in order is to look for a comprehensive graphical representation of logical relations that constitutes possible objects by their form. The first step in the formation of a semantic system of visual logic is the introduction of a sign that visually represents a comprehensive, unspecified range of logical relations for all kinds of possible objects. Perhaps one of the most creative moves in the design of his existential graphs was Peirce's solution of this first problem: Peirce introduced a two-dimensional surface – called the "sheet of assertion" or "Phemic Sheet" – as a graph that stands for each and every logical relation in a universe of discourse:

Figure 1. The sheet of assertion.

May be, you are irritated by the fact that the *empty* sheet can represent anything at all. Take a moment to think about it: Could you imagine a better image for all the possible relations that could be represented by a surface then the blank sheet? Therefore, the blank sheet is the crucial first move to start a graphical logic *ab ovo*. It is introduced as a sort of semantic axiom, stating that the representation of an unlimited number of specifiable relations in two dimensions is possible. Surely, on any two-dimensional surface we may inscribe any number of graphical forms. But the sheet in diagram 1 is introduced to represent those logical relations which hold between the objects represented.

> *The sheet itself is the first and most important logical graph. It is a graph of the semantic relation that all other logical expression build on*: It visually presents the form of the relation by which logical thought instantiated in the visual perception of the graph and in the universe of objects.

The sheet of assertion performs the visual representation and instantiation of logical relations in one step, combining them into one perceptual object. Fascinated by this invention, Peirce uses quite a number of impressive concepts to describe the semantic role of the epistemic unity between representation and instantiation: "[the sheet of assertion, H.P.] iconizes the Universe of Discourse, since it more immediately represents a field of Thought, or Mental Experience, which is itself directed to the Universe of Discourse, and considered as a sign, denotes that Universe. Moreover, it (is because it must be understood) as being directed to that Universe, that it is iconized by the Phemic Sheet" [Peirce, 1931-1958, 4.561].

This "iconic" role, the implicit epistemic and semantic function of visual properties in the sheet of assertion (ShA, for short), is crucial for the logic of the existential graphs. In what sense? The graphical approach connects the epistemology of reasoning about objects and their relation with the formal requirements of logic. Semantic relations are a case of ordered relations instantiated by visual representations of logical connections on the ShA:

> Whatever form G you draw on it is included by all the other visual forms that a) are on the sheet and b) are completely included in G.

Therefore, ShA visually represents the relation of logical inclusion and is able to signify the universe of discourse in which the logical relations represented are valid.

What is the analogy between this graphical logic and pragmatism? Pragmatism asks us to establish a goal-directed ordering of beliefs: Select those beliefs that connect us with practical consequences. But of course what makes up reality is the sum of all (even possible) practical consequences.

One of the main claims of this paper is now coming forward. I claim in the title that there is a pragmatic logic of the way in which representations are ordered. But how is it possible that the order, logical form and the form of relations is one and the same? As I said above, the concept that subsumes all three terms is the concept of a structured, transitive relation, close to what we call in formal logic a relation of half-order. It is involved when pragmatism asks us to construct a connection that holds between beliefs running from theoretical to practical beliefs. But the relation of transitive order may also be used regulatively, as a normative principle:

> For any sequence of signs, one interpreting the other, in every sequence of thoughts or other mental processes, in a sequence of visual signs, in movies or on TV, or the sentences in a book, some relation of order must be supposed to hold, in order to make the sequence readable – as a sequence of signs which, by the way they are connected, is related to a universe it represents.

And in fact we find that Peirce used the ordering requiring for relations as a normative principle for the construction and unification in his philosophical theories and proposals that specify the semantics of his graphical logic, when he says:

> [...] on the principle that logicians call "the Nota notae" that the sign of anything, X, is itself a sign of the very same X, the Phemic Sheet, in representing the field of attention, represents the general object of that attention, the Universe of Discourse. This being the case, the continuity of the Phemic Sheet in those places, where, nothing being scribed, no particular attention is paid, is the most appropriate Icon possible of the continuity of the Universe of Discourse – where it only receives general attention as that Universe – that is to say of the continuity in experiential appearance of the Universe, relatively to any object represented as belonging to it [Peirce, 1931-1958, 4.561].

This is a semantic requirement in terms of ordertheoretic properties assumed for all signs written on the sheet of assertion: It requires us to understand the semantical content of the sheet of assertion in terms of the relation of the quasi-order represented by all possible relations the *blank* sheet might contain. Because the blank sheet is a visual sign of our relation to the universe of objects – the blank sheet is and represents the "field of attention" – every graphical form written on it will necessarily *inherit* this semantical relation. In this way the semantic requirement is spelled out in terms of the transitivity property for syllogistic relations and the relation of universal predication in general: The represented form, the blank sheet, is related to every object in every individual universe of discourse – to which our attention is turned. Note, how Peirce combines here his epistemological phenomenalism with a formal account of experience in terms of relations.

What does the Aristotelean *nota-notae* principle mean here? This principle is sometimes also called *dictum de omni*. For us it is only important to note that the *nota-notae* principle is the concept in traditional logic that expresses an ordertheoretic property that the semantic relation of the sheet of assertion has: the transitivity and reflexivity of the relation between the visible sheet and its universe of discourse[1]. Peirce only claims, in traditional terminology, that the diagrams of his visual logic exhibit the same sort of transitively ordered referential relation which we for example assume in our everyday interpretation of photographs.

Let me state my main point in a slightly different way: the logical function of ShA is to represent iconically the form of the relation between the objects in its universe of discourse. In this way, a sort of implicit quantification is already present *before* we write any individual visual form on the sheet. It is defined as a two-dimensional surface that presents any number of specifiable geometrical inclusion relations between any logical signs that may be written on it and may represent the connected visual experience of any number of objects. And, because of the transitivity of the representation relation, it always represents the real objects of visual experience. In this way the order relation that holds between the real objects is represented by the blank ShA. The relation of the iconic logical form – of the two-dimensional suface of the sheet – to the possible objects in the real world invokes our experiential access to a universe of discourse only because the connection is transitive and reflexive[2].

---

[1] In fact, when Aristotle in his *Categories* formulates this principle he describes it as a semantic principle connecting predicated contents to an object. For what he says can be read as saying: if something is predicated of an object, than everything which is said about this predicate holds for this object too (Cf. Aristotle, Categories, 3.ch., 1b).

[2] Reflexivity of the logical relation, called "principle of identity" by him, Peirce required at other places, e.g. 4.348. There he said: "In order to form a system of graphs which shall represent ordinary syllogisms, it is only necessary to find spatial relations analogous to the relations expressed by the copula of inclusion [i.e. a sort of generalized conditional connective; H.P.] and its negative and to the relation negation". In the next sentence he formulate the crucial properties for the graphical logic: "Now all the formal properties of the compula of inclusion are involved in the principle of identity [i.e. reflexivity, in Peirce's notation, every X is an r to an X; H.P.] and the *dictum de omni* [i.e. transitivity, in Peirce's notation, if every X is r to a Y, and every Y is r to a Z, every X is r to a Z]" [Peirce, 1931-1958, 4.348].

Summing up Peirce's basic ideas for logic of the existential graphs we may say that the validity of graphical logic depends on the fact that the graphical forms express logical relations that represent, by a transitive relation between objects, experience and thought. How does Peirce's ordertheoretic semantics of iconicity apply to the relation of the graphical shapes and lines written on it? All graphical forms inherit the ordering that two-dimensional continuous manifold allows for. Second, the inclusion relation between graphical forms written on the sheet allows for a transitive ordering of logical signs which refer to objects by connecting them to the logical graph we look at. To see this in more detail, look at the logical graph representing a conjunctive assertion of two sentences visually:

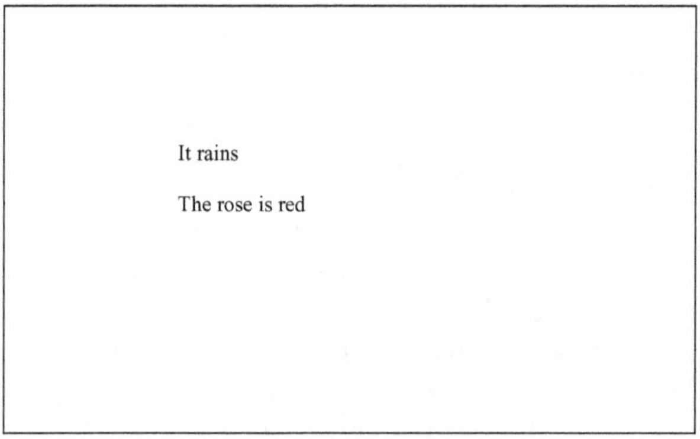

It rains

The rose is red

Figure 2. Conjunction.

Given the definition of the sheet of assertion by its representation of transitive ordering relative to its universe of discourse, every individual proposition written on the sheet inherits the transitive relation to that universe. Therefore, diagram 2 can now be read as saying that there is a universe of discourse represented by the sheet of assertion of diagram 2 such that the two assertions "It rains" and "The rose is red" are true in it. The writing of any number of propositions on the same sheet always means that they are true in the same universe of discourse. The mere graphical occurrence of two propositions is sufficient to express conjunction.

Given that the sheet of assertion is understood as the sign of an area of transitively structured relations, it is obvious that negation can be expressed simply by cutting the connection between a proposition and the sheet on which it is written. The graphical equivalent to negation Peirce therefore called "the cut" and it is for this reason, that the complete enclosure of a proposition symbolizes diagrammtically the negation of the conjunction of diagram 2, that is, of "It is not the case that it rains and the rose is red":

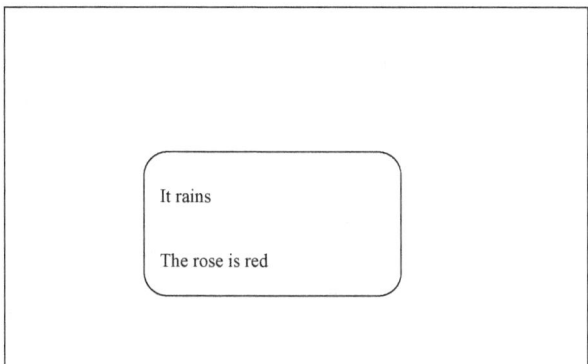

Figure 3. Negation.

Now, conjunction and negation taken together are sufficient to express all other logical connectives of the propositional calculus. Just to demonstrate the expressive power of the Existential Graphs which lets you *see* the equivalence between "If it rains, the rose is red" and "It is not the case that: it rains and the rose is not red", let me give you the symbolization of the material conditional:

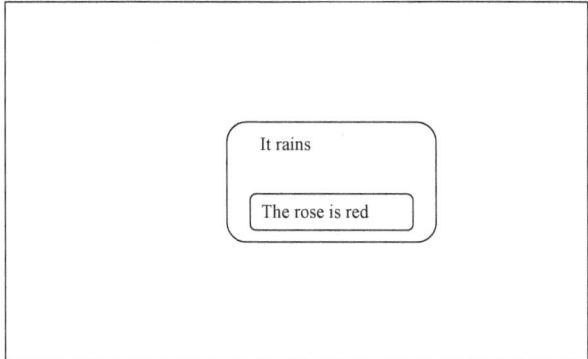

Figure 4. The material conditional.

3 The formal and the systematic connection between pragmatism and visual logic

Peirce's implicit use of ordering properties of relations between signs and their objects is the backbone of the semantics and syntax of his graphical logic. But how does the order relation apply to pragmatism? On a first glance, there seems to be no obvious way in which the order of relations are relevant to the methodological issues of pragmatism.

Pragmatism is many things to many people. But let us ask whether the relation between theoretical and practical beliefs must encorporate any formal properties that are of crucial importance for the task the pragmatist wants to achieve. Our point of departure is the principle of pragmatism that Peirce formulated in 1878, the Pragmatic Maxim (PM for short), which for him was a "rule for attaining [...] clearness of apprehension": "Consider what effects, that might conceivably have practical bearings, we conceive the object of our conception to have. Then, our conception of these effects is the whole of our conception of the object" [Peirce, 1931-1958, 5.402].

First, note that the issue PM talks about *is not* meaning or verification (or a theory of it) but how *objects* are related to or may be subsumed under various concepts. The natural habitat of concepts are for Peirce the beliefs we have about them. Every concept occurs in propositions and is explicated in arguments and inferences. According to the implicit reading of the meaning of theoretical concepts, PM tells us to connect a concept of "practical bearings" with a theoretical concept to clarify our conception of an object.

Second, concepts can be explicated by beliefs of different sorts and in some cases their difference is crucial for the way in which our thought can be clarified: A lot depends on the instances of "practical bearings" we use to explicate the implicit meaning of our theoretical notion of some objects. Therefore, what PM really gives us, is a methodological rule that calls for the construction of a very specific kind of relation between beliefs: We are asked to relate more general, theoretical beliefs to more practical ones about specific actions, individual objects and perceptions that are supposed to clarify the theoretical ones in relation to the objects that they apply to. For this reason, PM is a methodological rule that will establish a very specific kind of relation between *a) different types of beliefs*, namely, *b) theoretical*, fairly *general and therefore risky beliefs*, on the one hand, and *c) practical*, more concrete and rather *sound beliefs about perceptions and actions* on the other. My thesis is: this "specific relation between beliefs" must instantiate, to some extent, a transitive, reflexive relation. But what are the reasons for this ordertheoretic properties to apply here?

Note that any process of clarification is a goal-directed teleological one:

We want to achieve a predefined goal, namely clarity, precision in explicating concrete criteria or material conditions under which more information about an object can be known. To clarify a theoretical belief, you are advised to think about it in terms of its possible consequences and applications: If we can find logical relations that connect these two kinds of beliefs, then we have a good chance to clarify our high-risk theoretical thought in terms of the more concrete information about specific applications or other practical consequences. So this is the sort of connection between beliefs that P.M. asks us to construct: We need a series of knowledge-extending relations between different sort of beliefs that run in one direction: *From a more general notion of an object to more specific and applicable ideas about the same object.* But all relations of identity are ordered relations.

BIBLIOGRAPHY

[Peirce, 1931-1958] C.S. Peirce. *Collected Papers of Charles Sanders Peirce*. Harvard University Press, Cambridge, MA, 1931-1958. vols. 1–6, Hartshorne, C. and Weiss, P., eds.; vols. 7–8, Burks, A.W., ed.

[Peirce, 1960] C.S. Peirce. *The New Elements of Mathematics*. The Hague, Paris, 1960.

Helmut Pape
Forschungsinstitut für Philosophie
Universität Bamberg, Hannover
E-mail: helmut_pape@web.de

Agent-Based Abduction
Being Rational through Fallacies

LORENZO MAGNANI AND ELIA BELLI

ABSTRACT. In this paper we will try to describe the close relation between logical fallacies and abduction considered as inference to the best explanation. To this aim we will exploit the so-called *agent-based reasoning* framework, which adopts the perspective of a cognitive agent. Then we will analyze some types of reasoning that in the perspective of classical and informal logic are defined *fallacies*. We will describe how in an agent-based reasoning this kind of *fallacious* reasoning can in some cases be redefined and considered as a good way of reasoning. In treating this issue we will also refer to the concept of *abduction*; from the point of view of the classical logic, abduction is a kind of formal fallacious argument – the well-known fallacy of affirming the consequent – even if it is also recognized is a very precious method of explanation and discovery in science and in everyday reasoning (for example abduction is able to govern inconsistencies that arise in scientific domains, when anomaly is treated through the generation of new hypotheses. Finally we will illustrate how in a celebrated form of practical reasoning, moral deliberation, the concept of abduction can demonstrate a central cognitive attraction also permitting to overcome some sterile consequences of the old-fashioned assumption that in the tradition of moral philosophy prohibits to derive an "ought" from an "is".

1 Agent-based reasoning, agent-based logic, abduction

It is well-known that in classical logic a good argument is a sound argument and, from a semantic point of view, it is a valid argument based on true premises. Even if this conception of good inference is usually able to model many kinds of argumentation of real human beings, its appeal to true premises is ill suited to many contexts which are often characterized by the presence of hypothetical and uncertain beliefs, by great disagreement about what is true and false, by ethical and aesthetic claims which are not easily categorized as true or false, and, finally, by variable contexts in which dramatically different assumptions may be accepted and rejected.

We share with Gabbay and Woods [2005] the idea that logic can be considered a formalization of what is done by a cognitive agent. Starting from this perspective, logic is *agent-based* [Gabbay and Woods, 2001]. Agent-based reasoning consists in describing and analyzing the reasoning occurring in problem solving situations where the agent access to cognitive resources encounters limitations such as

- bounded information

- lack of time

- limited computational capacity.

Hence, the "beings-like-us" that Woods describes in his "Epistemic bubbles" [Woods, 2005] discharge their cognitive agendas under press of incomplete information, lack of time, and limited computational capacity. We can consequently say that cognitive performances depend on information, time, and computational capacity. An *agent-based logic*, as a discipline that furnishes ideal descriptions of *agent-based reasoning*, returns to be thought of as a science of reasoning and considered agent-centered, task-oriented, and resource-bound. Woods says:

> So, then, a principal function of reasoning is to facilitate cognition, this means the reasoning agent is also a cognitive agent. If logic is to press forward as a renewed science of reasoning, it would do well to reflect on what cognitive agency is like, on what it is like to be a knower [Woods, 2005, p. 732].

In dealing with these features we arrive to what has been called the "Actually Happens Rule" [Woods, 2005, p. 734] that states that "to see what agents should do we should have to look first at what they actually do and then, if there is particular reason to do so, we would have to repair the account". This rule is a particular attractive assumption about human cognitive behavior mainly for two reasons. The first is that beings like us make a lot of errors; the other is that cognition is something that we are actually very good at.

In the following section we will discuss the case of "fallacies" as errors that people make. These errors occur in ways of reasoning and acting that from some perspectives are good and from others are bad. In dealing with this matter we will try to give an account of fallacies seen from the viewpoint of agent–based reasoning. We will try to give some examples of fallacious reasoning treating both informal fallacies (such as the inductive ones like "hasty generalization") and formal fallacies (such as abduction). We will treat induction (in the case of hasty generalization), and abduction as fallacious ways of reasoning that in spite of their fallacious character are fruitful for the cognitive agent: a way of being rational through fallacies. Secondly, we will analyze the problem of the role of "reasons" in the perspective of abduction offering an account of what happens in the typical agent-based situation of moral reasoning, a cognitive situation particularly sensitive to abduction.

Abduction can be easily considered in the perspective of agent-based reasoning because in abductive reasoning [Magnani, 2001b] both the activity of guessing new explanatory hypotheses (or new "reasons", in the case of moral deliberation, see below section 4), and the activity of selecting already existing ones, is based on incomplete information. In this case we deal with "nonmonotonic" inferences: we draw defeasible conclusions from incomplete information.[1] From this perspective, abductive reasoning also represents a prototypical case of practical reasoning (like in the case of moral reasoning we will illustrate below in this paper): we adopt deliberations based on incomplete information and on particular abduced hypotheses – guesses – that serve as "reasons".

2   Beings-like-us as hasty generalizers

As already noted, people make errors in reasoning. This means that in analyzing the beings-like-us argumentations we have to face problems regarding agent's access to cognitive resources such as information, time, and computational capacity, and logical attributes such as truth-preservation. It is in this sense that we have previously said that agents discharge their cognitive agendas under press of bounded information, lack of time and limited computational capacity.

The successful use of fallacies into many kinds of reasoning can be fruitfully accounted for in the framework of *agent-based reasoning*. It is undeniable that in human reasoning mistakes are widespread. The peculiarity of fallacies seen in the perspective of agent-based reasoning is that mistakes that are actually committed are mistakes that do not seem to be mistakes to those who commit them. In some sense we can say that they are ways of reasoning that are felt truth preserving for the reasoner but are not considered truth preserving for the logicians!

A fallacy is a pattern of poor reasoning which appears to be a pattern of good reasoning [Hansen, 2002]. Fallacies are forms of reasoning and ar-

---

[1]It has to be noted that an explanatory reasoning can be causal, but explanations are also based on other aspects. Thagard [1992, chapter five] illustrates various types of explanations (deductive-nomological – so-called in the neopositivist tradition of philosophy of science – statistical, causal, analogical, schematic).

gumentation typical of organic agents and in this sense we can say they are suitably shaped by evolution. Simple inductions and abductions performed more or less consciously by both humans and animals are surely two great results of this evolutionary process. Two main disciplines respectively clearly illustrate different kinds of fallacies: formal logic, which recognizes and explains "formal fallacies", and informal logic, that describes the so-called "informal fallacies". First of all, we can say that the validity of a deductive argument depends on its form, consequently, formal fallacies are arguments which have an invalid form and are not truth preserving (for example the fallacy of the "affirming the consequent" and of "denying the antecedent"). On the other hand, informal fallacies are any other invalid modes of reasoning whose failing is not strictly based on the type of the argument (for example the "*ad hominem* argument" or the "hasty generalization").

Even if there is no agreement upon an established taxonomy, the fallacies discussed in *informal* logic contexts typically still include formal fallacies (of course also discussed in *formal* logic) such as the famous fallacy of affirming the consequent[2] and the fallacy of denying the antecedent, but also the proper informal fallacies such as "*ad hominem*" (against the person), "slippery slope", "*ad baculum*" (appeal to force), "*ad misericordiam*" (appeal to pity), and "two wrongs make a right".

From the point of view of classical logic a fallacy is a bad argument that looks good (cf. Figure 1). From the point of view of agent-based reasoning a fallacy is not an argument that looks good but is bad, but an argument that is bad in some aspects and good in some others. Let us consider the inductive case of the so-called "hasty generalization", that can lead the cognitive agent – in spite of its fallacious character – to fruitful outcomes.

This fallacy occurs when a person (but there evidence of it also in animal cognition, for example in mouses) infers a conclusion about a group of cases based on a sample that is not large enough. It has the following form:

- Sample S, which is small, is taken from the group of persons P.

- Conclusion C is drawn about the group P based on S.

It could take also the form of:

- The person X performs the action A and has a result B.

- Therefore all the actions A will have a result B.

---

[2] In the light of classical logic abduction coincides with the fallacy of affirming the consequent (cf. below section 3).

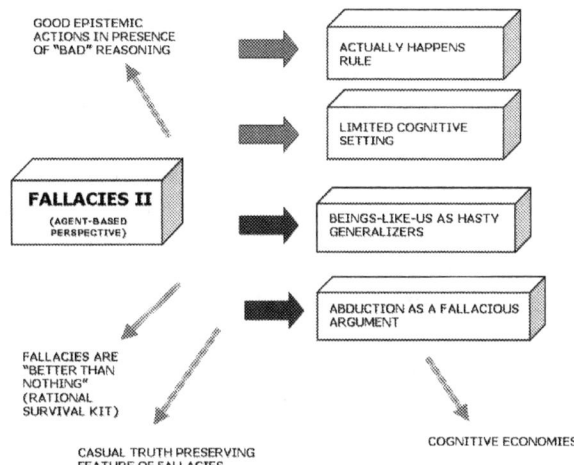

Figure 1. Fallacies from the point of view of classical logic.

The fallacy is committed when not enough A's are observed to warrant the conclusion. If enough A's are observed then the reasoning is not fallacious, at least from the *informal* point of view. Males, when driving their cars, have probably quarreled with a woman driving her car and, while quarreling, they have argued (when not shouted) "all woman are bad drivers!" This is our case of fallacious reasoning.

Insofar, small samples will be likely to be unrepresentative. Another simple case is the following. If we are asking one person that even recently met a lot of Italians what he thinks about the recently new established Italian proportional electoral system, his answer clearly would not be based on an adequate sized sample for determining what Italians in general think about the issue. This is because the answer given is based only on a reduced experience and that judgment can not be relevant in dealing with a generalization about the matter in question. This means that this fallacious argument implies that small samples are less likely to contain numbers proportional to the whole group of cases.

People often commit hasty generalizations because of bias or prejudice. For example, someone who is a sexist might conclude that all women are unfit to fly jet fighters (or to drive a car) because one woman crashed in either case. People also commonly commit hasty generalizations because of laziness or sloppiness. It is very easy to simply jump to a conclusion and much harder to gather an adequate sample and draw a justified conclusion. Thus, avoiding this fallacy requires minimizing the influence of bias and taking care to select a sample that is large and meaningful enough.

Moreover, we can recognize another important occurrence. We have said that people commit errors and are hasty generalizers because of prejudice, mindlessness, bias, and so on. What we are trying to underline is that the hasty generalization is not always a bad generalization for two reasons. The first is that, getting true conclusions, hasty generalizations might be good if the result of the generalization we made coincides with the *result* of a good generalization in the philosophical sense of induction, or in the sense of inductive logics. We call this case "casual" truth preserving feature of hasty generalization. The second reason is that, in some sense, even if we do not reach good conclusions, not exploiting the casual truth preserving feature, we can say that hasty generalization is good in some sense, obviously not in the classical logic one. We will now try to understand what it can be.

Think of a toddler that for the first time touches a stove in his kitchen [Woods, 2004]. His finger is now burned because the stove burns. Starting from this evidence, the hasty generalizer toddler thinks that all the stoves are hot and decides not to touch stoves anymore. This is obviously a hasty generalization:

- X of observed A are B (The stove *touched* burns).

- Therefore all A are B (*All* the stoves burn.)

Or:

- Sample S, which is too small, is taken from the group of persons P. (The toddler touches the stove and at a first touch the stove burns).

- Conclusion C is drawn about the group P based on S. (All the time the toddler will touch the stove, it will burn).

We can also say that this is a case of bad argument also from the formal point of view because it is not truth preserving, in the light of classical logic. However, in the perspective of agent-based reasoning the problem now is: can we say that this argument is good from some perspective? Indeed the hasty generalization is sometimes a "prudent" strategy. It also presents a cognitive economy: given the task of not being burnt for a second time, the hasty generalization is a kind of reasoning that is fruitful because, being a prudent strategy, it embeds the canons of *strategic rationality* in the sense of the "strife for survival". Moreover, it also involves a *cognitive success*.

Hence, it is clear that fallacies (hasty generalization in this case) have some relevant relations with strategic rationality. However, the prudent strategy of "not touching the stove" is obviously incorrect for at least two reasons: 1) the first reason is that it is not good to generalize from only one sample available and 2) the second reason resorts to the fact that the causal relationship between the fact that a stove burns if it is overheated is not discovered by the toddler. From applied natural physics, we can say that it is a state of affairs that a simple stove (not overheated) does not burn because a stove is made of iron or some other metals and metals burn only if they are overheated. So there is something "bad" in this kind of *fallacious* reasoning certainly – first reason – from an informal logic point of view but also – second reason – from the perspective of natural physical principles of heat, that constitute the only epistemologically satisfactory knowledge of the situation. But even if we acknowledge these wrong aspects, there is an idea of some rationality embedded in this example due to the fact that the toddler prevents himself from being burned. It seems that hasty generalization (like in the case of other fallacies, such as the fallacy of affirming the consequent can be considered resources that form a sort of *human survival kit* [Woods, 2004, p. 7]. As some unconditioned reflexes, hasty generalization is a response (in the form of a reasoning and then of a possible subsequent action) to something that the toddler is involved to. The cognitive result of a hasty generalization is bad but only in the sense that it is based only on one sample and because it does not *explain* the burnt stove. It is instead a form of good reasoning because it preserves the toddler from being burnt another time.

We have also to note that hasty generalization also allows the toddler to produce a new *successful cognitive information*. In the perspective of the logical tradition, this piece of information is "bad" because obtained through fallacious reasoning, but in agent-based terms we notice that the same information contributes to solve the toddler's problem and, in this sense, can be endowed with "good" cognitive relevance.

The goal reached by the toddler through the hasty induction is in some sense a success. In describing agent-based reasoning, we have said that agent-based logic (as a matter analyzing agent-based reasoning) returns to be thought of as science of reasoning and considered agent-centered, task-oriented and resource-bound. The task of the toddler was to prevent himself from being burnt: the task is reached with a bad performance, in the perspective of classical logic. We can say that his fallacious strategy "satisfies" his purpose of not being burnt; we can also say not only that hasty generalization halts upon the attainment of the good-enough, but also reaches a specific target that responds to a survival strategy. In some sense, the toddler's reasoning in hasty generalization terms can be seen as a very good strategy that satisfies his purpose in an optimizing way. From this point of view, the study of fallacies allows us to think about the interplay between reasoning, strategy, goal, and evaluation of goals (cf. Figure 2).

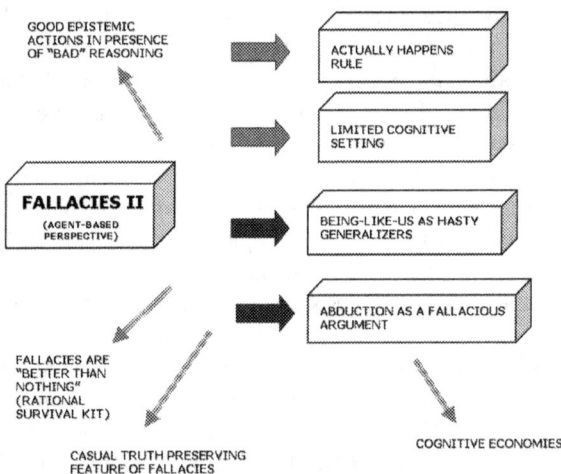

Figure 2. Fallacies from the point of view of agent-based reasoning. The big dark arrows show the examples of fallacious reasoning and the other two arrows underline the setting in which we are moving. The thin arrows display the possible implication of fallacious reasoning considered in agent-based term.

For a better understanding of the matter, we can follow the Kirsh and Maglio's dichotomy: they distinguish actions into two categories: *pragmatic actions* and *epistemic actions* [Kirsh and Maglio, 1994]. Pragmatic actions are the actions that an agent performs in order to arrive as close as possible to a goal. The environment in which the agent moves is modified so that it assumes a configuration that helps the agent to reach the physical goal. On the other hand, epistemic actions are the actions that an agent performs in the environment in order to discharge the mind of a cognitive load or to extract information that is hidden or that would be very hard to obtain only by internal computation (cf. Figure 3).

The toddler's action of touching the stove is a pragmatic action performed by the toddler in order to reach a goal. For example, we can think that the baby was trying to reach some biscuits on the fridge and in trying to reach

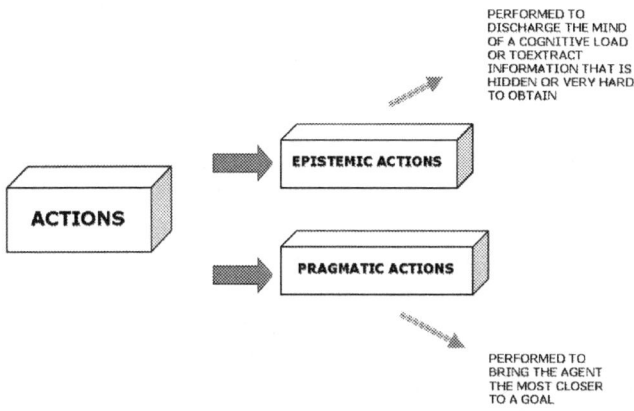

Figure 3. Epistemic and pragmatic action.

them, he touched the burning stove. But, on the other side, the toddler's action is also an unexpected but welcome epistemic action. In fact, it furnishes the basic information on which the inductive hasty generalization can be based. As we have seen in the perspective of the logical tradition the reached generalization is "bad" because it is a form of fallacious reasoning, but in agent-based terms the same information contributes to solve the toddler's problem of not being burnt for a second time and, in this sense, can be endowed with "good" cognitive relevance. In other words, we can say that, in this case, the action that is the basis of a fallacious reasoning is a good epistemic action, but also a good pragmatic action (in the case the toddler has previously reached the biscuits on the fridge).

We are trying to say that fallacious forms of reasoning can be good not only for practical purposes, such as *task-oriented reasoning*, in solving here-and-now problems. The question we are gathering is if action deriving from fallacious reasonings can receive an epistemic endowment. In fact we can say that, from the situation previously experienced through the "touching the stove" action, the toddler obtains new information and creates a strategy that prevent himself of not being burned more seriously for a second time. His fallacious reasoning (hasty generalization, in this case) suggests that he should not touch the stove for at least two reasons. Practically because this *touching* action has already given him bad results, and theoretically because he made a logical inference that governs his way of future acting, no matter if the inference is bad from a classical logical point of view. We maintain that fallacies also deal with epistemic action, in the sense that they can exploit them to extract new information.

We have contended above that fallacies are forms of reasoning and argumentation typical of organic agents and in this sense we have concluded they are part of a "survival kit" suitably shaped by evolution. We have also added that induction and abduction performed more or less consciously by both humans and animals are surely two great results of this evolutionary process. We know that in the last centuries humans have also characterized induction (and abduction) in various "ideal" philosophical and logical ways, so going beyond the spontaneous use of those kinds of thinking we have just illustrated. Already Mill provided "Methods" for Induction and Peirce integrated abduction and induction through the famous syllogistic framework where the two non-deductive inferences can be clearly distinguished: it has to be noted that Mill also said that what he called "institutions" rather than individuals are the embodiment of "inductive logics". Following this Millian perspective [Mill, 1843], Gabbay and Woods also add that it is typical of human individuals to function as practical agents and that it is typical of "institutions" to function as theoretical agents [Gabbay and Woods, 2005, p. 14]; moreover, agents tend toward enhancement of cognitive assets when this enables the achievement of cognitive goals previously unaffordable or anattainable.

In summary, organic agents like human beings are hasty generalizers and more or less naive inducers and abducers but are also the creators of external cognitive and logical representations that for example provide deductive and computational realizations of those reasoning performances. The interplay between these "external" tools and the already "internalized" templates of reasoning certainly realizes a continuous improvement of the internal templates themselves but also expresses the centrality of the hybrid exploitation of both levels in reasoning.

3 Abduction as a fallacious argument

It is well-known that abduction appears to be a formal fallacy that can be recognized from the classical logic point of view: the fallacy of affirming the consequent (cf. below) [Peirce, 1931–1958]. However, from the point of view of both everyday and scientific knowledge, abduction is an important kind of inference used to explain facts and invent hypotheses and theories [Magnani, 2001b].

Abduction – a distinct form of reasoning – is the process of *inferring* certain facts and/or laws and hypotheses that render some sentences plausible, that *explain* or *discover* some (eventually new) phenomenon or observation; it is the process of reasoning in which explanatory hypotheses are formed and evaluated. There are two main epistemological meanings of the word abduction [Magnani, 2001b]: 1) abduction that only generates "plausible" hypotheses ("selective" or "creative") and 2) abduction considered as inference "to the best explanation", which also evaluates hypotheses (cf. Figure 4). An illustration from the field of medical knowledge is represented by the discovery of a new disease and the manifestations it causes which can be considered as the result of a creative abductive inference. Therefore, "creative" abduction deals with the whole field of the growth of scientific knowledge. This is irrelevant in medical diagnosis where instead the task is to "select" from an encyclopedia of pre-stored diagnostic entities. We can call both inferences ampliative, selective and creative, because in both cases the reasoning involved amplifies, or goes beyond, the information incorporated in the premises.

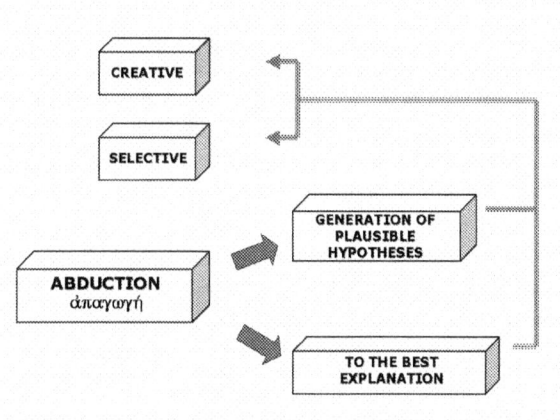

Figure 4. Creative and selective abduction.

Magnani [2001b] has introduced the concept of *theoretical abduction* as a form of neural and basically internal processing. He maintains that there are two kinds of theoretical abduction, "sentential", related to logic and to verbal/symbolic inferences, and "model-based", related to the exploitation of models such as diagrams, pictures, etc. (cf. Figure 5).

Theoretical abduction certainly illustrates much of what is important in creative abductive reasoning, in humans and in computational programs,

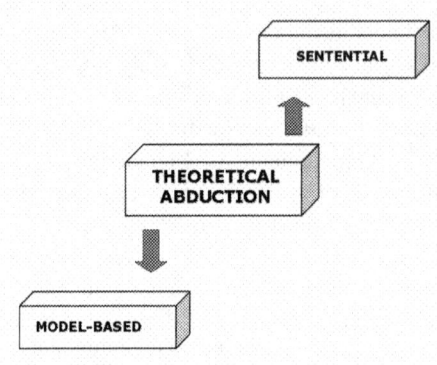

Figure 5. Theoretical abduction.

but fails to account for many cases of explanations occurring in science when the exploitation of environment is crucial. It fails to account for those cases in which there is a kind of "discovering through doing", cases in which new and still unexpressed information is codified by means of manipulations of some external objects called *epistemic mediators* [Magnani, 2001b]. The concept of *manipulative abduction*[3] captures a large part of scientific thinking where the role of action is central, and where the features of this action are implicit and hard to be elicited: action can provide otherwise unavailable information that enables the agent to solve problems by starting and by performing a suitable abductive process of generation or selection of hypotheses.

Let us now illustrated the fallacious character of abduction. Since the time of John Stuart Mill, the name given to all kinds of non deductive reasoning has been induction, considered as an aggregate of many methods for discovering causal relationships. Consequently induction in its widest sense is an ampliative process of the generalization of knowledge. Peirce distinguished various types of induction: a common feature of all kinds of induction is the ability to compare individual statements: by using induction it is possible to synthesize individual statements into general laws – inductive generalizations – in a defeasible way, but it is also possible to

---

[3]Manipulative abduction and epistemic mediators are introduced and illustrated in [Magnani, 2001a] and [Magnani, 2001b].

confirm or discount hypotheses.

Following Peirce, we am clearly referring here to the latter type of induction: abduction creates or selects hypotheses; from these hypotheses consequences are derived by deduction that are compared with the available data by induction. This perspective on hypothesis testing in terms of induction is also known in philosophy of science as the "hypothetico-deductive method" [Hempel, 1966] and is related to the idea of confirmation of scientific hypotheses, predominant in neopositivistic philosophy but also present in the anti-inductivist tradition of falsificationism [Popper, 1959].

Deduction is an inference that refers to a logical implication. Deduction may be distinguished from abduction and induction on the grounds that the truth of the conclusion of the inference is guaranteed by the truth of the premises on which it is based only in deduction. Deduction refers to the so-called non-defeasible arguments. It should be clear that, on the contrary, when we say that the premises of an argument provide partial support for the conclusion, we mean that if the premises were true, they would give us good reasons – but not conclusive reasons – to accept the conclusion. That is to say, although the premises, if true, provide some evidence to support the conclusion, the conclusion may still be false (arguments of this type are called inductive, or abductive, arguments).

All these distinctions need to be exemplified. To describe how the three inferences operate, it is useful to start with a very simple example dealing with diagnostic reasoning and illustrated (as Peirce initially did), in syllogistic terms:

1. If a patient is affected by a pneumonia, his/her level of white blood cells is increased.

2. John is affected by a pneumonia.

3. John's level of white blood cells is increased.

(This syllogism is known as Barbara).

By deduction we can infer (3) from (1) and (2). Two other syllogisms can be obtained from Barbara if we exchange the conclusion (or Result, in Peircian terms) with either the major premise (the Rule) or the minor premise (the Case): by induction we can go from a finite set of facts, like (2) and (3), to a universally quantified generalization – also called categorical inductive generalization, like the piece of hematologic knowledge represented by (1) (in this case we meet induction as the ability to generate simple laws, contrasted with induction as a way to confirm or discard hypotheses, cf.

above). Starting from knowing – selecting – (1) and "observing" (3) we can infer (2) by performing a selective abduction. The abductive inference rule corresponds to the well-known fallacy called *affirming the consequent* (simplified to the propositional case)

$$\frac{\varphi \rightarrow \psi}{\varphi}$$

It is useful to give another example, describing an inference very similar to the previous one:

1. If a patient is affected by a beta-thalassemia, his/her level of hemoglobin A2 is increased.

2. John is affected by a beta-thalassemia.

3. John's level of hemoglobin A2 is increased.

Such an inference is valid, that is not affected by uncertainty, since the manifestation (3) is pathognomonic for beta-thalassemia (as expressed by the biconditional in $\varphi \leftrightarrow \psi$). This is a special case, where there is no abduction because there is no "selection", in general clinicians very often have to deal with manifestations which can be explained by different diagnostic hypotheses: in this case the inference rule corresponds to

$$\frac{\varphi \leftrightarrow \psi}{\varphi}$$

It is now clear that in the light of classical logic abduction is a fallacy. Despite this fallacious character it is of fundamental importance in many agent-based reasoning situations like scientific explanation, scientific discovery, and also in moral deliberation, as we will illustrate below. We can furnish another reason that stresses the fruitfulness of abduction in agent-based reasoning: it is a powerful inferential process able to govern inconsistencies. For example, in the case of the formation of new scientific hypotheses and theories epistemologists have always recognized the role played by inconsistencies and anomalies that violate the paradigm-induced expectations derived from previously established conceptual frameworks. Logicians have in turn shown that inconsistencies generated by anomalies are difficult to be managed in deductive frameworks: they are unexpected facts that the rules of classical logic are not able to manage.

In fact, if we consider *deduction* in scientific reasoning, an anomaly in the system is a problem that logical hypotheses will explain with difficulty: *classical* deduction has no sense in giving reasons for unexpected or unknown facts. In a deductive system, all premises are given and new information that contradicts the previous one clashes against the classical law of *ex falso sequitur quodlibet*.

On the contrary, abduction starts from the philosophical assumption that information is incomplete and this retroductive method shows that it is a fallacy if considered in the framework of classical deduction. In other words, in a classical logic system, adding *new* information to the old one does not let us to revise previous assumed hypotheses. But if we start from incomplete information we will have to face the problem of inferring conclusions that are not foreseeable or already stated and that in the meantime contradict the previous ones. As a consequence of this we obtain that previous deduced conclusions can be invalidated and modified in presence of a new information. For these reasons logicians quickly recognized that abduction is a form of *non-monotonic* reasoning that only non-standard logic environments can grasp. Recently Gabbay and Woods [2005] have nicely reminded us that in the reasoning situations characterized by limited information reasoning cannot be anything but a fruitful *ignorance preserving reasoning*. In a logic able to deal with this fundamental aspect we have to clearly distinguish between the formation of hypotheses and the cogency of their validation[4].

Stating this, we can outline two different ways of perceiving abduction: 1) from the point of view of classical logic, abduction is a formal fallacy, not truth preserving; 2) from the point of view of everyday reasoning and epistemology, abduction is an important kind of reasoning able to discover new hypotheses and also to furnish explanations of scientific facts. In delineating the structure of a new agent-based perspective of logic able to encompass fundamental human reasoning devices like abduction Gabbay and Woods further point out that logic has to be considered an "account of how thinking agents reason and argue" [2005, p. 1]. As we have already mentioned their idea is that logic has to be considered the disciplined description of the behavior of real-life logical agents. Logic has to be thought of as an *agent-based logic*. From this viewpoint, abduction can be rendered as that kind of logical reasoning in which the fact of not being truth preserving (but *ignorance preserving*), has to live together with the fact that it is fruitfully used by real logical agents. Similarly induction is seen as *probability enhancing* and deduction as *truth preserving*.

The description of this new logical perspective is based on the assumption

---

[4]An example of ignorance-preserving reasoning is *diagnostic reasoning* in which abduction is useful plausible diagnoses and deduction for its validation.

that the reasoner reasons in a bounded information set typical of real agents and consequently on the redefinition of the significance of many forms of reasoning previously considered fallacious.

The new agent-based perspective on logic makes the logical system more open and flexible. In this sense the study of fallacies, as important tools for the human "kit" providing evolutionary advantages as well, allows us to maintain that, in some sense, a fallacious form of reasoning is better than nothing [Woods, 2004]. In the case of abduction, we can go beyond this definition and say that fallacious reasoning give us even more: in fact, the use of abduction is good for at least two reasons. Abduction is not only a simple formal fallacy, but also a specific case of ignorance-preserving *agent*'s reasoning that can be fruitfully idealized in theoretical logical agents; on the applicative side, abduction is a good process able to provide new hypothesis and govern inconsistencies.

4 The agent-based structure of reasons in moral deliberation

This section illustrates that "abduction", that is the reasoning to hypotheses, is central to the problem of "inferring reasons" in a fundamental kind of practical reasoning: decision making. Moral deliberation will be our example. We have seen that in abduction we usually base our guessing of hypotheses on incomplete information, and so we are facing nonmonotonic inferences: we reach defeasible conclusions from limited information, and these conclusions are always withdrawable[5]. It is in this sense that abductive reasoning constitutes a possible useful model of practical reasoning: ethical deliberations are always adopted on the basis of incomplete information and on the basis of the selection of particular abduced hypotheses which play the role of *reasons*. Hence, ethical deliberation shares some aspects with hypothetical explanatory reasoning as it is typically illustrated by abductive reasoning in scientific settings. To support this perspective on the "logical structure of reasons" we will provide an analysis based on the distinction between "internal" and "external" reasons and on the difficulties in "deductively" grasping practical reasoning, at least with the only help of classical logic.

In a previous work [Magnani, 2006b, chapters six and seven] devoted to introduce the methodological problems of ethical deliberation, Magnani contends, following Rachels, that morality is the effort to guide one's conduct by reasons, that is, to do what there are the best reasons for doing while giving equal weight to the interests of each individual who will be affected by one's conduct.

---

[5] For further specifications cf. above section 3, "Abduction as a fallacious argument".

"The logical structure of reasons" is the title of chapter four in Searle's book *Rationality in Action* [2001]. We plan to use Searle's conceptual framework to better understand what exactly are "reasons" in the specific case of ethics. Whereas Searle deals with rational decision making, many of his conclusions appear to be appropriate for ethical cases, too.

By criticizing the classical model of rational decision making (which always requires the presence of a desire as the condition for triggering a decision), Searle establishes the fundamental distinction between *internal* and *external* reasons for action: those that are internal might be based on a desire or on an intention, for instance, while external reasons might be grounded in external obligations and duties. When I pay my bill at the restaurant, I am not doing so to satisfy an internal desire, so this action does not arise from internal motivations; instead, it is the result of my recognition of an external obligation to pay the restaurant for the meal it has provided. Analogously, if an agent cites a reason for a past action, it must have been the reason that the agent "acted on". Finally, reasons can be for future action, and this is particularly true in ethics where they do not always trigger an action – in this case, however, they must still be able to motivate an action: they are reasons an agent can "act on".

Searle's anti-classical emphasis on "external reasons" does not have to appear strange: in [2006b] Magnani has often stressed the fact that human beings delegate cognitive roles (and moral worth) to external objects that consequently acquire the status of deontic moral structures. This also occurs when we articulate ideas in verbal statements – promises, commitments, duties, and obligations, for example – that then exist "over there", in the external world. Imagine the deontic role that concrete buildings (like for instance the ones whose shapes restrict routes people can follow) or abstract institutions (for example, constitutions usually compel us to consider equality of citizens as important) can play in depicting duties and commitments we can (or have to) respect. Human beings are bound to behave in certain ways as spouses, tax payers, teachers, workers, drivers, and so on. All these external factors can become – Searle says – reasons/motivators for prior intentions and intentions-in-action of human beings.

As it is illustrated in [Magnani, 2006b] many things around us are human made, artificial – not only concrete objects like a hammer or a PC, but also human organizations, institutions, and societies. Economic life, laws, corporations, states, and school structures, for example, can also fall into that category. We have also projected many intrinsic values on things like flags, justice rituals, or ecological systems, and as a result, these external objects have acquired a kind of autonomous automatism "over there" that conditions us and distributes roles, duties, moral engagements – that is, it supplies potential "external reasons". Non-human things (as well as so-to-say "non-things" like future human beings and animals, etc.) become moral clients as well as human beings, so that current ethics must pay attention not only to relationships between human beings, but also to those between human and non-human entities.

Moreover, we can observe how external things we usually consider to be morally inert can be transformed into those moral mediators which express the idea of a distributed morality. For example, we can use animals to highlight new, previously unseen moral features of other living objects, as we can do with the earth or with (non natural) cultural objects; we can also use external "tools" like writing, narratives, others persons' information, rituals, and various kinds of institutions to morally reconfigure social orders. Hence, not all moral tools are inside the head along with the emotions we experience or the abstract principles we refer to; many are over there, even if they have not yet been identified and represented internally, distributed in external objects and structures which function as ethical devices available for acknowledgment by every human agent. These delegations to external structures – thus transformed in moral mediators - encourage or direct ethical commitments, and, they favor the predictability in human behavior that is the foundation for conscious will, free will, freedom, and of the ownership of our own destinies: if we cannot anticipate other human beings' intentions and values, we cannot ascertain which actions will lead us to our goals, and authoring our own lives becomes impossible.

Let us return to the role played by reasons in ethical reasoning. Intentional states with a propositional content have typical *conditions of satisfaction* and *directions of fit*.

1. First, mental and linguistic entities have directions of fit: for example, a belief has a *mind-to-world* direction of fit. For example, if I believe it is raining, my belief is satisfied if and only if it is raining "because it is the responsibility of the belief to match an independently existing reality, and it will succeed or fail depending or whether or not the content of the belief in the mind actually does fit the reality of the world" [Searle, 2001, p. 37]. On the other hand, a desire (or an order, promise, or intention) has a *world-to-mind* direction of fit: "if my belief is false, I can fix it up by changing the belief, but I do not in that way make things right if my desire is not satisfied by changing the desire. To fix things up, the world has to change to match the content of the desire" [Searle, 2001, p. 38].

2. Second, other objects (not mental and not linguistic) also have a direction of fit similar to the ones of beliefs. A map, for example, which

may be accurate or not, has a *map-to-world* direction of fit, whereas the blueprints for a house have a *world-to-blueprint* direction of fit because they can be followed or not followed [Searle, 2001, p. 39]. Needs, obligations, requirements, and duties are not in a strict sense linguistic entities, but they have propositional contents and directions of fit similar to the ones of desires, intentions, orders, commitments, and promises that have a *world-to-mind, world-to-language* direction of fit. Indeed, an obligation is satisfied if and only if the world changes to match the content of the obligation: if I owe money to a friend, the obligation will be discharged only when the world changes in the sense that I have repaid the money.

When for example we apply the moral principle of the *wrongness of discriminating against the handicapped* to the a specific moral case (for example the recent famous case of Baby Jane Doe where the parents had to decide in favor or against a fundamental surgical operation), we resort to a kind of "external" reason that we have to "internalize" – that is, recognize as a reason worth considering for a possible deliberation. If we instead exploit strong personal feelings like pity and compassion to guide our reasoning, we would decided for or against the operation based on a completely "internal" reason. We have to note, of course, that external reasons are always observer-relative. It is only human intentionality that furnishes meaning to a particular configuration of things in the external moral or non-moral world. The objective fact that, say, I have an increased white blood cell level acquires a direction of fit that is a direction for action only if related to a human being's interpretation (for example only "in the light" of a diagnosed disease, that same fact can trigger the decision for a therapy).

Searle also discusses the so-called collective intentionality that enables people to create common institutions such as those involving money, property, marriage, government, and language itself, an intentionality that gives rise to new sets of "conditions of satisfaction", duties, and commitments. In our perspective we say these external structures have acquired a kind of delegated intentionality because they have become *moral mediators*, they have acquired a kind of moral "direction", as Magnani [2006b, chapter six] has illustrated. In those cases, when we have to deal with a moral problem through moral mediators, evaluating reasons of any kind immediately involves manipulating non-human externalities in natural or artificial environments by applying old and new behavior *templates* that exhibit some uniformities. This moral process is still hypothetical (abductive): these templates are embodied hypotheses of moral behavior (either pre-stored or newly created in the mind-body system) that, when appropriately employed, make possible what can be called a moral "doing" [Magnani, 2006b].

We contend that external moral mediators are a powerful source of information and knowledge; they redistribute moral effort by managing objects and information in new ways that transcend the limits and the poverty of the moral options immediately represented or found internally (for example exploiting resources in terms of merely internal/mental moral principles, utilitarian envisaging, and model-based moral reasoning – emotions, for example).

It follows from the previous discussion that many entities can play the role of deontic moral structures. This fact can lead to a re-examination of the concept of duty. In this perspective duties can be also grounded on trained emotional habits, visual imagery, embodied ways of manipulating the world, exploitation of moral mediators – as we have just seen, endowed with a sufficient ethical worth in a collective.

4.1 The ontology of reasons

What are these "reasons" that, following Searle, are the basis of rational actions and, in the Baby Jane Doe case, the basis of moral action? A reason answers the question "Why?" with a "Because"; it can be a statement, like a moral principle, as in the answer to "Why should we perform surgery on Baby Jane Doe?": "Because of the wrongness of discriminating against the handicapped". In reality, reasons are "expressed" by the statements-*explanations* in so far as they are *facts* in the world (the fact that it is raining is the reason I am carrying an umbrella). They are also represented by *propositional intentional states* such as desires (my desire to stay dry is the reason I am carrying the umbrella), and, finally, by *propositionally structured entities* such as obligations, commitments, needs, and requirements, like in the case of our moral "principle" of "the wrongness of discriminating against the handicapped". All good reasons explain and all explanations give reasons. Searle also distinguishes between reasons that justify my action and thus explain why it was the right action to perform, and the reasons that explain why in fact I did it.

1. First of all, in rational decision making, when we must provide a reason for an intentional state, we have to make an intelligent selection from a range of reasons that exist either internally or externally – in the latter case, we must take the external reason, recognize it as good, and internalize it. With respect to our ideal of an ethical deliberation sustained by "reasons", we can affirm that it is not unusual for the "deliberator" to have limited knowledge and inferential expertise at his or her disposal. For instance, she may simply not have important pieces of information about the moral problem she has to manage, or she may possess only a rudimentary ability to compare reasons and

ascertain data. Ethical reasoning is so abductive and defeasible: because it is impossible to obtain all information about any given ethical situation, every instance of moral reasoning occurs without benefit of full knowledge, so we must remember that any reason can be rendered irrelevant or inappropriate by new information. Generally speaking, as illustrated above, these reasons can take three different forms: external *facts* in the world, such as empirical data; internal *intentional states* such as beliefs, desires, or emotions; and *entities* in the external world like duties, obligations, and commitments with the direction of fit upward (world-to-mind). External facts must be internalized and "believed", while external entities must be internalized and adopted ("recognized") as good and worth of consideration. The same happens in the case of rational moral deliberation.

2. Second, we must remember that maintaining a flexible, open mind is particularly important when we lack the ethical knowledge necessary to confront new or extreme concrete situations.

When evaluating an ethical case, we have at hand all the elements of rational moral decision making: the problem we face, the "reasons", and the agents involved. Every reason, Searle says, contributes to a "total reason" that is ultimately a composite of every good reason that has been considered – beliefs, desires, obligations, or facts, for example. As already observed, first, rationality requires the agent to recognize the facts at hand (I have to believe that it is raining) and the obligations undertaken (I have to adopt the principle of the sanctity of human life) without denying them (which would be obviously irrational) [Searle, 2001, p. 115]. Second, reasons can be more than one, indeed I need at least one motivator, but in some cases there are many, and these reasons often conflict with one another; it then becomes necessary to appraise their relative weights in order to arrive at the prior intention and the intention-in-action.

In abductive reasoning, this kind of appraisal is linked to evaluating various inferred explanatory hypotheses/reasons, and, of course, it varies depending on the concrete cognitive and/or epistemological situation. In the section 3 below, we have illustrated that epistemologically using abduction as an inference to the best explanation simply requires evaluating competing hypotheses (that express competing "reasons" in the ethical case). The best [total] reason would be the one that creates prior intention and intention-in-action.

What criteria can we adopt to choose the reason(s) that will become the motivator(s)? Thagard [2000] has proposed a framework in terms of coherence, in which ethical deliberation is seen as involving conflicting reasons (deductive, explanatory, deliberative, analogical) that can be appraised by testing their relative "coherence". This "coherence view" is terrifically interesting because it reveals multidimensional character of ethical deliberations. The criteria for choosing the most coherent "reason/motivator" represent a possible abstract cognitive reconstruction of an ideal of "rationality" in moral decision making, but they can also describe the behavior of real human beings. Human beings usually take into account just a fraction of the possible knowledge when performing ethical judgments. For example, when making judgments, it is common for utilitarians to employ only what Thagard calls "deliberative" coherence or for Kantians to privilege principles over consequences. Psychological resources are limited for any agent, so it is difficult to mentally process all levels of ethical knowledge simultaneously in an attempt to calculate and maximize the overall coherence of the competing moral options. The "coherence" model accounts for these "real" cases of human moral reasoning by showing they fit only "local areas" of the coherence framework: in general, real human beings come to immediate conclusions through one moral aspect (for instance, the "consequentialist" one) and disregard the possible change in coherence weight that could result from considering other levels (for instance, the "Kantian" one).

Searle interprets rationality in decisions naturalistically: "Rationality is a biological phenomenon. Rationality in action is that feature which enables organisms, with brains big and complex enough to have conscious selves, to coordinate their intentional contents, so as to produce better actions than would be produced by random behavior, instinct, tropism, or acting on impulse" [Searle, 2001, p. 142]. We agree, but we would add that rationality is a product of a hybrid organism. This notion obviously derives from the fact that even the external tools and models we use in decision making – an externalized obligation, a computational aid, and even Thagard's "coherence" model described above – are products of biological human beings, but at the same time these tools constitutively affect human beings, who are, as we already know, highly "hybridized"[6].

4.2 Abduction in practical agent-based reasoning

Searle considers "bizarre", and we strongly agree with him - that feature of our intellectual tradition, according to which true statements that describe how things are in the world can never imply a statement about how they ought to be: in reality, to make a simple example, to say something is true is already to say you ought to believe it, that is other things being equal,

---

[6] Searle [2001] calls the means and ways of performing an action (for instance to fulfill an obligation) "effectors" and "constitutors". An obligation to another person is an example. I know I own you some money: "I can drive to your house" and "give you the money" – effector and constitutor.

you ought not to deny it. Also, logical consequence can be easily mapped to the commitments of belief. Given the fact that logical inferences preserve truth, "The notion of a valid inference is such that, if p can be validly inferred from q, then anyone who asserts p ought not deny q, that anyone who is committed to p ought to recognize its commitment to q"[7]. This means that normativity is more widespread than expected.

Certainly, theoretical reasoning can be seen as a kind of practical reasoning where deciding what beliefs to accept or reject is a special case of deciding what to do. The reason it is difficult to "deductively" grasp practical reasoning is related to the intrinsic multiplicity of possible reasons and to the fact that we can hold two or more inconsistent reasons at the same time[8]. The following example illustrates how practical contexts are refractory to logical modeling. Given the fact that we consider it a duty to do p and that I also feel committed not to do p, we cannot infer that I am committed to do (p and not p). I am a physician committed to not killing a patient in a coma, but at the same time my compassion for the patient commits me to the opposite duty. This does not mean that I want to preserve the life of the patient and, at the same time, I want to kill him – that would lead to an inconsistent moral duty. All this represents an unwelcome consequence of the fact that commitment to a duty is not closed under conjunction [Searle, 2001, p. 250].

In practical reasoning, we are always faced with desires, obligations, duties, commitments, needs, and requirements, etc., that are at odds with one another. Moreover, even if I consider it a duty to do p and I believe that (if p then q), I am not committed to do q as a duty: I can be committed to killing a patient in a coma and at the same time believe this act will cause pain for his friends, but I am not committed to causing this pain. *Modus ponens* does not work for duty/belief mixture [Searle, 2001, pp. 254–255].

The examples above illustrate the difficulties that arise when classical logic meets practical reasoning. They further stress the importance we attribute to abductive explanatory inferences in agent-based practical settings, where creating, selecting, and appraising hypotheses are central functions.

---

[7] Cit., p. 148.
[8] Searle "reluctantly" declares that it is impossible to construct a formal logic of practical reasoning "adequate to the facts of the philosophical psychology" [2001, p. 250]. We think that many types of non-standard logic (deontic, nonmonotonic, dynamic, ampliative, adaptive, etc.) reveal interesting aspects of practical reasoning by addressing the problem of defeasibility of reasons and of their selection and evaluation.

5 Conclusion

In this paper we have described the relation between logical fallacies, induction, and abduction considered as inference to the best explanation. We have exploited the so-called *agent-based reasoning* framework, which adopts the perspective of a cognitive agent, to show how the so-called strategic rationality is important in many human reasoning performances. We have described how in the perspective of agent-based reasoning *fallacious* reasoning can be redefined and considered as a good way of reasoning. A quick illustration of the main aspects of abduction has been introduced showing its character from the point of view of classical logic, which depicts it as a kind of formal fallacious argument – the well-known fallacy of affirming the consequent. We have stressed the strategic role of abduction as a method of explanation and discovery in science and in everyday reasoning (for example abduction is able to govern inconsistencies that arise in scientific domains, when anomaly is treated through the generation of new hypotheses). Finally we have illustrated and furnished examples of practical reasoning where abduction exhibits its main cognitive appeal: moral deliberation. In this case abduction nicely accounts for the choice of internal and external "reasons" in decision making processes; we also believe it can help us to overcome some sterile consequences of the old-fashioned assumption that in the tradition of moral philosophy prohibits to derive an "ought" from an "is".

BIBLIOGRAPHY

[Gabbay and Woods, 2001] D. Gabbay and J. Woods. The new logic. *Logic Journal of the IGPL*, 9(2):141–174, 2001.

[Gabbay and Woods, 2005] D.M. Gabbay and J. Woods. *The Reach of Abduction*. North-Holland, Amsterdam, 2005. Volume 2 of *A Practical Logic of Cognitive Systems*.

[Gabbay and Woods, 2006] D.M. Gabbay and J. Woods. A formal model of abduction. In L. Magnani, editor, *Abduction, Practical Reasoning, and Creative Inferences in Science*, volume 14(2), pages 189–220, 2006. Special Issue of the *Logic Journal of the IGPL*.

[Hansen, 2002] H.H. Hansen. The straw thing of fallacy theory: the standard definition of "fallacy". *Argumentation*, 16(2):133–155, 2002.

[Hempel, 1966] C.G. Hempel. *Philosophy of Natural Science*. Prentice-Hall, Englewood Cliffs, NJ, 1966.

[Hintikka, 1998] J. Hintikka. What is abduction? The fundamental problem of contemporary epistemology. *Transactions of the Charles S. Peirce Society*, 34:503–533, 1998.

[Kirsh and Maglio, 1994] D. Kirsh and P. Maglio. On distinguishing epistemic from pragmatic action. *Cognitive Science*, 18:513–549, 1994.

[Magnani, 2001a] L. Magnani. *Philosophy and Geometry. Theoretical and Historical Issues*. Kluwer Academic Publisher, Dordrecht, 2001.

[Magnani, 2001b] L. Magnani. *Abduction, Reason, and Science. Processes of Discovery and Explanation*. Kluwer Academic/Plenum Publishers, New York, 2001.

[Magnani, 2002] L. Magnani. Epistemic mediators and model-based discovery in science. In L. Magnani and N.J. Nersessian, editors, *Model-Based Reasoning: Science,*

Technology, Values, pages 305–329, New York, 2002. Kluwer Academic/Plenum Publishers.
[Magnani, 2005] L. Magnani. Creativity and the disembodiment of mind. In P. Gervás, A. Pease, and T. Veale, editors, *Proceedings of CC05, Computational Creativity Workshop, IJCAIO2005, Edinburgh*, pages 60–67, 2005.
[Magnani, 2006a] L. Magnani, editor. *Abduction, Practical Reasoning, and Creative Inferences in Science*, 2006. Special Issue of the *Logic Journal of the IGPL*.
[Magnani, 2006b] L. Magnani. *Knowledge as a Duty. Distributed Morality in a Technological World.* Cambridge University Press, Cambridge, 2006.
[Mill, 1843] J.S. Mill. *A System of Logic. Reprinted in: The Collected Works of John Stuart Mill.* Routledge and Kegan, London, 1843. ed. by J.M. Robson.
[Peirce, 1931–1958] C.S. Peirce. *Collected Papers.* Harvard University Press, Cambridge, MA, 1931–1958. 1–6, ed. by C. Hartshorne and P. Weiss, 7–8, ed. by A. W. Burks, 1931-1958.
[Popper, 1959] K.R. Popper. *The Logic of Scientific Discovery.* Hutchinson, London, New York, 1959.
[Searle, 2001] J. Searle. *Rationality in Action.* The MIT Press, Cambridge, MA, 2001.
[Thagard, 1992] P. Thagard. *Conceptual Revolutions.* Princeton University Press, Princeton, 1992.
[Thagard, 2000] P. Thagard. *Coherence in Thought and in Action.* The MIT Press, Cambridge, MA, 2000.
[Woods, 2004] J. Woods. *The Death of Argument.* Kluwer Academic Publishers, Dordrecht, 2004.
[Woods, 2005] J. Woods. Epistemic bubbles. In S. Artemov, H. Barringer, A. Garcez, L. Lamb, and J. Woods, editors, *We Will Show Them: Essays in Honour of Dov Gabbay*, volume II, pages 731–774. College Publications, London, 2005.

Lorenzo Magnani
Department of Philosophy and
Computational Philosophy Laboratory
University of Pavia, Pavia, Italy; and
Department of Philosophy, Sun Yat-sen University
Guangzhou (Canton), P.R. China.
Email: `lmagnani@unipv.it`

Elia Belli
Department of Philosophy and
Computational Philosophy Laboratory
University of Pavia, Pavia, Italy.
Email: `elia.belli@unipv.it`

www.ingramcontent.com/pod-product-compliance
Lightning Source LLC
Chambersburg PA
CBHW071223230426
43668CB00011B/1284